Handbook of Computer Animation

Springer
London
Berlin
Heidelberg
New York
Hong Kong
Milan
Paris
Tokyo

John Vince (Ed)

Handbook of Computer Animation

Springer

John Vince
Bournemouth Media School
Bournemouth University
Poole, UK

British Library Cataloguing in Publication Data
Handbook of computer animation
1.Computer animation
I.Vince, John, 1941–
006.6'96
ISBN 1852335645

Library of Congress Cataloging-in-Publication Data
Handbook of computer animation / John Vince (ed.).
p. cm.
Includes bibliographical references and index.
ISBN 1-85233-564-5 (alk. paper)
1. Computer animation--Handbooks, manuals, etc. I. Vince, John (John A.)
TR897.7.H365 2002
006.6'96--dc21 2002030648

ISBN 1-85233-564-5 Springer-Verlag London Berlin Heidelberg
a member of BertelsmannSpringer Science+Business Media GmbH
http://www.springer.co.uk

Typesetting: Digital by Design Ltd, Cheltenham, Gloucestershire, England
Printed and bound at the Cromwell Press, Trowbridge, Wiltshire, England
34/3830-543210 Printed on acid-free paper SPIN 10856356

This book is dedicated to the memory of Rob Edwards

Preface

The UK's National Centre for Computer Animation (NCCA) is based at Bournemouth University and has been associated with teaching, research and consultancy in computer animation for over 15 years.

Currently, there are approximately 250 students studying computer animation on a three-year undergraduate course and three one-year postgraduate courses. The overriding philosophy behind these courses is based upon the marriage of art and science. Students come from a variety of creative and technical backgrounds and during their stay at the NCCA they are immersed in a multidisciplinary approach to computer animation.

To gain access to the undergraduate course, students must possess strong creative skills, but also the potential to cope with mathematics, programming and computer technology. During the three-year course, students develop strong animation skills, and become expert in animation systems such as Maya, Houdini, RenderMan and OpenGL. At the same time they learn to program in C, C++ and Java, and are taught a range of mathematical topics that include vectors, transforms, matrices, parametric curves and surfaces. Students also become familiar with computer architectures, operating systems, computer hardware and networks.

As one would expect, graduates from this course are in great demand by industry. This multifaceted education, together with strong problem-solving skills, make these graduates ideal candidates for a wide variety of jobs in the computer animation sector. Perhaps the most relevant role is that of a Technical Director within a special effects company.

The postgraduate courses comprise: MA 3D Computer Animation, MA Digital Effects and MSc Computer Animation. Students on these courses are exposed to the same philosophy of marrying art with science, which is manifested in the wide range of masters projects they undertake in their final semester.

Running such an ambitious range of courses requires a highly qualified academic team, which the NCCA has managed to acquire and maintain. Members of the team come from a variety of backgrounds that include computer science, computer animation, traditional animation, fine art, engineering and media production. This wealth of knowledge provides the academic framework to support the above courses and is the key to maintaining a vibrant and stimulating student environment.

Like many other computer-based disciplines, computer animation is in a constant state of flux. New techniques are constantly being developed; software products come and go; increased processor speeds transform working practices; and new GUIs improve performance and project complexity. Anyone working in such a rapidly changing subject must constantly update their knowledge and remain abreast with state-of-the-art developments. In computer animation, this means following international events such as SIGGRAPH, IMAGINA and CGI, monitoring recent research papers and books, as well as attending film festivals and watching key films. It can also mean that personal research projects are pursued to develop an individual expertise.

Staff at the NCCA are engaged in all of the above activities. There is an active research group, which is recognized as the UK's center of excellence for computer animation. Academic staff are also engaged in a variety of personal research projects, computer animation productions and the writing of books, which brings us to this particular volume.

Two years ago, we decided to write a book on 3D computer animation. Each member of staff would contribute a chapter on their particular area of expertise. The challenges associated with such a task were threefold: the first was agreeing on the level of technical competence expected of the reader; second, maintaining a level of technical continuity across the book; and thirdly, keeping a flowing literary style across the chapters.

We agreed that the technical expertise expected of our readers would be that associated with a typical NCCA 3rd-year undergraduate or postgraduate student, and that the subjects addressed in the book would correspond to key topics associated with their academic studies. This provided the freedom to address what we believed to be central topics within computer animation, but which were not necessarily linked directly to one another. We also wanted to address the breadth of diverse subjects associated with computer animation such as computer games, special effects, curves and surfaces, and genetic algorithms. As far as we knew, no such book existed, and writing such a book presented a unique opportunity to provide the student community with a book that addressed a collection of major topics relevant to anyone studying computer animation.

The book is divided into seven chapters – each one is written by an individual member of staff. I will introduce each chapter in this preface and attempt to describe the *raison d'être* behind its inclusion and the expertise of the author. The chapters are organized in the form of a journey through the world of computer animation. We begin with computer game design and finish with rendering and shading. During our journey we stop at some very interesting points that include genetic algorithms, shooting live action, digital effects, cubic polynomial curves and surfaces, and subdivision surfaces.

Chapter 1 explores various issues associated with computer game design and is entitled *Computer and Video Game Design Issues*. It is written by Lee Uren who was a graduate of the NCCA's BA (Hons) Computer Visualization and Animation course. After graduating, Lee joined the NCCA as a Demonstrator and because of his unique knowledge of computer games joined the lecturing staff and aligned himself with the teaching of this subject. Like many other graduates of the NCCA's BA course, Lee enrolled for the MA Computer Animation and was awarded a distinction in 1999. Lee is now employed as an animator at Industrial Light & Magic, Inc. Fortunately, before Lee left, he finished the manuscript for this chapter where he traced the history of computer games and documented the rapid technological developments that occurred over the past two decades. The chapter concludes by addressing recent events such as the launch of the Playstation II and finally he gazes into the future to see what is coming in over the horizon.

Chapter 2 is entitled *Evolutionary Algorithms in Modeling and Animation* and is written by Anargyros Sarafopoulos. Anargyros was a graduate of the NCCA's undergraduate course and is now a Senior Lecturer. Anargyros lectures on computer programming and he has recently completed a PhD where he has been investigating the role of genetic algorithms in computer animation. Much of this chapter is drawn from his research and is truly state-of-the art. The chapter updates the reader's knowledge with the fundamental principles of genetics and quickly moves to the digital domain where genetic algorithms are starting to get a foothold in computer graphics and computer animation. He explains how evolutionary algorithms can be used as an artistic meta-tool, and even though still images are used to demonstrate the technique, the potential for animation looks very promising. Anargyros' PhD supervisor at University College London was Professor Bernard Buxton, who also contributed to the chapter.

Chapter 3 is written by Professor Mitch Mitchell, who is a Visiting Professor at the NCCA. Mitch was a cameraman and founding video effects supervisor at the BBC before going freelance. He later joined The Moving Picture Company in London and as Director of Special Effects developed new techniques and technologies as well as supervising visual effects on many commercials and broadcast programs. Freelance again he has worked as visual effects supervisor on amongst others the BBC's award winning promotional film *Next Generations*, the Orange car chase commercial "Hold Up", a 10 h mini-series *The Tenth Kingdom* and is currently completing effects and editing of a TV movie as well as writing a Media Manual on effects. Throughout his career he has written about and spoken at many conferences on new technologies in motion picture and television production and has lectured extensively to students on effects photography. In his chapter *Shooting Live Action for Combination with Computer Animation* Mitch shares with the reader a wealth of practical knowledge acquired from a professional life devoted to special effects.

Chapter 4 is called *Elements of Digital Effects* and is written by Steve Hubbard. Steve completed both the BA (Hons) Communication and Media Production, and MA Computer Visualization and Animation courses at Bournemouth University, before staging a ten-year journey into computer animation and digital effects production. He then returned to the NCCA in 2001 to communicate his discoveries and experiences to a new generation of prospective digital effects practitioners, by running the NCCA's MA Digital Effects course.

Whilst in industry, he worked on commercials, simulator ride films, cinema titles, feature films and documentary series. These included the BBC's landmark series *The Human Body* and *The Planets* and Channel 4's *What Happened to the Hindenburg*? Largely freelance, he worked at many London companies such as Lost in Space, Cell Animation, Cinesite and The Mill. He was also a founder of three computer animation companies.

Steve's chapter takes over from the previous chapter by explaining the techniques of compositing live action with synthetic images. The chapter begins by defining what is meant by digital effects, continues with a brief history of analogue and digital effects, and then explains the terminology and techniques associated with modern compositing systems.

Chapter 5 is written by Professor Peter Comninos who founded the NCCA in 1989. Peter's knowledge of computer animation is deep, as it is broad. Prior to coming to Bournemouth University in 1987 he was working at Teesside University where he had developed his CGAL computer animation system, which was the result of his PhD research. CGAL was the central software system at the NCCA for several years, and many students acquired an understanding of computer animation using CGAL before securing key jobs in industry. Apart from directing the research programme at the NCCA, Peter also teaches Advanced Animation Systems and Mathematics on the NCCA's undergraduate course and on the MSc Computer Animation course. He is one of the few people in the world who have developed a complete computer animation system single-handed, and although this could have been an exciting topic for the book, he decided to write about another subject with which he is an authority: *Cubic Polynomial Curves and Surfaces*. This chapter takes the reader into the mathematical world of curves and surfaces, and although this is the first chapter to introduce mathematics, the reader will find that he has made every attempt to communicate some fundamental mathematical ideas in a clear and simple manner.

Chapter 6 – *Smooth Surface Representation over Irregular Meshes* – follows on from the previous chapter by exploring the subject of subdivision surfaces. The author is Professor Jian Zhang who joined the NCCA in 1996, and is now Professor of Computer Graphics. Professor Zhang has published a number of papers in computer graphics, computer animation and surface modeling, and he is responsible for the research projects of the NCCA. One of his research interests is in the modeling of surfaces of irregular topology. Such surfaces have become increasingly important and widely used in computer animation, as they complement the ordinary parametric surfaces where they are generally restricted to have a quadrilateral topology. In this chapter he introduces three such surface modeling methods: Cutmall–Clark subdivision surfaces, Doo–Sabin subdivision surfaces and loop triangular patch-based spline surfaces. Although there are a number of other modeling methods, which are equally capable to handling irregular topology, these three methods are widely used in animation applications. For each method, he introduces both the theoretical background and the algorithms. This will help the reader both understand the principles and be able to implement these methods.

Chapter 7 concludes the book with *Rendering and Shading*. This chapter is written by Dr Ian Stephenson who joined the NCCA in 1997. Ian's PhD research work was in *Massively Parallel Computer Architectures*. He has worked in industry and before pursuing an academic career he was a systems programmer at Cambridge Animation Systems where he was working on the *Animo* system. At the NCCA he teaches on a variety of technical subjects that include Operating Systems Techniques, Computer Architecture and RenderMan.

Ian takes a system's approach to rendering by considering the implementation of a simple renderer, and showing how the stages interact with one another. The RenderMan standard is used as the basis of the discussion, which allows real practical problems to be considered rather than imaginary ones. The chapter begins by considering some of RenderMan's features and then explores in greater detail the concepts of shading engines. The chapter concludes by looking at more advanced features such as spectral colors, motion blur, deferred shading and occlusion culling.

We wish to thank staff and students at the NCCA for their support during the writing of this book and their permission to use their images. On behalf of the above authors, I hope that you will find the book useful, interesting to read, and that you will discover something new that will increase your understanding of computer animation.

John Vince
Bournemouth, UK
2002

Contents

Contributors

Professor Bernard Buxton
University College London, London, UK

b.buxton@cs.ucl.ac.uk

Professor Peter Comninos
Bournemouth University, Poole, UK

peterc@bournemouth.ac.uk

Steve Hubbard
Bournemouth University, Poole, UK

shubbard@bournemouth.ac.uk

Professor Mitch Mitchell
Bournemouth University, Poole, UK

mitch@mitchell.co.uk

Anargyros Sarafopoulos
Bournemouth University, Poole, UK

asarafop@bournemouth.ac.uk

Dr Ian Stephenson
Bournemouth University, Poole, UK

istephen@bournemouth.ac.uk

Lee Uren
Industrial Light & Magic, Marin County, USA

lee_uren@hotmail.com

Professor John A. Vince
Bournemouth University, Poole, UK

jvince@bournemouth.ac.uk

Professor Jian Zhang
Bournemouth University, Poole, UK

jzhang@bournemouth.ac.uk

1 Computer and Video Game Design Issues

Lee Uren

1.1 Gameplay

Throughout the history of Digital Entertainment Systems, game designers have aspired to create more and more engaging and immersive experiences, which grab the player's attention and refuse to let go.

Early arcade games generally tended to be fast, frantic, action-based shooting games, for the single reason that more money could be extracted from a player if the game pulled them in quickly with simple game mechanics and clearly defined rules, but often a steep difficulty curve. (The almost universal use of pixellated violence and destruction probably stems from the fact that excitement and drama are most easily borne from conflict.) The initial ease of play followed by a fast increase in difficulty results in a challenge being set – "Why did I die there? I'll do better next time!" – and a high frustration level, urging enthusiastic players to take "just one more go" and get that little bit further into the game.

Typically, games with a high-action content, reflex-based play systems or fixed-sequence obstacles score highly on initial interest, as they can be exciting and easily learned. This is typical of almost all arcade or "coin-op" gameplay systems. Games that are more strategic and require analytical thought, or are of a more varying structure, tend to have greater life spans in terms of longer lasting appeal, as the experience can be different and more complex each time it is played. However, complex controls and play systems often mean that the game may initially be played in shorter bursts as the player comes to terms with learning the game system, and sometimes reading a hefty instruction manual. Generally speaking, this kind of game is more commonly associated with use on home computers.

The desire to see carefully allocated rewards such as new "power-up" features, different types of enemies and impressive graphic differences in newly discovered levels are more incentives for gamers to attempt to progress through a game, as well as the obvious high-score glory: risk should offer reward.

One of the key originators of exactly this type of experience was Eugene Jarvis. Working with Williams, he produced an impressive quantity of innovative, polished and highly successful games in the industry's infant years, including *Defender*, *Stargate* (*Defender II*), *Blaster*, *Robotron: 2084* and *Smash TV*. He describes his goal behind this as being "to create a new generation of video-game addicts".

He says of *Robotron*, with it's novel twin joystick controls, "When I design a video-game, my philosophy is to give the player as much freedom of movement and firing as possible. Why restrict motion to a single line, plane, or direction? In reality there are no such limitations". Yet few of the early games could be called "realistic". Hardware was primitive by today's standards, so artists and designers had to resort to using a bold, bright, iconic approach to represent the game worlds on screen. Realism and immersion came almost entirely through creating a compelling, absorbing gaming experience. The game system's rules and interaction represented far more than the simple graphics ever could, relying on a player's imagination to fill in the visual gaps. *Robotron*'s exponential reward system for saving more and more humans on a single level adds the final essential ingredient – an exploitation of the player's greed for points, and thus a coveted entry in the high score table. This is played off against an exponential increase in the difficulty of attaining that reward – putting your own character in ever more dangerous situations, which makes for a truly winning combination: the greater the risk, the greater the reward.

Many of the other early, classic games still stand up well today in terms of game mechanics. *Defender* is one such example, with no pre-set attack routes or simple paths for enemies to follow. Every foe has it's own artificial intelligence (AI) system, and so the game plays slightly differently every time. Jarvis says, "I wanted to design a game where the character's behavior was not merely deterministic, but incorporated higher orders of game strategy – a rudimentary artificial intelligence as opposed to artificial stupidity". This enhanced replay value ensures that *Defender* is still an enjoyable, challenging game today, while the graphically similar but far more repetitive *Scramble* is a less compelling experience.

Another strength of both *Robotron* and *Defender* is that they ask the player to think as well as simply target and react. Both games feature hordes of fast, relentless enemies to constantly fend off, but both also require strategic decisions regarding the protection of defenseless humans. The sheer speed of play and ever-changing pyrotechnic colors feed the adrenaline rush, whilst the strategic elements feed the brain.

Sinistar, also from Williams, requires still more decision making on defending your ship from the constant attack of smaller threats, whilst having to prepare for the larger threat of the *Sinistar* weapon which is gradually being constructed by your opponents from pieces of mined ore that you let go. All of these gameplay features are represented in 2D, in very limited resolutions and low color depths, but it is the strong game mechanics that shine through the rudimentary graphics.

This tradition started to influence the transition into true 3D with the release of Dave Theurer's *I, Robot* through Atari, a true classic in terms of combining reflexes, careful timing, spatial thinking and quick decision making.

Today, writing certain types of games could in fact be seen to share just a few similarities with film script writing, although the two media are fundamentally very different – an interactive entertainment medium versus a purely staged, choreographed entertainment medium. Just as there are rules for script writing – the three-act paradigm, for example – there are some basic laws of gameplay design. At least, for certain types of games.

Japanese companies such as Nintendo, Sega, Konami, Capcom and Namco have the art of designing the "pure game" down to a fine art, keeping the play mechanics simple and easily accessible, but gradually expanding upon the *use of* the game system or features as the player progresses through the game. Classic examples of this are *Super Mario World* and *Legend of Zelda: Ocarina of Time*, where the player is gradually taught how to gain more control of the lead character through a series of progressively more challenging "exercises". These developers also currently seem to have the monopoly on games featuring content which could be most likened to film, as in the *Metal Gear Solid* and *Resident Evil* series. Both these franchises shine with cinematic use of camera, strong characterization and enthralling plot development. By way of contrast, US companies such as Westwood Studios, Electronic Arts and LucasArts often tend to have much more complex game systems with a multitude of different controls and outcomes possible right from the outset. In these cases, a good deal of the player's time is spent learning the actual control systems and interfaces (this can sometimes even become part of the challenge in itself). A good example of this type of game is *Command & Conquer*, with its many arrays of units, and icons to direct the player's forces as they move around over the battlefield.

Ultimately though, rules can be broken, and some of the best work is produced when this is true. Original game design can thus largely consist of one of the following.

- New features being added to tried-and-tested game systems. For instance, team-based or "tag-team" fighting added into SNK's fighting games and Capcom's *X-Men vs. Street Fighter*. Or the use of a user-selectable weapon powerup system, pioneered many years ago in Konami's *Gradius*.
- A cross-fertilization of genres, such as strategy games brought into a real-time combat environment (*Command & Conquer*, *Battlezone* – 1997 version).
- Borrowing ideas straight from other, slightly different games within the same genre, such as the energy refuelling lane taken from *F-Zero* and transplanted into *wipEout: 2097*.
- Inspiration! Just as there are techniques to script writing, but no easy formula for creating the spark of original ideas, game design can often be a process of instinctively knowing what will work interactively in a given situation. A sound knowledge and experience of playing both classic and contemporary titles is invaluable here, as it gives an understanding of what can work and what does not. This is why so many game testers go on to become level and game designers, as having spent most of their waking hours playing games they gain an instinctive knowledge of gameplay mechanics.
- Having a real passion for playing games is obviously one of the best incentives of all. As David Braben, co-creator of the legendary *Elite* says, "You have to be creating the game for yourself, not for some imaginary market". Although with such huge development budgets at stake these days, many developers do of course take great care to ensure their products will be suited to a sizable target audience.
- Rarely, but sometimes, even ignorance can lead to a successful game. The often criticized but commercially successful *Myst* seems to have been borne from this approach, as Richard Vanderwende of Cyan attests: "One of our strengths is that we don't play games, so we aren't influenced by them".
- It is also worth mentioning that not all good games are based on original concepts, and many "best of genre" titles are the latest release in a long line of clones or sequels. In such cases, very gradual, incremental improvements can be made on each successive release of essentially the same idea. One of the best examples of this evolutionary process is Capcom's incredibly long-running *Street Fighter* series.
- Finally, and this is an interesting point, many games are created out of a direct link with an emerging technology or technique. Sometimes the idea can drive the use of the technology, and sometimes vice versa. It can thus often seem as though a game has been created out of a drive to conquer something new, much as mountain climbers often explain their drive to conquer a mountain: "because it's there".

1.2 Technology: an Enabler or a Limiter

Just a few years ago players were truly amazed by newly emerging, highly detailed 3D environments appearing in games for the first time. These were only made possible by the development of hardware accelerator cards for home PCs, offering high-resolution displays and advanced features such as texture filtering. The period of time between then and now has seen an exponential development curve for real-time 3D graphics, with noticeably more detail and effects becoming possible seemingly every month that passes by.

As a matter of history, the types of technologies available have always influenced the types of games created. It is interesting to note that consistently, as soon as a new computer technology was available, a game had made use of it. *Space War* was created just after the first computers entered universities, and even id's *Doom* and *Quake* made use of Intel's increasing domestic PC processor speeds, pushing the hardware to the limit to create new game environments as soon as the power was available to do so. In the case of Doom, this happened even before the power was really available to render "true" textured-mapped 3D worlds, by using tricks to create an environment that only appears to be 3D.

One school of thought is that a game should be designed independently of any target hardware. This is a view shared by Nintendo's Shigeru Miyamoto, who fairly recently expressed his concern that games seemed to be far too reliant on new technologies.

But the co-dependence of gameplay and hardware technologies goes right back to the very beginning of the industry. Ultimately, the graphics and sounds experienced during the game are often designed as a direct consequence of the specification of the target system. Nintendo have a track record of this (*F-Zero* and *Mario Kart* taking advantage of the SNES' bitmap stretching effects, and *Starfox* making famous use of the *Super FX* chip, the world's first 3D games accelerator), as in fact do most hardware developers who also produce games.

Sega pioneered sprite scaling technology in arcade-based games (*Space Harrier, Outrun, Afterburner*), Atari produced custom hardware to accommodate vector graphic 3D games (*Battlezone, Star Wars*) and then eventually the first fully shaded polygonal 3D game, *I, Robot.*

In fact, a large percentage of titles now deemed classics have made use of new hardware or software technologies to a lesser or greater extent.

- Hardware sprite animation and simple starfield/ground parallaxed hardware scrolling – *Defender.*
- Use of a subtly defined 16-color palette to give the impression of light-source-shaded objects in 2D bitmap form – *Sinistar, Joust.*
- Use of synthesized or sampled speech to enhance audio dynamics – *Berzerk, Bosconian, Sinistar.*
- A blitter chip to move blocks of memory very quickly, enabling animation of very large numbers of objects on screen – *Robotron 2084.* (A similar chip later featured prominently in Commodore's Amiga computers.)
- Character set display combinations used to produce an apparent 3D image ("block graphics") – *3D Monster Maze.*
- Analogue vector graphic displays (2D or 3D rendered) – *Asteroids, Battlezone.*
- Use of many bit-planes to give more detailed parallax scrolling effects – used extensively on literally hundreds of 2D games, such as *Bounder, Super Mario World, Battle Garegga,* etc.

- Techniques producing bands of moving color to simulate traveling across a ground plane (raster-scan interrupts or similar methods) – see many early racing games including *Pole Position*, *Roadblasters*, *Electraglide*.
- 2D algorithms used to generate illusions of 3D grids in a single point perspective – *Ballblazer*, *Space Harrier*.
- Use of a simple Z-buffering technique to hide 2D bitmap images behind others in sensibly sorted depth orders – *Crystal Castles*, *Knight Lore*, *Marble Madness*.
- Sprite scaling techniques, to give an illusion of fast 3D movement – Sega's arcade systems, including *Buggy Boy*, *Hang On*, *Space Harrier*, *Galaxy Force* and *Outrun*.
- Bitmap perspective transformations to give an illusion of textured planes moving in 3D space – *F-Zero*, *Mario Kart*, *Wolfenstein 3D*. Later used to produce multiple animated planes in *Darius Gaiden* and *Radiant Silvergun*.
- Two-point perspective "fake" 3D games combining combinations of sprite-scaling, bitmap perspective distortions and simple polygonal rendering – *Wolfenstein 3D*, *Doom* and their many clones to follow.
- Flat-shaded true 3D polygonal rendering – *I, Robot*, *Virtua Racing*, *Virtua Fighter*. Then later, Gouraud and similar variant smooth shading algorithms applied to blend edge discontinuities.
- Texture mapping techniques used to enhance polygonal graphics – *Virtua Fighter 2*, *Tekken*, *Quake*. And soon afterwards, games designed to look best with filtered textures – *Quake 2*.
- 3D polygonal effects used to improve 2D games – *Raystorm*, *Einhander*, *Radiant Silvergun*.
- Transparency combination algorithms to layer bitmaps which appear translucent – hundreds of games including *Axelay*, *Alien 3* and *Thunderforce V*.
- Variable 2D bitmap distortions producing rotating curved surface approximations – *Axelay*, *Darius Gaiden*.
- Chips used to speed up specific custom 3D rendering effects, e.g. a 3D math accelerator – *Starwing*, via use of Argonaut's *Super FX* chip onboard the game cartridge, later the *3DFX* card and the like for PCs.
- Algorithmically generated landscape geometry – *Rescue On Fractalus*.
- CD audio digital soundtracks, notably licensed or commissioned music from well-known bands in games such as *Quake* and the *wipEout* and *Rollcage* games. (On CD-ROM-based systems.)
- Laserdisc-based video games – *Dragon's Lair*, *Space Ace*.
- Combined Laserdisc video and digital animation overlays – *Firefox*.
- CD-ROM used as a mass data storage medium for large game content – *Final Fantasy VII*, *Resident Evil*. Also for multimedia concept "point and click" games – *Myst*, *Riven*.
- Fast CD-ROM data retrieval and decompression algorithms to give "full motion video" in-game effects or cut-scenes – *Novastorm*, *Voyeur*.
- Worldwide Internet data transfer used for multiplayer real-time 3D gaming – *Quake*.
- High-resolution 3D computer-generated images (CGI) rendered as bitmaps, used either as 2D sprites or background images – *Donkey Kong Country*, *Killer Instinct*, *Resident Evil*.
- Video-captured footage used as sprite frames – *Mortal Kombat*. Also stop-motion animation used in precisely the same context – *Primal Rage*, *Clay Fighter*.

- Simple motion-blur techniques possible in real time – *Metal Gear Solid, Rollcage.*
- Bump-mapping effects possible in real time – modern PC 3DFX-accelerated games.
- Curved surface representations used for storage of geometry – *Quake III Arena, Evolva.*
- Various other high-end 3D graphics techniques coming into very widespread use. Including texture filtering algorithms, depth fogging, alpha channel transparency effects, etc. used on almost all 3DFX-accelerated, Nintendo 64 (N64) and Dreamcast games.
- Advanced use of depth buffers for rendered images becoming possible to simulate camera focal lengths. Also use of voxel-based engines and other more unusual computer graphics (CG) techniques.

So has all this hardware and software innovation widened the gaming experience, or has it led us down a narrow path all this time, the types of games being produced being purely dependent upon new technologies? Clearly the answer is somewhere in the middle. Technology is a tool to be used by the designer, not a master to the end product. It is only very recently that Mr Miyamoto made his intentions known to get back to basics and produce simpler games. This is an interesting change of position for Nintendo, the SNES being one of the most technology-driven consoles yet released with it's extensively used Mode 7 bitmap processing, transparency effects and the inclusion of accelerator chips on games such as *Starfox*. This change of attitude will no doubt have been influenced by the amount of resources that *Zelda 64* tied up at Nintendo over its three-year production cycle, culminating in the involvement of almost every artist and designer in the company over the final six months of its production. However, this in no way means that designers will go back to producing purely 2D games and leave 3D gaming behind. The advantages of an immersive 3D environment for a game are clear, as it more closely represents the world that we are used to interacting with all the time.

One of the first things to change in game design and production when the move to 3D was underway was the size of a development team. Gone are the days of the one-man software producer who could program, create graphics and bring it all together with sound single-handedly. Designers such as Dave Theurer, Matthew Smith, Bill Hogue and Jeff Minter all quickly produced many remarkable games in the 1980s, but nowadays, large teams of individuals produce games over development scales measured in years. Modern game systems have advanced visual and sonic abilities, meaning that specialists are needed to produce material for every aspect of a game. It takes time to produce detailed, textured 3D graphics and to write efficient engines to drive them. In Japan in particular, it is not uncommon to find a team of over a hundred staff working on a big production at companies such as Square and Konami.

There is always the danger that the time and effort necessary to produce ever-improving graphics will detract from the effort spent designing the game system or level designs, and this is indeed often the case. Market research also rears its ugly head, as do sales figures, and large proportions of games released conform rigidly to the most popular genres. Attractive screenshots on the back of the box help to shift games that have just not had the time spent on level design or depth of play.

But can small teams of game developers cope with the demands being asked of them to produce fully immersive, interactive experiences that can compete with the level of visual excitement that we are becoming used to? Most high-budget special-effects-driven films only have a few minutes worth of digital effects embedded within the live-action footage. And now game artists are being asked to produce visuals that will suspend the player's disbelief for hours at a time. Clearly, the feeling of believability and "realism" in a game cannot be carried by the graphics alone.

1.3 Realism, Immersion and Believability

One of the most noticeable trends in recent years has been the thrust toward "realistic" looking and feeling games. Graphically speaking at least, it is now possible to represent a reasonable approximation of the real world – or an entirely imaginary one – within a game system.

Since its inception, computer graphics has always seemed to be obsessed with creating realistic images. The entire concept and structure of a 3D package dictates this bias – using perspective, color theory, lighting models and shading algorithms that are all based on real-world models. This is quite unlike older art forms, which often actually set out to challenge and confront our perceptions of reality, and is perhaps a telling sign of how young the medium of CG still is. Computer animation and games are media with their own inherent qualities and subtleties. By way of analogy, a painter rarely sets out to produce a photograph (the obvious exception being a matte artist).

Any framed view of a world (be it a painting, film, television or a piece of theater) already presents a scene in a consciously filtered way. *Edward Gordon Craig* (1872–1966) was a theatrical designer who started a movement away from naturalism. His theories and reasons for this amounted to what is now known as *media integration*, *concentration* and *action*.

Craig found that *media integration* (bringing together actors, sets, props, lighting, music, etc.) became more cohesive when not trying to just replicate reality. This is of course now widely accepted in cinema, with staged use of framing, camera movement, lighting, editing and so forth altering the perception of a scene for artistic effect. A reduction to selective or necessary elements, or a non-naturalistic setting, can also affect the viewer's *concentration* – enabling a director or designer to highlight specifics within a scene. Extraneous detail detracts from what the designer intends to be seen. Craig likened the theater set to a human face, which always has the same features recognizable, but which can be altered to produce different "expressions" to communicate ideas or emotions. This defines his third idea of *action*. Dynamic lighting, moving props, camera (viewpoint, movement, lens, focus, etc.) and interactive scenery are all extensions of these theories, and all can be found in computer games. (Although sometimes reduction of detail may of course actually come from necessity rather than planned concentration, with selective inclusion of props due to limited available processing power.)

Research into the social and behavioral effects of using virtual environments (VEs) has resulted in some useful theories and terminology, which can equally be applied to game theory. With specific regard to virtual environments, the *theory of cognitive immersion* defines four variables that contribute to the sense of immersion: experience, absorption ability, interactivity and avatar. The former two are attributes of the user and vary from person to person, while the latter two are attributes of the virtual environment and vary in quality between VEs (Tromp 1995).

- *Immersion* is described as the sense of "being there", of being absorbed within a given medium. This could be a book, a film, a play or a 2D or 3D virtual environment.

- An *avatar* is the incarnation (the embodiment of the awareness and identity) of the user within a virtual world.
- *Space* describes the spatial representation of a 3D world – its geometry, texturing, lighting, etc.
- *Place* describes a space that has been endowed with an implication of some social, cultural or behavioral understanding. This can then provide *behavioral framing*, shaping our actions within that place. An example of the difference between space and place would be a house compared to a home – "house" implies a generic weatherproof structure made from stone and wood, a "home" implies a similar structure but endowed with warmth and personal value. The way that we act and behave in our homes will be different to the way that we behave in other houses and other types of buildings. Almost all games include spaces, but the difference that an effectively endowed sense of place can make to an environment is profound. (See Harrison and Dourish (1996) for an in-depth discussion of the concepts of space and place.)

Most importantly of all though, it is important to remember that playing a game is all about having fun. If a gaming experience isn't enjoyable, then it is simply not worth playing. Games are also about escapism. They are at their very least a momentarily entertaining distraction, and at the very most a perfect escape from our everyday lives. This then raises the question, "should games be truly realistic at all?" For if we crave an escape from reality, then what we probably really need from a game is an *alternate* reality, in some way quite different from our own.

One of the most commercially successful types of game today is the racing game, arguably one of the genres most attempting to capture the sense of realism – of "being there". But while the graphics strive to provide realistic renderings of built-up and natural environments, the essential difference of this virtual reality (VR) is that the player is in a situation that they can only dream of in real life. Very few of us will ever get to take part in a real life professional race. So, even if a game appears to be fairly realistic, there will almost always be some level of escapism involved, some kind of unrealistic situation, at least as far as the average player is concerned. And even for the experienced professional, a realistic simulation is not always what they would like from a game.

Often when we say a game is *realistic*, it is actually more likely that we mean we find the experience believable. Our interface with the real world around us consists of five senses – sight, hearing, touch, taste and smell. Therefore it might stand to reason that to make an environment seem truly realistic we would have to fool every one of these senses to achieve a genuine effect. 3D graphics stand in for sight (even possibly with stereoscopic vision), surround sound systems such as Dolby Digital AC-3 can provide accurate placement to sound effect sources, and force feedback controllers can provide some level of tactile response. The last two senses, taste and smell, are obviously of some lesser importance, although some games have made (usually disastrous) attempts at providing a sensation of smell. The infamous *Leather Goddesses of Phobos* text adventure provided a "scratch and sniff" card that the player was instructed to activate at relevant points throughout the game.

In practice, however, we can fool the mind with far less. Cinema has been manipulating people's perceptions of reality for years through careful use of staging, plot, music, lighting, special effects and all manner of trickery, even deliberately depicting events off-camera for different results. Watching a film can be a believable experience even when the narrative is not explicitly represented sequentially or in real time, or even on camera at all. One of the most important stages of the film making process is editing, where decisions are made on what to leave in and what to remove from the narrative. Shiny Entertainment's Dave Perry supports this philosophy: "Making games is about knowing what the player needs to see and feel, and more importantly, what they *don't* need to see and feel". Perry goes on to say that *MDK* purposefully presents the action in third person so that the player can actually see *Kurt Hectic* getting injured, in a similar fashion to watching a character in a film, rather than try to convey this feeling in first-person mode.

There are even some issues central to interactivity that genuinely question a need for absolute realism. Any human–computer interface needs to be designed around metaphors to aid understanding and ease of use. These metaphors can be quite abstract and not necessarily related to a specific element of the real world. An interface also needs to be designed for ease of use. Imagine trying to access a piece of knowledge in a virtual environment. In an *absolutely* realistic simulation of our own world, the user would have to travel to a library, find a book or CD-ROM containing the information, then flip through the pages with a virtual hand or type on a virtual keyboard to access the files. If this information is required quickly then it would be far more practical to be able to access the information via an abstracted interface – for instance, a HUD (head-up display) system overlaid onto the rendering of the virtual scene, or some kind of all-knowing virtual PDA (personal data assistant).

In certain instances, a game can indeed be far more enjoyable if it purposefully presents *un*realistic situations. Some racing games have convoluted AI routines that frequently "cheat", for instance repositioning cars at strategic points along the track to provide a continual stream of opponents for the player to compete with (the original SNES incarnation of *Super Mario Kart* being a fine example of this). This can sometimes be a far more exciting experience than being able to quickly assume the pole position at the front of a race, and thus not have the opportunity to interact with another vehicle for many more laps to come.

This points to one of the most effective ways of enhancing the believability of a game – behavior routines, or *artificial intelligence*. If we can truly "buy" the believability of a computer-controlled character's behavior, we will probably find the experience of interacting with them a realistic one. The enemies in *Half Life* and *Unreal* react to the player's presence in a radically different way from those previously seen in *Doom* or *Quake*. The newer breed of first-person shooters actually depicts opponents appearing to communicate with each other, using recognizable tactics to defend and attack. The monsters of *Doom* merely awaken when the player is heard or seen, and then automatically attack via the route of least resistance. Of course, this apparently new-found intelligence is just another programming trick, based upon a series of rules and attributes, but it can be a very effective one.

More advanced AI routines might even eliminate the common "skill setting" choices that can be found on many of today's games. How is a player expected to know whether they will get the most fulfillment from choosing "easy", "normal" or "hard" from a menu before they have even tried the game? A more advanced AI, which reacts to how well the player is doing, will enable opponents to respond at a level judged challenging enough to give the best experience. This is not a new idea, being used very effectively in Konami's *Gradius* many years ago. *Gradius*' level of aggression toward the player varied radically based on how well the player was doing in terms of powerups collected and scoring potential. Some fighting games can "learn" from a given player's style of attack and respond accordingly. But this sort of system has a long way to go before games can really react quickly to advanced tactics, or tricks such as sudden changes in the style of playing, or even multiple players taking turns on the same game.

1.4 Player Control and Freedom of Choice

The degree of control a player feels to be exerting over the situations in a game can also have a strong effect on the believability. Even a 2D platformer that allows complete freedom of movement in all scrolling directions will seem to be far more immersive than one with purely right-to-left, fixed-speed scrolling.

As games have developed, the idea occurred of offering multiple routes through a game, sometimes resulting in a different ending sequence. Sega's *Out Run* and the *Darius* shooting games from Taito both pioneered the use of a simple branching system where a choice of "left" or "right" at the end of each stage would allow traversal through an expanding choice of alternative routes through to the end of the challenge. A little later, Treasure's *Guardian Heroes* had over 50 distinct pathways through the game, each player choice unlocking a slightly different version of the story.

Nowadays of course, many games are far less linear still, with a player's choices and actions apparently writing the story of the game as progress is made. Modern role-playing games (RPGs) have expanded on this kind of idea the most, featuring plots that seemingly unfold before the player's eyes. Charles Cecil of Revolution Software (famous for the *Broken Sword* RPGs) says that this kind of interactive experience consists of three components:

- immediate appeal (based on conventions);
- primary gameplay (mechanics); and
- narrative structure (content).

Very often, narrative does not mix well with freedom of control (interactivity). Narrative tells a sequence of story events in an artistically authored form, whereas freedom of control (by definition) dictates that these decisions are instead up to the player. Thus, most modern games featuring narrative are made up from "knots", which literally tie up plot threads at critical points throughout a game's structure. In this way, all available branches and choices ultimately meet back up again at the next central plot point, then branch back out again offering more alternatives that converge again at the next point. In this way, an overall pre-defined narrative is possible, with many different "sub-plots" being written by the player.

One of the most noticeable stumbling blocks has always been achieving realistic interaction with other characters in a game, beyond merely trying to kill them. Handling of conversations has long been a problem, with solutions often resulting in menu-driven "multiple-choice" question and answer systems (as featured in *Monkey Island*, then handled slightly differently in Oddworld Inhabitants' *Abe* games). The days of truly conversing with a computer-controlled character are still some way off and largely depend upon relatively undeveloped or emerging technologies such as speech recognition, language interpretation and higher orders of AI.

1.5 Cinematic Influences

The fusion of film and game is often compared to water and oil. The view is that [they] never mix – they are of different properties and different dimensions. As proof of such thinking, most games out there go only as far as linking the movie and game portions. However... by breaking down film elements (story, theme, direction, camera work, lighting, etc.) digitally,

incorporating them in the game media, and unifying them, the fusion c
becomes complete.

Hideo Kojima, Konami, director o

Many modern games have looked to the cinema for inspiration and expressi...
The obvious film tie-ins abound, often poor licenses living off the strength of the film's name. (The biggest exception to this being Rare's excellent *GoldenEye*.) Some games even make the journey the opposite way, onto the big screen from interactive origins, although the first one of these to actually make a profit at the box office was *Mortal Kombat*.

Original game titles such as *Metal Gear Solid*, the *Resident Evil* and *Final Fantasy* series, *Parasite Eve*, *Dino Crisis* and *Silent Hill* have all been well received both commercially and critically. Some of these games are actually a lot like interactive blockbuster movies with carefully storyboarded, cinematic use of camera, exciting set pieces and spectacular effects-laiden graphics.

Metal Gear Solid is one of the most notable titles to lead the player through a singular fixed plot, featuring incredible attention to detail in the environments and level designs, precisely because it does not have a freely explorable structure. The limits set on the narrative and design have allowed the careful fine-tuning of the tightly storyboarded, highly directed gameplay. In fact, many entirely free-roaming games can get tiresome at times, with the player uncertain of where to go or what to do next, continually retreading old ground and exhausting every possibility until a route of progression can be discovered. Thus, the level of freedom can sometimes be limited in favor of a narrower, tighter sense of cinematic or plot. In other words, in cinematically planned or narrative-driven, authored games, the player is often given only the impression of control.

In sharp contrast to these, games like *Quake III Arena* deliberately eschew any trace of storyline, cinematic, plot or even a true single-player "story" mode in favor of a purely multiplayer interactive experience. Even in single-player mode, the game will simulate other players in traditional *Deathmatch* style using newly developed AI routines.

Films and film-like games are manipulative and controlled. Games based on fully interactive systems are manipulable and controllable.

1.6 Emotion

Getting an emotional reaction from players pays off big ... Any game that doesn't evoke reactions simply will not work.

John Romero, ex-id Software and co-founder of Ion Storm.

One of the most important aims when designing a game is to make it an emotional experience. After all, if it is not enjoyable to play, then no one will want to play it, regardless of how pretty the graphics might be (although spectacular graphics can of course be used to generate excitement). The type of enjoyment will need to vary depending on the intended game experience. A fast action-based game needs to be exciting, giving the player an adrenaline rush, while a turn-based strategy game needs to present a deeply engrossing mental challenge.

The *Final Fantasy* series has been particularly successful with regard to portraying character and getting the player involved in a very personal sense. Killing off a well-developed character deep into the story of *Final Fantasy VII* was a bold move, akin to Hitchcock's decision to kill off the lead character in *Psycho* – something that the audience was certainly not expecting at that point in the history of cinema. This is, however, fundamentally different to a player getting killed in a less character-oriented game such as *Quake*, where they can then just restart from the last saved point with no real sense of loss and an infinite supply of lives. Reality is not as forgiving.

This is just one of the as yet unsolved questions in games – if death is indeed trivial, if we can always have another chance, then how do we make life a valuable commodity? We also never have any real consequences for our actions, so the lives of other in-game characters are likewise instantly devalued.

It is well worth realizing that visceral emotions (e.g. aggressive excitement through a violent game) are considerably easier to evoke than profound or subtle emotions. Too few games have the power to move a player in the same sense that an emotionally charged, dramatic film could.

1.7 Character

You don't want to think about what your character might say first thing in the morning, but you have *to think about it, and give them an attitude about everything in life.*

Dave Perry, Shiny Entertainment.

One of the most effective ways of holding the player's attention is to create an appealing character. Just as we care more about characters in films if we admire them or can connect with them in some way, we can become more immersed in the role of a character on screen if we truly empathize with them.

1980s *Pac-Man* was possibly the first easily recognizable game character, quite a feat given that he was initially represented in only a single color – bright yellow. Pac-Man's creator at Namco, Toru Iwatani, was allegedly inspired by a pizza eating experience not only to make a game about eating, which he felt could appeal to a wider audience than shooting games, but also to base the very design of his character on his partly eaten pizza.

Shigeru Miyamoto was the next to bring a widely accepted game character to the public in the form of a middle-aged plumber from Italy. *Mario* (initially intended to be named Jumpman) was born as the hero in *Donkey Kong*. This was released only a year after Pac-Man, and although he is much more complex than his predecessor, Miyamoto says that his design was largely based on solutions to graphical display problems. His distinctive blue and red dungarees were simply to give his arms a clearly different color to the rest of his body, so he would be easier to animate in so few pixels. The mustache clearly separates his nose and mouth, and the addition of a hat hides difficult hair detailing. Nintendo went on to develop Donkey Kong, as well as his extended family as fully fleshed-out characters through a long series of games, and later added a new hero, *Yoshi* the dragon. Yoshi was initially limited to helping Mario in his fourth starring game (*Super Mario World*), but then became the star of his own successful games across Nintendo formats.

Part animal, part human *Fox McCloud* debuts in Nintendo's *Starfox* (*Starwing/Lylat Wars*) games along with several other human–animal hybrids. Also from Nintendo, the elf-like *Link* appears in the hit *Zelda* games. It should be noted, however, that even Nintendo are not infallible when it comes to character design. *F-Zero* introduced *Captain Falcon*, intended to be Nintendo's new flagship character for their brand new Super Famicom console, but for some reason we never saw the blue-suited racing hero again.

Sonic the Hedgehog, designed by Yuji Naka of Sega Japan, is another strong character. Sonic's creator describes him as "a reliable friend to young people". Sonic is 16 years old, his running shoes suggest speed (as does his name) and his spikes are styled to become both speed lines and defensive weapons during the game. The inspiration for Sonic was allegedly Felix the Cat crossed with Mickey Mouse.

A somewhat similar marsupial appeared on the PlayStation in 1996 in the form of *Crash Bandicoot*. Extremely popular as a character in Japan, the Western-designed bandicoot also has a relaxed attitude as clearly expressed by his casual jeans and tennis shoes. Crash was among the first of the polygonal characters to have distinct facial expression animations, hence his oversized head to let them be easily seen. Jason Rubin is credited with designing Crash in conjunction with Warner Interactive.

Core Design's *Tomb Raider* gave us a new type of heroine – *Lara Croft*. Intended to be clearly upper class and British, and based upon "Indiana Jones, Tank Girl and Neneh Cherry", Lara became very famous very quickly. Toby Gard, Lara's designer, says that a man will feel as though he is protecting her when guiding her through the caverns, whereas a woman will assume her role more literally. He freely admits that her chest is her "chief asset", and the shades give a toughened look. Lara's backpack not only helps to define the exploring side of the character, but also helps hide flaws in the earlier versions of the 3D model.

Other famous characters worthy of mention are *Gordon Freeman* from *Half Life* (whom we interestingly never see), *PaRappa*, *Duke Nukem*, and *Cloud* and *Squall* from *Final Fantasy VII* and *VIII*. Also a myriad of beautifully designed individuals from many fighting games including the *Street Fighter*, *Tekken*, *Mortal Kombat*, *Virtua Fighter* and *Soul Calibur* series.

Most of these are now available to buy in action figure or cuddly toy incarnations.

1.8 The Virtual 3D World

One of the biggest leaps in gaming has been the move into the third dimension. Sometimes this is only a graphical move, with the game itself being played out in a 2D space, and sometimes we can interact with a true 3D virtual world.

Many modern games are still 2D in terms of gameplay despite the 3D polygonal graphics. *Pandemonium*, *Einhander* and *Street Fighter EX Plus Alpha* are all 2D games underneath the 3D rendered images. One reason to keep the 2D gameplay intact is not to change a successful formula. Another reason is that interaction with a 2D projection of a 3D world (i.e. a flat rendered image) is not an easy thing to implement successfully.

In the case of the first-person shooting genre, a lot of the problems inherent to 3D game design are avoided. First, there is no need to worry about camera issues such as framing or keeping the camera focused on the player's character, as the camera *is* the player's character, and is implicitly controlled by the user. Precise, analogue mouse control is most often used to mimic the turning of the player's head, while cursor keys provide intuitive and independent movement for the feet.

Racing games are one of the most logical genres to have been fully updated into 3D. The design of detailed outdoor landscapes has been pushed forward a lot since the first 2D, sprite-based racers. Again, handling of the camera can be fairly straightforward, as the action is usually viewed either through the eyes of the driver (first person), or from slightly above and behind the vehicle, looking forward across the roof of the car itself (third person). The only concern in third-person mode is to keep the car and the oncoming road adequately in frame, possibly even anticipating turns slightly before they arrive. Specialist analogue control systems are also now fairly intuitive, with steering wheel add-ons available at a price.

Space- and air-based first-person or flying games (such as *Elite*, *Colony Wars*, *Starfox* and all flight simulators) also have the luxury of being largely viewed through the eyes of the pilot, or having definable, fixed relative position "chase" cameras. Modern analogue controllers make for a good input device with diving, climbing and banking all being controlled quite effectively, and roll orientation possible on "shoulder" buttons.

Fighting games have pioneered the representation of the human figure more than any other type of game, mainly because the whole power of the system is often utilized to render just two highly detailed characters on the screen. The first game to do this in true 3D was Sega's *Virtua Fighter*, featuring polygonal characters on a sparse, enclosed, flat arena surface. (Sega did in fact have a pseudo-3D fighting game called *Dark Edge* in arcades before this, but it was based on their sprite scaling technology and not polygonal geometry.) Detailed surroundings that can be fully interacted with are only just starting to be produced in 3D, as in Capcom's recent *Power Stone*. Early attempts at an (almost) first-person fighter proved exciting with *Punch Out* but since then, most fighting games present the action from a side-on view of the two combatants. (*Tekken 2* does have a secret mode with a similar viewpoint system to Punch Out, but this is really only a novelty feature). In almost all fighting games, the camera will circle, cut or dolly whenever necessary to keep the characters framed from a side viewpoint, and the reason for this is immediately obvious. In real life we have two eyes, and rely on our binocular vision to judge distances, thus it is very difficult from a first-person view to judge spatial relationships quickly or accurately enough for this type of game, when viewing a single flat rendered image. Most of the newer fighting games do however allow circling of the opponent, making use of the 3D space.

Perhaps the genres that have struggled the most to make use of 3D have been the platform and action–adventure formats, most often viewed in third person with the character presented fairly large in the middle of the screen. As Tony Lloyd of Argonaut Software recently said, "I'm positive that 3D platformers are in their infancy".

Toby Gard defines a 2D platform game as a combination of agility-based timing, puzzles, secrets, traps, pickups, an intuitive control system and enemies, and that all that can be retained in 3D but needs a lot of re-thinking, as it's no longer presented in a plain side-on scrolling view.

Almost every 3D platform game made to date has some kind of flaw with its camera system, and this is not surprising. Camera handling becomes very difficult when a character can be surrounded by enemies on any side and chased around. How can a camera intelligently cope with this situation, when it becomes necessary to convey information on the area in front of the player (to see where you are going), as well as framing enemies attacking from behind or the sides in complex built-up environments? In addition, when navigating through tight tunnel spaces the camera may often move through walls, which can obscure the character from view. The addition of a collision detection system for the camera can go some way toward improving this, but the character can still get lost behind an item of scenery, and the line of sight between character and be player broken. A fixed-camera view from above and behind is clearly not adaptive enough, so most games provide a manual override system whereby the player can force the camera to circle left or right around the character. In addition, the option is sometimes included to momentarily move in to assume the character's point of view to assess the surrounding area (e.g. *Tomb Raider*, *Metal Gear Solid*).

Super Mario 64 takes this one step further again with the inclusion of a controllable camera operating character, but Shigeru Miyamoto says he would have preferred not to have any manual camera control at all, if it had really been possible to automate the process effectively.

Jason Rubin credits *Crash Bandicoot's* success to its simple camera, viewing the simple play mechanics (a limited view of a limited world). Other games such as *Resident Evil* have entirely fixed views, cutting between cinematically framed, locked off cameras for both artistic and technical reasons.

Control systems too are varied, and range from rotational controls (*Tomb Raider*, *Resident Evil*) to camera orientation-specific directional controls (*Spyro the Dragon*, *Mario 64*). If a game requires speedy navigation through complex pathways, rotational controls can prove too cumbersome. And if a situation calls for precise movement along a straight line when using camera-specific controls, an automated camera which starts to circle the moving character will require the player to constantly adjust the angle that the character needs to travel on screen, in order to keep the trajectory straight. Also, some games will require the player to spot an item or a ledge to jump to at a certain point, so the camera needs to be equipped to let the player see this.

Another problem is that judging spatial relationships is often far more crucial in platform games than any other genre. Jumping across gaps in scenery can be very tough in a third-person 3D view, especially if different strengths of jump are needed to accurately travel the correct distance to the next platform. *Zelda 64* introduces a kind of semi-automated jumping system to help the player out in this respect. Accurate jumping becomes harder still in a first-person game (see *Turok*'s later levels) as the position of the character's usually invisible feet is not at all apparent.

1.9 2D into 3D

The worldwide marketplace has evolved toward the more impressive qualities of 3D space. To some extent, the industry magazines have helped usher the end of 2D products by terming 3D to be the cutting edge. Shaping consumer perceptions that 2D is less sophisticated has effectively limited their commercial appeal.

Scott Steinberg, Crystal Dynamics

It is now quite rare to see a purely 2D game released, but just as in other media, the game designers of today often look back in time for their inspiration.

Quite a few of the most popular classic games have been translated into 3D with varying degrees of success. Often, the flat 2D gameplay does not sit well within a 3D environment. Many recent updates of classic games have been poorly received both critically and commercially, perhaps because the gameplay has either remained too rigidly fixed to the originals, or the redesign has not actually added anything worthwhile. And sometimes even because the polygonal graphics are all too obviously an attempt to cash in on a previously successful name.

Other updates have been much more successful, notably Shigeru Miyamoto's *Mario* and *Zelda* games on Nintendo 64, both adapted from already acclaimed 2D incarnations running on the SNES. Both games were instantly and widely embraced as the very best of their genres.

Many 3D updates of 2D games are in the pipeline, or have appeared over the last few years: *Super Mario, Zelda* and *Donkey Kong 64, Metal Gear Solid, Gauntlet Legends, Robotron X, Pitfall, Battlezone, Sinistar Unleashed, Castlevania 64, Lemmings 3D, Xevious3D/G, R-Type Delta, G-Darius,* the *Street Fighter* and *Mortal Kombat* fighters, *Sonic Adventure, Solar Assault* (*Gradius*), *Tetrisphere, Speedball, Q*Bert, Centipede, Asteroids, Prince of Persia 3D, Space Invaders,* even *Pong.* Some take precisely the same 2D gameplay and retouch it with modern looking 3D graphics, others look far removed from the originals, taking the basic concept and completely reworking it to suit a full 3D environment.

Compare the amount of reworking done on *Mario* or *Zelda 64, Metal Gear Solid* or *Battlezone* to the work done on the *Frogger, Mortal Kombat, Asteroids* or *Robotron* updates. Clearly some fundamental changes need to take place, not only for the general public to accept it as a worthwhile title, but for the gameplay to work properly in a full 3D environment. A thoughtful and cohesive design process will rework the title from the ground up, not just bolt on polygonal graphics.

All of these are just the officially licensed updates – ideas for 3D games often come from 2D ones. *Block Out* was obviously *Tetris* in a 3D cube, but just imagine *Gauntlet* played from a first-person view and you might visualize something not completely unlike *Doom.*

1.10 A Case Study: Comparing the 2D and 3D *Super Mario* Incarnations

Shigeru Miyamoto's two platform games are probably the most successful platformers to date in terms of fully exploiting the possibilities of 2D and 3D gaming, respectively.

1.10.1 *Super Mario World* – Side-scrolling 2D Platform Adventure for the Super Nintendo Console

- The game objective is to rescue the Princess who has been kidnapped by the evil Bowser.
- Mario can move anywhere within the boundaries of each level, which are typically rectangular and many times the area of the TV screen. Most need to be negotiated from left to right, and very occasionally the scrolling happens at a forced, non-controllable rate (used as a specific gameplay feature).
- Mario can walk, run, jump (both to negotiate platforms and crush opponents), spin-jump, swim, pick up and throw some objects and enemies, climb fences and punch whilst climbing. After collection of various special items, Mario can throw fireballs, fly around in a limited fashion, float through the air and ram the ground to defeat multiple enemies. An extra character, Yoshi the dragon, can be found at many points throughout the levels and can be utilized to eat enemies, spit them back out (sometimes as fireballs), freely fly around and crush stronger enemies that Mario cannot defeat by himself.
- Coins and stars can be collected, leading to bonus points and extra lives being obtained. Boss characters lie at the end of certain levels and must be defeated to progress further.

- Level access is generally quite linear, with some shortcut or selectable routes that are initially kept secret from the player. Once any area is completed, the player can return to it to search for hidden areas or replay for extra points and coins. An island map contains all the various stages, with extra pathways opening up to Mario after completion of each stage.
- Levels are often themed and this can affect the gameplay – for instance, icy levels have Mario sliding around with greatly increased inertia.
- Early stages are simple in design and require a minimal amount of skill from the player. As the courses progress, more and more of Mario's moves and special abilities need to be learned and exploited to conquer the challenges. The player is often guided or given hints on how to perform these new moves the first time they are required.
- Very precise control is needed over Mario to negotiate many of the platform-based stages. Learning to control the character accurately is of prime importance to doing well in the game. Flying around as Super Mario is one of the trickiest elements to master but yields the greatest rewards, allowing travel to parts of levels that are not otherwise accessible. These areas often contain bonus items or large numbers of coins, which can be collected for extra lives.
- Several special stages called Switch Palaces are hidden within the game, and completion of each will unlock special blocks, which when smashed allow further features to be found in the standard levels, so that they may be played over again with more content there to be enjoyed.
- On most levels the scrolling window can be moved left or right by the player to view a little more of the environment immediately surrounding Mario. This allows more of the playfield (either ahead of or behind Mario) to be seen.

1.10.2 *Super Mario 64* – Third-Person 3D Platform Adventure for the Nintendo 64 Console

- The objective of the game is to once again defeat the evil Bowser and rescue Mario's friends.
- Mario has full directional control within the boundaries of each contained level. He has more abilities this time, including: walk, run, jump, crouch, backflip, bounce off walls, climb, dive, slide, swim, pick up and throw or jump onto enemies, perform superleaps, smash the ground to disable multiple enemies, and grab larger opponents and swing them around. Special abilities can be obtained which include flying, walking underwater or through solid objects.
- Coins can be collected for bonus points and lives, and there are multiple objectives for each stage, including collecting stars or negotiating a path through each environment to defeat a boss character. Further, more specific objectives are announced by the various characters within each level and include races that need to be won and the retrieval of specific items or other characters.
- Access to levels is not at all linear, however a number of items need to be collected from the initially available stages in order to unlock the more advanced courses. A castle provides the central "hub" for the courses, which branch off from various locations within the castle, the entrances to which are sometimes hidden.

- Each world has a graphical theme, which also determines different gameplay characteristics for the level: water, volcanic, snow, caves, haunted mansion, etc.
- Early stages are simple in design and require a minimal amount of skill from the player. As the courses progress, more of Mario's moves and special abilities need to be learned and exploited to conquer the challenges. The player is again often guided, or given hints on how to perform these the first time they are required from other characters or by reading information in the levels.
- Very precise control is needed over Mario to negotiate many of the more platform-based stages. Learning to accurately control the character is of prime importance to doing well in the game.
- Several secret stages are hidden within the game, and completion of each will further unlock features in the standard levels, so that they may be played over again with even more features to be enjoyed.
- Camera control is a mixture of automated and user-controlled, offered in selectable near and far modes, each with switchable rear, left or right side perspectives.

1.10.3 *Mario World* to *Mario 64* – Changes Made to Facilitate the Move to 3D

As we can see from the above descriptions, the 3D version of Mario is conceptually very similar to the 2D version. However, there have been a few noticeable changes, other than the obvious differences in graphics.

The use of analogue control is the first noticeable difference. On a 2D screen, digital left and right motions on the digital pad (D-pad) are appropriate, with an action button held down to run faster. This makes perfect sense when presented with the task of moving precisely left or right through a level. In the 3D version, the eight possible directions on a digital pad have given way to 360° analogue control, and the further the pad is pushed in any direction, the faster Mario will run. This generally makes for a much more natural, intuitive and precise control system for negotiating a ground surface presented in perspective. Note that the directional control is always dependent upon the camera position and orientation, with all movements for Mario effectively working in eye space coordinates. This is a good choice of control system, but can occasionally present some difficulties when the camera is moving or rotating and the player needs to keep Mario's trajectory constant within world space.

Complete freedom of movement within each level has been maintained, but the shapes of the levels now vary much more. Levels are no longer just rectangular in design, and of course all three dimensions are made use of. Platforms now have a width, rather than just a length and height. This makes for a feeling of much greater freedom of movement, but also means that absolute mastery of the 3D control system is required. For example, when jumping to attack an opponent, you have to deal with aiming to land on a 3D target rather than a 2D one. This also means that exploration is now much more exciting with objects often obscured by others until a pathway corner is turned, dramatically revealing a new and often expansive area.

The variable camera control makes judging spatial relationships much easier than if the player were limited to a fixed view. By rotating the camera around Mario to one side, jumping distances can be assessed from a good angle, and the zoom feature presents either a close-up view for precision movement, or a wider angle for keeping an eye on much more of the surrounding area.

The addition of different missions within the levels also adds more variety to the game system. Ideas are borrowed from other genres such as racing games, but crucially these are presented as integrated parts of the main game rather than disjointed, separate levels. Some of these missions also make good use of the potential of 3D, such as launching Mario from the cannons which makes use of a first-person perspective view for aiming purposes.

Characterization is also brought to the fore with many different speaking characters within the levels, who can either be friendly (giving advice and hints) or aggressive.

But the real reasons that Mario can succeed as a 3D platformer where so many others fail, is that it retains the most important aspects of a 2D platformer and adapts them to sit in a 3D environment: simplicity of the central play system, speed and responsiveness, and the challenge of precise control being absolutely necessary to progress through the game, but (very importantly) without being unfair to the player. It could perhaps be viewed as quite an easy game to upgrade, as the original version was the fourth in a series that was already very finely tuned, and the top of its class. However, all new changes have also clearly been very finely tuned and tested, with good compromises made to the control and camera systems, in particular. Despite all the extra missions and features, the most important aspects of a good platform game are still there – ease of learning, responsiveness, precision control, simplicity of the central design and the variously themed terrains determining the subtle or pronounced variations in gameplay.

Mario 64 was also very carefully designed and planned out, not just in terms of level design but also down to the actual units of the game world. Mario's length of jump, speed of movement and the precise scale of the scenery were all based on specific measurements against his height to ensure it would play consistently, with all jumps and walkway paths mathematically planned out. (The same process was applied to a much simpler, but nonetheless successful platform game, Naughty Dog's *Crash Bandicoot*. Crash is precisely one "Crash Unit" wide and deep, and two units tall, while all the other objects in his world relate to a carefully planned proportion of this.)

Shigeru Miyamoto and Nintendo's attention to gameplay is clearly visible in all their other games: *F-Zero* **X** successfully transcends the entirely faked 3D perspective from the SNES version into fully polygonal 3D tracks for the N64. The designers understood that responsiveness and speed were the two main ingredients that made the SNES original successful, so the design of the N64 version is deliberately sparse in terms of 3D graphical detail. This enables incredibly fast rendering, and thus it is one of the few games on the N64 that runs at an arcade-like 60 fps, which was seen as far more important to the game's handling and feel than including attractive scenery.

Miyamoto's N64 title *Legend of Zelda Ocarina of Time* was also adapted from a 2D incarnation on the SNES. However this 3D version is much more of a complete redesign, and exhibits possibly the finest examples of attention to detail, characterization and feeling of place yet seen within a videogame.

1.11 A second case study – *Robotron X*

By way of contrast, compare all this care and attention to detail to the conversion work done on a less successful 3D update such as Midway's 1996 PlayStation version of *Robotron X*, and we see a huge difference. Superficially, it still looks like *Robotron*, but several important aspects that made the original so playable have been lost.

First, the frame rate has dropped drastically, making it difficult to stay aware of multiple fast moving targets. On busy levels it is not uncommon to see characters or missiles moving as much as around 20 pixels or more between frames. This ultimately leads to a feeling of unfairness when an enemy appears next to your character without warning, usually meaning a life is lost in an instant.

Secondly, having extra waves of enemies "warping-in" throughout the duration of a level (an idea ironically used first in *Defender*) serves little purpose, and seems in this case more of an attempt to control the on-screen polygon count rather than being a tactical gameplay feature. (This is a big change from the first incarnation of *Robotron*, which shows the player complete information about the current level as soon as it is encountered.)

Typical of such an update (see recent versions of *Asteroids*, etc.), the simplicity and purity of the central game design has been compromised with the inclusion of token power-up weaponry, which only serves to tip the balance of play.

Finally, neither of the two camera views offered to the player works very well. The scrolling three-quarter 3D view does not show enough of the playfield on screen at once to keep the player informed on the state of the action (the original sensibly had the entire playfield on a single screen). And as the whole game mechanics is based around eight-way digital control, the 3D perspective foreshortening actually detracts from the intuitiveness, further hampered by constant camera rotation as the character is guided from side to side. The second view offered, an overhead plan view, works better but makes *Robotron X* feel like a version of the original that is running far too slowly. (And what is the real point of having a 3D world if it is only to be viewed in a plan orthographic projection?) Being a 3D projection, this view also makes identifying other characters slightly harder – remember, in the original, even though the game is presented in plan view, we see side elevations of the character sprites which makes them instantly recognizable.

While certainly not being a bad game by any means, this particular 3D update is certainly a missed opportunity.

Other attempts at *Robotron*-like games include *Llamatron*, Jeff Minter's reworking of the *Robotron* theme for the PC (which is available to download for free). This remains 2D and offers a far more exciting challenge than Midway's 3D update, integrating several new features whilst remaining true to the original's minimalist game mechanics.

Finally, Eugene Jarvis' own *Smash TV* takes the same premise and introduces full color bitmap graphics, simultaneous two-player co-operation, bosses and non-linear progression through a branching map grid of single-screen levels. The plot concerns a game show as featured in *The Running Man* film, with cash and prizes standing in for *Robotron*'s humanoids that needed to be saved in the original version of Jarvis' concept.

1.12 Approaches to Game Design I – Pre-production and Planning

1.12.1 The Design Document

Once an idea is found and approved for pre-production, it is nowadays usual for a team to produce a *design document*. The degree to which this document is actually used during the production phase varies wildly between companies. For larger productions involving dozens of team members, it can be an invaluable design "bible". Smaller projects with fewer team members can much more easily evolve continually throughout the actual production cycle, so for this kind of environment the design document can often be more loosely adhered to.

Whatever the practice of the company, to make further use of the document, the first purpose it will be put to is to gain funding or approval for production of the game. In today's market, games are very expensive and time consuming to produce, so often a smaller developer will use the design document to secure backing from a larger publisher. The games industry is not completely unlike the film industry in this respect, with treatments being pitched to studios for the same purpose – to secure funding and distribution for the finished product. For this very reason, most documents end up being quite thoroughly planned out, polished and well presented before being handed to a prospective publisher. Often this can even happen well into the actual production cycle of the game, if the developer has funded the project by themselves thus far. And in a larger company that develops its own games in-house, the design document serves as a pitch from the team to the management.

The exact structure of the document can vary between companies, but typically a modern design document contains the following information.

Summary

Containing:

- **Introduction.** Briefly describing the scenario and any USPs (unique selling points).
- **Programming, Art and Sound Briefs.** A brief outline of the requirements within each of these categories. For example, whether existing engines and editors are to be used or new ones need to be written, the graphic style and the types of music to be included.
- **Risk assessment.** Describing anticipated problems or resource issues. For instance, if experienced programmers have not yet been hired to make up a complete team, or technical unknowns such as writing for a new platform, which has not yet been commercially released.

Game Overview

Describing:

- **Storyline.** Up to the point whereby the player takes the lead role.
- **Player characters.** Visual designs and biographies.
- **Game structure.** A breakdown of the levels and the way these are to be accessed (linear or non-linear progression). Usually containing flow diagrams depicting the traversal possibilities between the different areas.
- **Camera and characters.** Control systems, interaction and combat techniques, and a list of NPCs (non-player characters) together with their attributes or special abilities.
- **Level descriptions.** Overviews of aims or objectives, themes, architecture (tunnel/room/exterior-based), lighting and any special gameplay features. Also which characters or creatures inhabit the levels, any bosses to be defeated at the end, and a list of items to be found on the level together with their uses.
- **Multiplayer modes.** Co-operative, deathmatch, head-to-head, etc.

Appendices

Explaining specific aspects of the game in detail as necessary. For instance:

- **Camera control.** Strategies for getting the most from the camera system based around the gameplay requirements, cinematic and architectural features. Also the specifics of automation or user control.
- **Production schedule.** Outlining the amounts of time allotted to each aspect of the production. Such as prototyping, modeling, animation cycles, engine, tools and editors, AI, sound effects and music, collision detection, damage systems, documentation, level design and production, project reviews and final testing and approval stages.

Other headings included in the appendices can be specific to a particular genre; for instance, a detailed description of the combat system if the game is a fighting game.

1.13 Approaches to Game Design II – Production Issues

1.13.1 The Target System

There are still plenty of titles that are developed exclusively for one hardware system, but most games are now developed for multiple platforms to maximize financial returns. Whilst the development teams can make efforts to ensure that all versions look, sound and play as close to each other as possible, the differences in the hardware technologies and input devices of the different target systems will inevitably mean that compromises will have to be made at some point. Conversions (or ports) between systems are now commonplace, and are sometimes handled by a different studio or developer; for instance, after a game has already proven financially successful on one particular platform. The first game to be ported was *Miner 2049er*, which originally traveled from the Atari 400 to the other systems of the time.

PC and Dreamcast titles are often reproduced in cut-down graphic form on the PlayStation or N64. Reduction in the detail of polygonal models is obviously one of the major compromises. The PlayStation is said to realistically handle somewhere around 2500 rendered polygons on screen in a full game (roughly the same as the Saturn and N64), although this does vary a great deal depending upon the game requirements, the exact frame rate required and the skill of the production team. A typical PC can now handle well over ten times this number, and the PlayStation 2 and Microsoft consoles can cope with even more.

Reduction is also normally required on texture resolution and color depth, as current consoles have only a fraction of the memory of a reasonably equipped modern PC. Conversions to the Gameboy handheld almost always require a drastic redesign of the entire game format, as it is not equipped to cope with 3D at all.

1.13.2 Tricks of the Trade

All videogame programming is about pulling the wool over people's eyes.

Mark Atkinson, Computer Artworks.

A game *engine* will normally be optimized for a specific kind of game and environment – racing track, first-person indoor corridor maze system, third-person exterior hill landscapes, deep space battle simulation, and so on. The reason for this is to achieve the best performance possible, as a more generic approach will undoubtedly waste computation time running standardized pipelines. Making a renderer as efficient and as fast as possible may ultimately require making use of specifically available hardware features, making conversions to other platforms even more difficult. But there are usually ways to enhance performance by taking shortcuts or "cheating". This can be fairly cheap and obvious, such as only modeling the parts of the environment that the player will actually see. This obviously depends very much on the type of camera system to be employed and will not work for games that feature complete camera freedom within a 3D environment.

Keeping an engine efficient, that is able to calculate up to 60 frames per second, means keeping the amount of information being processed every frame as small as possible. Models that seem to be intricately detailed are often represented in a series of *levels of detail* (LODs). These are then switched between according to the model's distance from the camera. Objects far away, which are rendered as only a few small pixels on screen, are usually extremely simple shapes. As they get closer to us, more and more complex models can be substituted until the highest resolution mesh is presented filling the whole screen.

Just as geometry is usually presented at different resolutions, so too are texture maps. Several versions of the same bitmap texture are usually stored, and switched based on the distance from the viewer. This technique is known as *MIP-mapping*.

Racing games commonly use about five resolutions of both cars and texture maps. In the very far distance, cars are usually just cubes with a very simple texture applied to each face. *Metal Gear Solid*'s Solid Snake has more than twice this LOD, with many models being utilized depending on the type of camera view being used.

Rendering can be made more efficient by using techniques such as potentially visible sets. Lighting too is often a mixture of real and pre-calculated, using bitmap texturing and vertex coloring to simulate environment shading. Even in-game physics can now be pre-calculated to some degree.

1.13.3 Design Tools

All games with a visual appearance on screen will require a 2D package in order to produce artwork. In addition, 3D games obviously require a 3D package to enable production of models, application of textures and of course animation sequences.

In a typical production environment, a range of unique tools, designed in-house, augments the off-the-shelf software systems. This is sometimes borne out of necessity because the tools do not exist commercially, and sometimes because a custom-designed tool can make a task easier or more efficient for the designers. For instance, most commercial systems have export filters to convert 3D geometry files to a variety of formats (including real-time libraries) but developers still often write their own tools. This makes it possible to maximize the efficiency and range of features when converting models from a commercial system to an in-game element. This may be due to the type of file format or animation system being employed within the game, or to make special or innovative uses of model attributes. With games technology advancing far more rapidly than large commercial packages can be modified, tested and released, this situation may well continue for some time to come.

Other more complex tools are often used for level design. First-person shooters are well known for this, with level editors being given away on the game CD for players to use at home, enabling new levels and characters to be designed, made and then put into the game. Such systems are often based around a technique of carving a tunnel system out of solid building blocks of space using Boolean operations. This can provide the game engine with much more detailed information about the boundaries of the environment (for essential uses such as collision detection and visibility systems) than level geometry imported from a more generalized 3D package could.

Racing games also typically have very specialized track editors, usually based around curve editors and a cross-section of geometry which can be varied at points around the curve. Track-side geometry can then be imported as objects and placed around the curve. Again, this type of system can more readily contain useful information for the game to utilize, such as which scenery objects needs to be visible at specific points around the track. With a good editor of this type, environment texturing can also be very quick to perform.

Some commercial tools are becoming available that are designed not only to create graphical content, but to cope with interaction issues as well. *MultiGen* and *GameGen* have been used by some developers, mostly for prototyping. *Motivate* and *Nemo* are two newer products that aim to address character interaction issues. Alias now directly provide several modules for their Maya software (developed mostly by third-party companies) tailored specifically toward games development, such as Granny, Alienbrain and RenderVision. Criterion Software supply the multi-platform *RenderWare* for real-time graphics rendering, and another system for speech interaction. Argonaut are working on advanced in-game lighting systems, and even real-time physics can now be more readily prototyped with the *MathEngine* system.

1.14 Superconsoles and Modern Game Design

The past couple of years have seen both radical changes and also a stabilization of sorts in the way that games are produced. The advances in graphical power continue to accelerate in line with Moore's law, if not faster still, and regardless of their advancing power, new systems are being released to the public at around the same initial cost each time. The number of platforms in popular commercial use at any one time has not really changed since the beginning of the home games market, with a small handful of consoles and the PC compatible dominating. Some of the players in the market have changed though. Sony has of course been the biggest company to enter the games market in the last decade, and has helped transform the medium from essentially a schoolboy's arcade activity into a huge, home-entertainment-based commerce with trendy mass-market appeal. Sega, one of the industry's founding giants and traditionally centered on the interests of the "hardcore gamer" has recently decided to opt out of the hardware market and concentrate solely on software development. Nintendo are still going strong, with their hopes pinned on the success of their new Gamecube console, as well as their staple handheld market with the Gameboy Advance. Microsoft, who have profited greatly from the PC software market for some time, are now after a slice of the hardware market too with their X-Box console, essentially a fixed-specification, low-cost PC in a single off-the-shelf box. Clearly, they are hoping to enter the market and repeat some of the success that Sony found with their original PlayStation. They certainly have the marketing resources, but only time will tell whether the games playing public will choose a PC-based system for a console, or continue to choose either a PC or a console for their recreational machine of choice.

Overall there has been no real degree of convergence in the number of hardware platforms as some people may have predicted, although there has been convergence in some types of technologies – many PCs and games systems now double up as a DVD player, for instance. Most games continue to be released on at least two or three different platforms. Essentially, this means that either they need to be developed to be as machine-independent as possible to allow rapid compiling to workable versions on several platforms, or they are written specifically for one format, and then ported across to one or more others. Both approaches have their advantages and disadvantages, of course.

A game written using a generic engine and a fairly standard 3D renderer will require a relatively small amount of effort to compile for a new system, but consequently will not make use of any machine-specific attributes that can be used to its advantage. This means that using standardized techniques can very often compromise the graphical performance. In today's market, with several popular machines being relatively equal in terms of their computing power, many developers choose this route. Their hope is that the majority of gamers will not notice if the frame rate is not quite up to speed when compared with some other titles on the system, or the texture resolution is not quite as high as it could be. One argument that does support this approach is that, with systems having finally become built primarily upon 3D technology, rather than 2D with "tricks" to fake a 3D look, it is less necessary to take advantage of platform-specific trickery. It is also clearly more prudent and ultimately cost-effective to develop with multiple platforms in mind, as profits can escalate with the successful release of a new version. In addition, it can make a good deal of sense for a smaller developer to make use of a third-party engine or renderer so as to minimize the length of the production schedule, and hence the cost of the investment.

Conversely, games written for one specific platform can take advantage of all the system's strengths and thus often have a much higher degree of graphical performance when compared with "ported" software. They also tend to be written by developers who work closely with (or are sometimes partly owned by) a hardware manufacturer. This can also lead to a benefit for the system manufacturer – having a degree of exclusivity for a good game, at least for a set period of time after its release, can greatly help boost sales for the system. Of course, in later months the same game can then be released for other platforms after the initial sales level out. Nowadays porting is sometimes undertaken by an entirely different developer from the original one, and can be done with varying degrees of success. One of the most technically successful ports recently was Bullfrog's PlayStation 2 version of *Quake III*, which runs at a fairly consistent 60 Hz frame rate and looks very close indeed to it's parent PC incarnation (albeit without the network play). This type of high standard is very rare indeed for a converted title.

With several large competitors still fighting for the biggest slice of the games market and even new manufacturers entering the scene, the multi-platform situation looks like it will continue for the foreseeable future. In many ways this is very healthy for the games playing public, with competition keeping hardware prices down and each company working hard to develop better systems than the others. It may mean that each new console continues to have a life span of five years before it is deemed obsolete, but those worried about this will continue to stick to their PC systems and their more incremental upgrades.

The biggest problem for developers actually lies not with the multitude of formats (it's always been this way, after all), but with the growing complexity of the formats. As stated before, this allows for increasingly more complex graphics and sound, which are expected to be backed up with more complex game content. And for "more complex", read "more time consuming" and "more expensive": games are now taking longer to produce than ever before, and the investment needed to make a quality title is getting higher all the time. Of course this means that the profits required from sales need to become bigger as well, to sustain the company's running costs. It is a difficult situation, and one that does encourage the use of modular or third-party tools and engines, and multi-platform releases. It is also no wonder that so many titles being released belong to the most popular genres. Luckily of course, it is still possible to be artistically and commercially successful at the same time.

1.15 The Future of Games

We can no doubt expect the ever-accelerating development of technology to continue at an even faster rate in years to come.

In 1997 id's John Carmack predicted that within five years, machines would be able to process over 10 million polygons per second. This has come about even quicker than he expected with the recent release of the PlayStation 2. This is expected to be the fastest single polygonal rendering machine ever made available, although polygon counts in ideal test situations are of course much easier to handle than in practical interactive situations. Argonaut's Jez San says that as a matter of history, these figures are usually five to six times higher than the rates that can be used practically in a real game.

But other developments are expected, other than just raw processing power. Dynamic mesh tessellation is being tested on games currently under development such as Shiny's *Messiah*, where extra levels of detail are added or removed from polygonal meshes on the fly, eliminating the need for rigid LOD model swapping.

Carmack himself is one of the pioneers writing the next generation of games which are beginning to use higher orders of surfaces to represent geometry. Bézier patches can be utilized on selected parts of level geometry and then tessellated into polygons before rendering. The distance from the viewer could even be used to determine how detailed the polygonal mesh needs to be to provide the optimum speed whilst maintaining a smooth visual quality. NURBS representation is another possibility, which is being looked into for use in real-time rendering.

Carmack also foresees a day when even curved geometry can be widely supplemented with digitally sampled objects, just as bitmap textures are digital samples of patches of 2D color: "There's a huge benefit to going to a digitally sampled environment. There are so many calculations that are so much easier there. The engines have a whole lot of elegance to them where you don't have to solve the level of detail separately. They kind of fall out of a raytrace engine into a voxel world. I do think that is where the future will be". Although voxel engines do exist, and have been used in games for some time, the truly widespread use of high-quality 3D sampled geometry may be out of reach for the moment, as no commercial hardware currently exists that is capable of accelerating voxel models in the same way that 3DFX cards accelerate polygon rendering.

Other recent PC accelerator cards such as the Nvidia *GeForce*-based models actually take some of the processing load away from the CPU. As well as rendering polygons to the screen, it contains its own 3D transforms and lighting systems onboard. This could free up more CPU time for curved surface math, mesh tessellation, physics or AI routines.

Texturing enhancements are appearing such as MPEG2-like decompression to save on memory and dynamically adjustable maps, enabling effects such as gunshot wounds, aging or rusting to happen in real time. But textures that have a truly dynamic level of detail are another possibility, which would enhance game graphics, eliminating the over-softening of textures when viewed up close (this would also alleviate MIP-mapping concerns).

Perhaps the fastest area of gaming to develop in the near future is the multiplayer experience. As Internet connections become faster, more data can be sent between computers, opening up faster and more accurate real-time gaming opportunities. Enhanced Internet speeds and widespread broadband connectivity could mean that most gaming experiences become multi-user.

Descent, *Hexen* and *Heretic* all featured support for VR technology, but such stereoscopic eye devices have proven unwieldy and uncomfortable over the past decade, with poor resolutions and display quality. Recent advances in screen miniaturization and enhancements may improve this, as well as making them more affordable. But there is also the problem of motion sickness to overcome, with the body's motion and balance sensors in the ears receiving conflicting information to the eyes.

Conversing with other characters could become a much more natural process with advances in voice recognition software. Of course, there is a vast difference between sampling and generating speech and actually understanding the language being used, and this is where current technology is at its weakest. Equally, the character's AI routines need to be up to the challenge.

Sometime in the future, display technologies may even advance to the point where detailed three-dimensional holograms can be rendered in real time, either on a 3D "screen" or eventually all around the player, making playing games a *truly* immersive experience. Possibly, one-day even display technology will be redundant, with devices being used to directly stimulate images within the brain. This technology could be used to produce visualizations of synthetic worlds that appear every bit as real as our own. Games made in this time will surely be very different from the ones we play today.

Bibliography

Harrison, S. and Dourish, P., Re-placing space: the roles of space and place in collaborative systems. *Proceedings from the Conference on Computer Supported Cooperative Work (CSW)*, 1996.

Tromp, J.G., Presence, telepresence, and immersion; interaction and embodiment in collaborative virtual environments. *Proceedings of FIVE'95 Framework for Immersive Virtual Environment, London, UK*, 1995.

2 Evolutionary Algorithms in Modeling and Animation

Anargyros Sarafopoulos and Bernard F. Buxton

2.1 Introduction

Generating computer animation involves two interwoven components. The first component is the *set of tools* (*tools*) used to generate and render computer animation. Tools consist of software and hardware that allow the creation of abstract geometric models, modification of these models over time, as well as their rendering. The second component is the *sequence of instructions to be carried out by the software and hardware tools*, to generate and render a specific animation sequence, here referred to as an *execution plan* or *execution*. Execution usually resides in the thoughts, story-boards and drawings of the director and animators that have to carry out the task of creating a specific sequence. Most modern software and hardware focus on the creation of *tools* that allow for the generation and rendering of realistic looking models and motion. Modeling and rendering tools are very important as they provide the materials used by the computer animator. Novel tools/materials often create new avenues for visual exploration. However, the quality of an animation sequence depends equally on the quality of the work carried out by the director and animators in terms of using the tools available, i.e. the quality of *execution*.

Sometimes the vision of the director and animator exhausts the capacity of the tools at hand so new tools and materials need to be employed or invented to assist execution. Execution as noted above, formed in the mind of the director and animators, often reduces to a chain of sampling of data (scanning of images or film), button presses, menu selections, and/or evaluation of scripts. However, an execution plan is not conceived in an instant or in a moment of inspiration. The director/animator will usually test and research a large number of forms and motions until they decide on how the final work will be done. The making of an execution plan often proceeds by trial and error, searching through the many options open to the animator in terms of creating a sequence. Animators often compose using a process of progression through trial and error. Models and animation are initially created using a given set of tools. The newly created temporal and spatial forms are typically tested against goals set at the beginning of the development process. The models or animation are then modified in order to fulfill preset requirements more closely and the process is repeated until the development time has been exhausted or the goals have been reached. Often there is a recurrent production cycle, where the fitness of execution is tested against the given set of goals. This results in a gradual improvement of execution plans through a "survival of the fittest" approach to creative design.

A parallel can thus be drawn between organic evolution and the human creative design process. Based on this analogy, new software is starting to emerge that focuses on assisting creative design in engineering (Furuta et al. 1995; Koza et al. 1999) and in the arts (Todd and Latham 1992; Lund et al. 1995; Bentley 1999; Bentley and Corne 2001). This software is being developed as a tool to assist with the invention of new algorithms, or the exploration of complex spaces generated by mathematical or procedural tools and can therefore be seen as a *meta-tool*. Such an evolutionary algorithms (EAs) applied to computer animation are often used as a *meta-tool* to assist in the second important component of computer animation mentioned above, the generation of *execution plans*. Evolutionary algorithms are an interdisciplinary research field that connects computer science, artificial intelligence and biology. EAs are based on simulation of natural evolution on the computer and form a computational search technique. "They combine survival of the fittest with structured yet randomized information exchange to form a search algorithm with some of the innovative flair of human search" (Goldberg 1989). Steven Dawkins describes his seminal work on generating two-dimensional (2D) branching forms (bioforms) using interactive evolution on the computer (Dawkins 1986, 1987). Dawkins' idea was extended and applied to the field of computer graphics by Todd and Latham (1991, 1992) and Sims (1991, 1992, 1993a,b, 1997). Latham and Sims use evolutionary algorithms for the generation of procedural models and animation. Reynolds (1992, 1994a,b,c,d,e) uses EAs for the evolution of the behavior of artificial agents in 2D environments, whilst other researchers present work for the evolution of behavioral animation (Zhao and Wang 1998; Griffiths and Sarafopoulos 1999; Lim and Thalmann 1999). Gruau, Gritz, and Kang describe the evolutionary design of controllers for articulated structures (Gruau 1996a; Gritz and Hahn 1995, 1997; Gritz 1999; Kang et al. 1999). Sims describes the evolution of behavior and topology using EA (Sims 1994a,b). The evolution of shaders and textures is investigated by Ibrahim (1998) and Wiens and Ross (2000, 2001). Using high-level commercial animation tools (such as Houdini and Maya) to evolve models and textures is described by Lewis (2000). Today there are numerous (Bentley 1999; Lewis 2001) interactive systems available for the evolution of models and animation.

This chapter provides some biological background and describes the main evolutionary algorithms as general optimization and search tools. We then concentrate on presenting two case studies applied to computer graphics and animation based on research carried out by one of the authors (Sarafopoulos 1995, 1999, 2001). The first case study involves interactive selection and the evolution of 2D procedural textures and the second focuses on the evolution of iterated function systems.

2.2 Biological Background

2.2.1 Evolution

Evolutionary algorithms are based on the simulation of a model of organic evolution formulated by Charles Darwin (1895). The Darwinian theory of evolution explains the emergence of complex living organisms as the result of gradual improvements through a recurrent process of *selection* and *reproduction* with *variation*. Selection is the processes of assigning *fitness* to individual organisms under a given environment. Individual organisms better equipped to survive and breed are said to have higher *fitness*. For living organisms to persist through the aeons, characteristics of parent individuals are transmitted to offspring through the process of reproduction. Reproduction is not a procedure of exact replication from parent to offspring. A controlled amount of variation is involved during the process of reproduction so offspring are not identical to their parents. Variation leads to random and undirected changes between *phenotypes*, i.e. the behavior and embodiment of individual organisms.

Natural selection favors individual organisms that are better adapted to their environment; in environments with finite resources better-adapted individuals exploit scarce resources more effectively and therefore survive to produce offspring. Individuals that are not adequately adapted constantly die off. Over many generations of individuals, natural selection continually filters out and eventually kills off unfit organisms. Since more individuals are born than survive to breed, natural evolution progressively amplifies fitness. Individuals with variations that are favored by the environment gain an *adaptive advantage*. Advantageous variations accumulate over time leading to highly adapted populations. A "*natural selection*" occurs resulting in populations of highly adapted living organisms.

The modern theory of evolution known as the *synthetic theory of evolution*, *modern synthesis*, or *neo-Darwinism* (Ridley 1996) unifies the Mendelian and molecular theory of inheritance with the Darwinian theory of natural selection. According to the Mendelian theory of inheritance, characteristics from parent to offspring are transmitted in the form of discrete particles called genes. Genes control the development of living organisms and encode their phenotype. The genetic make up of an organism, that is, all phenotypic characteristics encoded in genes, is referred to as the *genotype*. It is well known today that DNA (deoxyribonucleic acid) stores the genetic information of living organisms. The *synthetic theory of evolution* describes evolution in nature, as a process of *selection* and *reproduction* with *variation*, where selection is a process that operates on phenotypes and variation is a process that operates on genotypes. That is, where selection acts on an individual's physical embodiment and behavior, variation operates on the molecular level of DNA strands. Variation is a stochastic process that results from random errors during reproduction (*mutations*), and by the shuffling of genetic information passed from parent to offspring during sexual reproduction (*recombination*). Selection, however, is predominantly deterministic. To survive and procreate individuals depend on specific phenotypic characteristics (traits) that render useful skills in exploiting resources available under a given environment. Hence, selection can be seen as a function of the traits of an individual organism.

2.2.2 Fitness Landscapes

Through the process of natural selection populations evolve over time that become better adapted to their environment. The most common way to visualize evolutionary change is by depicting individual organisms as points in a multidimensional space, with one axis corresponding to each biological trait and an additional dimension is used to depict fitness. This abstraction is usually referred to as a *fitness landscape*, and was originally coined by the biologist Sewall Wright (1931) in the context of population genetics. In the simple three-dimensional (3D) case, individuals are represented as points on a 3D surface, having two trait coordinates and a fitness value. Such a 3D surface resembles a natural landscape with peaks and valleys depicting areas of high and low fitness (see Figure 2.1). However, from the biological point of view the concept of a static fitness landscape is often not realistic.

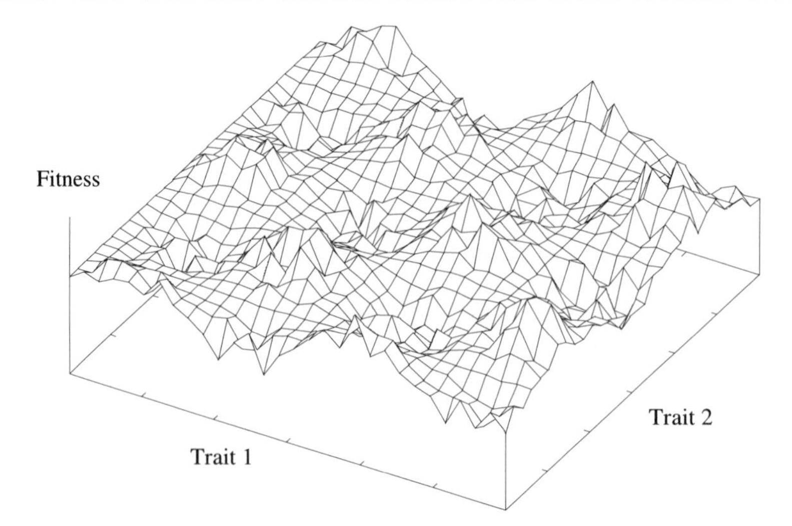

Figure 2.1 3D fitness landscape schematic. Evolution guides populations along the fitness landscape in certain ways; for example, adaptation makes a population move towards local maxima, i.e. peaks in the landscape. According to Sewall Wright's *shift balance* theory, genetic drift can cause sub-populations to step across ridges from a low fitness end of the landscape to another.

There are several reasons why this might be so. First, fitness in nature can be measured only indirectly (by the propensity of individual organisms to survive and procreate), since the exact biological traits and how they affect fitness are usually not known. The fitness of phenotypes may depend on their frequency (Ridley 1996). Secondly, populations often modify the resources of their environment, and consequently the mapping from traits to fitness may also vary as a function of time. Such environment–population interactions are a process that is inadequately understood.

Apart from natural selection, there are other factors that can change the way populations evolve (Ridley 1996); variation in small populations and changes of the state of the environment due to chance can cause the increase or decrease of certain traits, a phenomenon known as *genetic drift* (Provine 1986). Motto Kimura proposed the neutral theory of molecular evolution (Kimura 1983), where he argues that "most evolutionary changes at the molecular level are due to the random process of genetic drift acting on mutations, rather than natural selection. While recognizing the importance of selection in determining functionally significant traits, he holds that the great majority of the differences in macromolecular structures observed between individuals in a population are of no adaptive significance and have no impact on the reproductive success of the individual in which they arise. This contrasts with the orthodox neo-Darwinian view that nearly all evolutionary changes have adaptive value for the organism and arise through natural selection" (Martin and Hine 2000).

2.3 Evolutionary Algorithms

The problem of how to solve problems is unsolved – and mathematicians have been working on it for centuries.

Evolution and Optimum Seeking, Hans-Paul Schwefel

2.3.1 Artificial Evolution in Search and Optimization

The problem of finding solutions to problems is itself a problem with no general solution. In computer science and artificial intelligence finding the solution to a problem is often thought of as a search through the space of possible solutions. The set of possible solutions defines the *search space* for a given problem. Solutions or partial solutions to a problem are viewed as points in the search space. In engineering and mathematics finding the solution to a problem is often thought of as process of optimization. Problems at hand are often formulated as mathematical models expressed in terms of functions, or systems of functions. Hence, in order to find a solution, we need to discover the parameters that optimize the model or the function components that provide optimal system performance. There are several well-established search/optimization techniques in the literature. These are usually classified into three broad categories: enumerative, calculus-based and stochastic (Langdon 1998).

Enumerative methods are based on the simple principle of searching through potentially all points in the search space one at a time. In the field of artificial intelligence enumerative methods subdivide into two categories, *uninformed* and *informed* methods. Uninformed or blind methods such as the *minimax* algorithm (used in game playing) search all points in the space in predefined order. Informed methods such as Alpha–Beta and A* perform a more sophisticated search using domain-specific knowledge in the form of a cost function or *heuristic* in order to reduce the cost of the search (Russell and Norvig 1995, Chapters 3 and 4).

Calculus-based techniques are also often classified into two sub-categories: direct and indirect. Indirect or analytic methods stem from the origins of mathematical optimization in calculus and are based on discovering values that take function derivatives to zero. Direct or numerical methods such as the Newton–Raphson and Fibonacci methods are usually iterative techniques that navigate the fitness landscape using gradient/function information to move in the direction of the solution. Like a sightless climber who feels his way to the highest peak, they are also known as *hill-climbing* strategies. For an overview of hill-climbing strategies, see Schwefel (1995).

Stochastic methods are iterative methods that navigate through the search space using probabilistic rather than deterministic rules. A popular random method is *simulated annealing*. It was developed in explicit analogy with natural annealing, the process whereby, in order to harden steel, low-energy crystals are formed in copper by heating it to a liquid state and then gradually cooling it until it freezes (Kirpatrick et al. 1983).

In the 1950s and 1960s several researchers introduced stochastic search algorithms based on a simulation of Darwinian theory of natural selection, in order to solve search and optimization problems. Evolutionary algorithms (EAs) are stochastic search techniques based on a computer simulation of the genetics of natural selection. EAs operate on a population of points in the search space. The population is able collectively to learn better solutions by a process of selection and reproduction with variation. Individuals in the population gradually accumulate advantageous variations through a selection pay-off that leads the search to, or close to, a solution. Evolutionary algorithms allow us effectively to search spaces in which other traditional calculus or enumerative-based techniques fail. This is because EAs often operate on a coding of the problem not the problem itself, they use selection pay-off, as opposed to other quality information (such as derivatives), and they also operate a parallel search by simultaneously sampling many points of the search space (Goldberg 1989). There are four main variations of evolutionary algorithms: genetic algorithms (GAs), genetic programming (GP), evolution strategies (ES), and evolutionary programming (EP). The main difference between the above variations of evolutionary algorithms are on the encoding of an individual, and therefore in the representation or definition of the nature of the search space. A different encoding also implies a different method to modify individuals stochastically. Thus, each paradigm has a set of dedicated operations that allow for mutation and recombination of individuals in the population.

2.3.2 Genetic Algorithm (GA)

Holland first conceived of genetic algorithms as a theoretical framework for investigating artificial and natural evolution (Holland 1992). The theoretical framework specified by Holland was based on adaptation of a population of structures described as strings made out of characters of a discrete alphabet. A GA probably resembles natural evolution more closely than other evolutionary algorithms. GA genotypes are defined as strings made out of a binary alphabet, analogous to the four-letter alphabet made out of A (adenine), C (cytosine), G (guanine) and T (thymine) nucleotide bases that make up the genetic code of living organisms (Sedivy and Joyner 1992). The genotypes of individuals are defined as fixed-length binary strings. In order to encode a problem, the free variables have to be represented as fixed-length binary substrings of the string that represents the genotype. The analogy between natural evolution and a genetic algorithm may be described as follows.

- A chromosome or genotype in the context of genetic algorithms thus refers to a binary string that is a coding of some aspect of a candidate solution to a problem.
- A gene is a bit or a short sequence of adjacent bits in the chromosome that encodes for a particular feature or features of a problem. (In the context of function optimization, for example, genes usually encode the parameters of the function to be optimized.)
- A locus is the position of binary digit or the position of adjacent binary digits along the chromosome.
- Alleles are all the possible configurations of binary digits in a locus, i.e. the alleles of a two-bit string segment are 00, 01, 10 and 11.
- The phenotype or candidate solution to a problem is the decoded structure. (For example, in the context of function optimization, the decoded structure is often the function parameter set.)

Variation on binary strings as originally conceived by Holland operates in three modes.

- Recombination that is based on the exchange of genetic material between two parent individuals and is modeled in explicit analogy to the homologous recombination in nature.
- Inversion, modeled by analogy with inversions of DNA during replication, acts on a segment of the chromosome by inverting the sequence of bits along that segment.
- Mutation is a random change of the state of a digit along the chromosome.

The selection scheme used by Holland was one where each individual was selected probabilistically in proportion to the observed performance (i.e. fitness proportional selection). Fitness is calculated by evaluating an *objective* or *fitness function*, which assesses the performance of the candidate solution (see Figures 2.2 and 2.3).

The procedure that is the basis of most modern incarnations of the genetic algorithm is detailed in Goldberg's textbook *Genetic Algorithms in Search, Optimization, & Machine Learning* (Goldberg 1989), and it is usually referred to as the *simple genetic algorithm* (SGA).

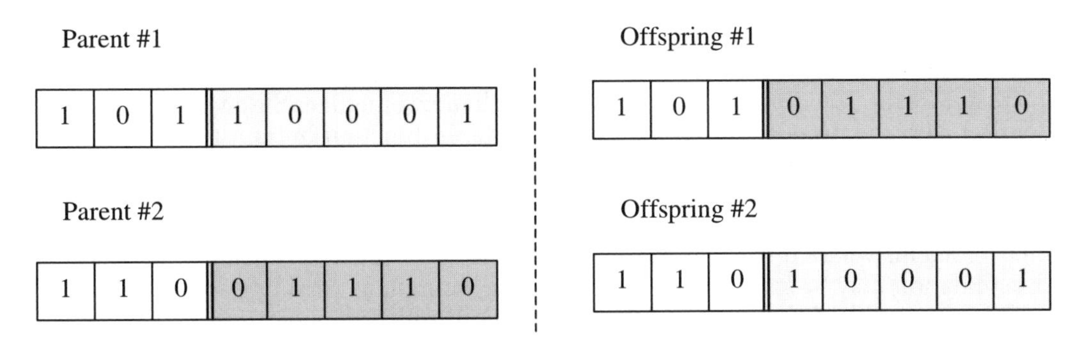

Figure 2.2 One-point crossover in SGA.

The outline of SGA is shown in Figure 2.4.

Given a decoding function d, a fitness function f, a set of control parameters $\{n, l, P_c, P_m$ – whose meaning will become clear below$\}$ and, a termination criterion t, the SGA can be described (more precisely) as follows:

(1) *Randomly create an initial population of n l-bit chromosomes.*

The algorithm starts by initializing the *gene pool* (all genes in the population) randomly, in order to scatter (using a uniform distribution) individuals across the landscape of all possible (2^l) binary string configurations.

(2) For each individual chromosome c in the population first decode and then calculate its fitness by evaluating $f(d(c))$.

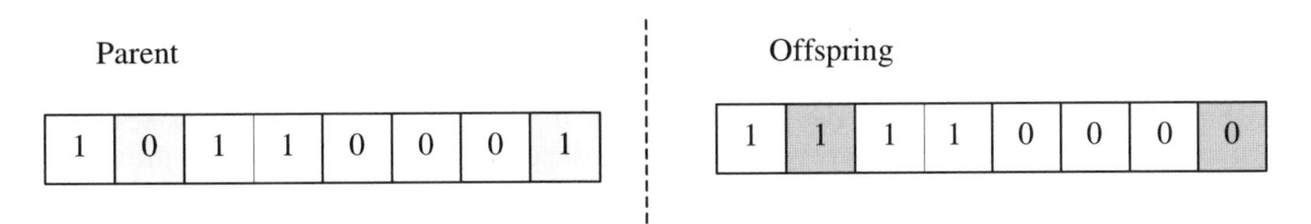

Figure 2.3 Mutation in SGA.

Simple genetic algorithm (SGA) outline

(1) *Randomly create an initial population of fixed-length binary string chromosomes, set the generation count to zero.*

(2) *For each individual chromosome in the population first decode and then calculate its fitness.*

(3) *Select the fittest individuals in the current population in order to create the new population through reproduction and variation using recombination and mutation operations.*

(4) *Replace the current population with the new population, and increment the generation count.*

(5) *If termination criteria are satisfied stop, otherwise go to step 2.*

Figure 2.4 Pseudo-code for the outline of simple genetic algorithm.

Assigning fitness involves the decoding $d(c) \rightarrow i$ and the evaluation of the fitness function $f(i)$, for each individual structure i. The nature of the encoding (*problem representation*), and therefore the decoding, depends on the task at hand. Sometimes the encoding is almost identical with the decoded structure. For example, in function optimization the binary coding is often a sequence of equal-sized binary strings that represent the parameter set of the function to be optimized. The representation of the parameter set might be Boolean or it might use the so-called Gray code (Hollstien 1971; Caruana and Schaffer 1988). In other problems where it is not just a set of parameter values that is being optimized the representation may be less direct. One interesting example is encoding solutions for the artificial ant problem (Collins and Jefferson 1991) in which the objective is to evolve an agent (ant) that is able to traverse a terrain which contains food pellets and gather these pellets in optimal time.

(3) *Select the fittest individuals in the current population in order to create the new population through reproduction with recombination probability P_c, and mutation probability P_m.*

The selection scheme used, referred to as *roulette wheel* or *fitness proportional* selection, is meant to be analogous with selection in nature where fitter individuals have a greater chance to survive and breed. Hence, the probability $p(x)$ of an individual structure x to be selected for reproduction depends on its fitness in relation to the fitness of other individuals in the population and is given by

$$p(x) = \frac{f(x)}{\sum_{i=1}^{n} f(i)}.$$

The simple genetic algorithm uses two operations to introduce variation in the gene pool: recombination or *crossover* (see Figure 2.2), and mutation (see Figure 2.3). Inversion, originally proposed by Holland, is typically excluded from the SGA. Fitness proportional selection of two parent individuals with replacement (that is, a parent individual can be selected more than once from the current population) is followed by recombination with probability P_c to produce two new offspring. If no recombination takes place (with probability $1-P_c$) the offspring are replicated from the parents. This is followed by mutation of the offspring at each locus with probability P_m. The new offspring are subsequently copied into the new population. The process repeats until the new population contains n new chromosomes. In Goldberg's original description of the SGA, the genetic operations of crossover and mutation are applied probabilistically on chromosomes during reproduction with crossover probability $P_c = 1$ (i.e. crossover is always performed), and mutation probability $P_m = 0.001$. The probabilities of crossover and mutation can vary but it is important that there is high crossover and low mutation probability. When the crossover probability is less than one, for example $P_c = 0.8$, it is possible for individuals to be copied verbatim (*reproduced*) to the next generation, thus implicitly gaining a longer life span.

(4) *Replace the current population with the new population.*

This is often referred to as generational GA, that is, progress is achieved through a series of well-defined and separate populations of individuals. However other algorithms exist such as "steady-state" algorithms where there are no distinct population intervals (Syswerda 1991).

(5) *If the termination criterion t is satisfied stop, otherwise go to step 2.*

The algorithm iterates several times through steps 2 and 5, and a single iteration is referred to as a *generation*. Typically we need many generations (e.g. from 50 to 500) in order to find an individual structure that is a solution to the problem. A complete sequence of generations through to termination is usually referred to as a run. The termination criterion is usually specified as a fixed number of generations, or an execution time limit. When the algorithm terminates one hopes to arrive either at a solution(s) or at a near solution to the problem, as the population tends to converge in exploiting a specific area of the search space that appears to provide optimal solutions or near-optimal solutions for that run. In order to solve a problem we usually perform several runs, typically executed by starting the search anew from a fresh set of uniformly distributed individuals in order to minimize the effects of random initial conditions and *premature convergence*. Premature convergence occurs when the population is trapped in a local minimum, by becoming fixated on particular gene combinations and virtually losing almost all other gene variations. (Premature convergence can be regarded as analogous to the phenomenon of niche pre-emption in nature where a biological niche tends to be dominated by a single species (Magurran 1988; Koza 1992, pp. 191–192).)

GAs rely mostly on recombination to provide improvements in fitness, whilst mutation is used mainly to maintain variation within the population. The emphasis on recombination is based on Holland's schemata theorem. Holland argues via the schemata theorem (Holland 1973) that, in certain cases, GAs provide near optimal use of the information provided by the search so far in order to guide the search in the next generation. By using an analogy with natural evolution, the notion of schemata can be thought of as a collection of certain configurations of genes. A collection of configurations of genes that combine well together to effect an increase in the performance of an individual is known as a "building block". The supposition that recurrent crossover, and sampling through selection pay-off, of short and fit schemata leads to strings of high fitness, is known as the *building block hypothesis*. Given the existence of building blocks, the genetic algorithm will progressively evolve, from a random initial population of mostly unfit structures, individuals with chromosomes that contain exponentially larger numbers of useful building blocks, thus evolving fitter structures.

Many of the early practical studies of GA investigated problems of function optimization (Hollstien 1971; De Jong 1975; Bethke 1981) with applications to engineering. Today the GAs have being applied to a wide range of fields, from biology and engineering, to sociological sciences (Mitchell 1996; Gen and Cheng 1999; Man et al. 1999). Several extensions of the SGA exist such as messy-variable length GA and hierarchical GA, which include new operations, selection methods and representations (Mitchell 1996).

2.3.3 Genetic Programming (GP)

Genetic programming is an extension of the genetic algorithm employed for the automatic generation of computer programs. Several researchers have investigated the induction of computer programs using evolutionary algorithms via different representations (Cramer 1985; Fogel 1999). However it was John Koza who systematically tested and formalized the use of LISP symbolic expressions (S-expressions) as the representation of choice for "the programming of computers by means of natural selection" (Koza 1992). Koza claims genetic programming to be the most general search paradigm in machine learning (Koza

1992). Perhaps the most important and most characteristic feature of genetic programming is the fact that solutions to problems are encoded directly as computer programs via the use of hierarchical, LISP-like, symbolic expressions. This feature is responsible for much of the generality of the genetic programming paradigm (Banzhaf et al. 1998, pp. 21–22) and, in fact, it can be shown that under certain conditions genetic programming is computationally complete (Teller 1994).

In order to demonstrate the nature of the encoding in GP, consider the task of program induction where we are asked to discover a program that calculates the area of a circle given its diameter. We could trivially write such a program using C, or LISP as follows:

```
/* calculate the area of a circle in C */
double area(double diameter) {
   double Pi = 3.14;
   return (Pi*(diameter*diameter))/4;
}
;;; calculate the area of a circle in LISP
(defun area (diameter)
   (setf Pi 3.14)
   (/ (* Pi (* diameter diameter)) 4))
```

Apart form the syntactical differences between the two languages, the important part of the program is the fragment where the calculation `(Pi*(diameter*diameter))/4` in C, or `(/(*Pi (*diameter diameter)) 4)` in LISP, takes place. In both cases the above two fragments of code perform the same task and in both cases the order of execution of the calculation involved is the same and can readily be visualized using a hierarchical tree graph (see Figure 2.5). This graph is, in fact, equivalent to the data structure that most compilers generate internally to represent computer programs before translation into machine code and is usually referred to as a *parse tree*. LISP S-expressions, because

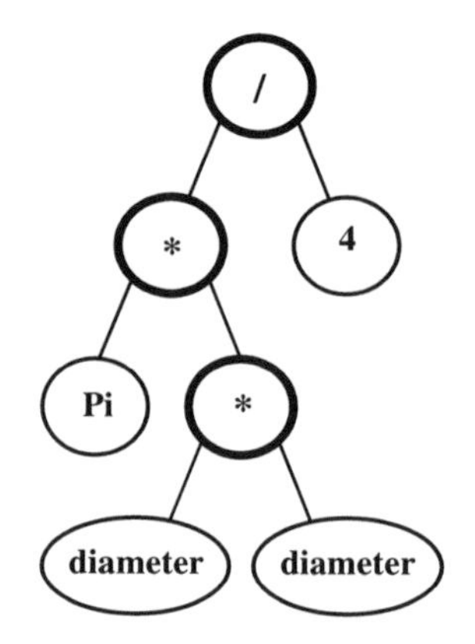

Figure 2.5 The parse tree for the S-expression `(/(*Pi(*diameter diameter))4)` used to calculate the area of a circle. The set of functions used in internal nodes of the tree constitutes the non-terminal function set (or *function set*) $F = \{/, *\}$ for this S-expression, whilst the functions used as leaf nodes compose the terminal function set (or *terminal set*) $T = \{2, 4, \text{Pi}, \text{diameter}\}$. Discovering a function that calculates the area of a circle, can therefore be thought of as search through the space of all possible S-expressions generated by composition of functions and terminals that belong to the combined set $C = F \cup T$.

of their simple prefix syntax, can be directly depicted as parse trees. The internal nodes of the tree are referred to as *non-terminal functions* (functions that accept arguments), and the external nodes or leaves as *terminal functions* (functions that accept no arguments or constants) such as *Pi* and the variable *diameter* in Figure 2.5. The root of the tree is the function appearing first after the leftmost parenthesis of the S-expression. Execution of the tree is carried out in a recursive, depth-first way, starting from the left. In genetic programming the terms *parse tree* and *S-expression* are used to mean the same thing. In GP the encoding of a solution, and hence the gene pool is made out of parse trees. The nature of the encoding is inherently hierarchical and of variable length in contrast to the SGA where the encoding is linear and of fixed length. It is interesting to note that, at least, in the case of program induction there is no distinction between genotypes and phenotypes, i.e. between the encoding and the decoded structures.

The algorithm that forms the basis of most modern incarnations of GP (including Koza's own, ongoing research on GP (Koza 1994; Koza et al. 1999)) is outlined in Koza (1992) and it is referred to here as standard genetic programming (*standard* GP).

Given (the provision of five ingredients) a terminal function set T, a non-terminal function set F, a fitness function f, a set of control parameters $\{n, P_c, P_m, P_r, D_c, D_i$ – the necessity of which will become clearer below$\}$, and a termination criterion t standard GP is outlined in Figure 2.6.

Possibly the most important decision that needs to be made (apart from the selection of the fitness function) before starting a GP run is choosing the functions and terminals that are required for the representation of a problem. The choice of functions and terminals is often referred to as the architecture or representation of a GP run. In standard GP, the composite set C, made out of the union of the function and terminal sets, has to meet two requirements. The first requirement is that the set C should be adequate to solve the problem – a property known as *completeness*. The second requirement is that the set is closed; that is, all functions and terminals (and all their compositions) return values and/or accept arguments of the same data type – a property known as *closure*. The first requirement ensures that the search space contains solutions to the problem and the second ensures that genetic operations (as described below) produce legal parse trees. For example, if we assume that the variable *diameter* is a positive real number other than zero, the composite

Standard genetic programming (standard GP) outline

(1) *Set generation count to zero. Generate an initial population of n S-expressions of initial maximum depth D_i made out of random compositions of functions from the combined set $C = T \cup F$, where T is a set of terminals and F is a set of non-terminal functions.*

(2) *For each individual S-expression in the population evaluate its fitness.*

(3) *Select the fittest individuals in order to create a new population of symbolic expressions through reproduction with probability P_r, recombination with probability P_c, and mutation with probability P_m, where D_c is the maximum allowed depth size of S-expressions created during the run.*

(4) *Replace the current population with the new population and increment the generation count.*

(5) *If the termination criterion t is satisfied stop, otherwise go to step 2.*

Figure 2.6 Pseudo-code for the standard genetic programming algorithm outline.

set C as specified in Figure 2.5 is both closed and complete.

A more precise description with discussion of particular aspects of the algorithm follows.

(1) *Generate an initial population of n S-expressions of initial maximum depth D_i made out of random compositions of functions from the combined set $C = T \cup F$.*

In order to spread the population across a wide variety of parse trees of various sizes and shapes the generation of the new population may be initialized, according to Koza (1992), using either "*full*", "*grow*" or "*ramped half and half*" methods.

The full generation method is based on creating the initial population out of S-expressions with each non-backtracking path between a leaf node and the root equal to maximum depth D_i. The grow method involves generating parse trees with branches extending at variable depths from the root, but with no path between a leaf and the root allowed to exceed depth D_i. Finally, according to the *ramped half and half* method, the population is divided into equal-sized groups. Each group has a unique associated depth value that belongs to an interval ranging from a minimum depth to the maximum specified depth. Half of the members of each group are created using the grow method and the other half are created using the full method. For example, a population made of 1000 S-expressions, where the minimum depth value is 2 and the maximum depth value is 6, will be divided into five groups, each group comprising of 200 members and each group being associated with maximum depth values of 2, 3, 4, 5 and 6, respectively. One hundred members of each group will be generated using "grow", and the rest using the "full" method, where the maximum depth for the full and grow methods is the group's associated depth value.

(2) *Calculate the fitness of each S-expression x in the population by evaluating $f(x)$.*

Assignment of fitness is explicitly provided by a problem-dependent user-defined fitness function. Fitness is often evaluated over a set of fitness tests. For example, consider devising a method that tests the performance of evolved S-expressions for the problem of program induction mentioned above (the calculation of the area of a circle given its diameter). The fitness function has to be representative of the problem domain as a whole. We cannot simply test the quality of evolved programs against the area of one circle of specified diameter. Instead we have to test newly generated programs against circles of varying diameters in order to estimate the quality of evolved solutions. Each such comparison, against observed data, is usually referred to as a *test case* or *fitness case*. Sometimes a predefined number of fitness cases are adequate for assessing the quality of a program. For example, we can sample 50 (x, y) pairs so that $y = \pi \cdot x^2/4$, each pair acting as a fitness case. In situations where the problem domain is vast or infinite, new fitness cases may have to be sampled for each generation of a GP run. In our example, this would imply sampling a different set of (x, y) pairs at each generation of the GP run.

A common method of measuring fitness in GP is the so-called *standardized fitness*, according to which the fittest individual is assigned a value of zero. In effect, standardized fitness measures error, the less erroneous an individual the better. For example, we could specify the standardized fitness f_s of an S-expression f calculating the area of a circle, for each (x, y) pair described above, as follows:

$$f_s = \sum_{i=1}^{n} |y - f(x)|,$$

where n is the number of fitness cases, x is the diameter and $f(x)$ is the area returned by the S-expression f. This problem can be seen as a *symbolic regression* problem; that is, given

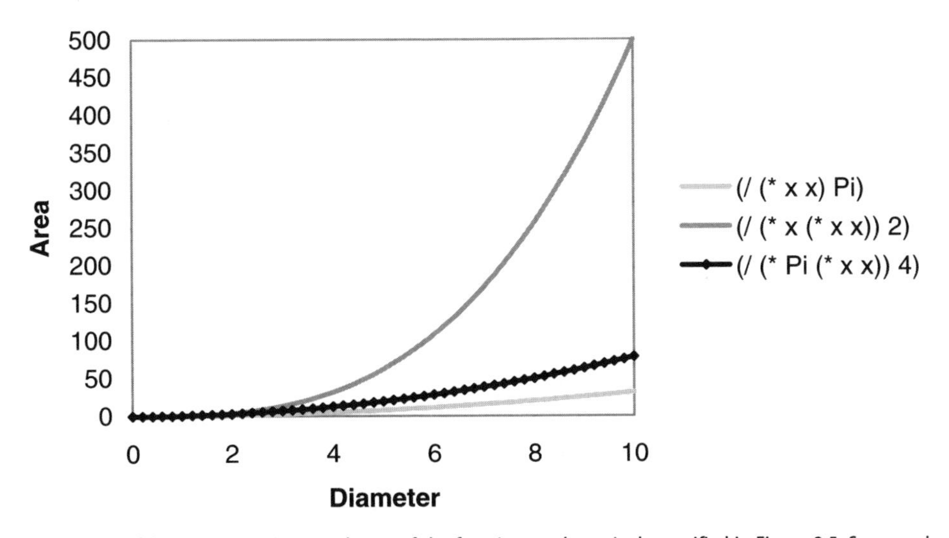

Figure 2.7 Plots of three S-expressions made out of the functions and terminals specified in Figure 2.5. S-expression `(/ (* Pi (* x x)) 4)` is a 100% correct solution to the problem of calculating the area of a circle given its diameter, and therefore has standardized fitness 0. The S-expression `(/ (* x (* x x)) 2)` has standardized fitness 5157.175 thus performing much worse than the S-expression `(/ (*x x) Pi)`, for which the standardized fitness is 801.029. The dots of the `(/ (* Pi (* x x)) 4)` plot represent the fitness cases generated by sampling 50 equally spaced (diameter) values between 0 and 10.

a sampling of data points we are asked to find a function (in symbolic form) that matches the given data, in this case a series of data points generated by the well-known relation $y = \pi \cdot x^2/4$ (see Figure 2.7).

(3) *Select the fittest individuals in order to create a new population of symbolic expressions through reproduction with probability P_r, recombination with probability P_c, and mutation with probability P_m, where D_c is the maximum allowed depth size of S-expressions created during the run.*

Selection in *standard* GP is performed mainly using fitness proportional selection with replacement, as in the SGA, actually the user can, if desired, specify other selection methods, such as *tournament* selection (Koza 1992, pp. 604–606). Tournament selection is based on competitions or tournaments between a small number of individuals in the population. The number of individuals taking part in a tournament is referred to as the *tournament size*. Tournament individuals are chosen at random and the best individual in a tournament is selected for reproduction. In the simplest case, two individuals are selected at random (a tournament size of two) and the best of the two is kept for reproduction. Selection proceeds with replacement, i.e. parents can be reselected. The fitness pressure can thus be adjusted when using tournament selection, by modifying the tournament size – the larger the tournament size the greater the selection pressure.

Creation of the new population proceeds by choosing one genetic operation amongst the group of available genetic operations, where operations are chosen stochastically depending on their associated probabilities. If reproduction or mutation is chosen, one individual is selected from the current population to create one new offspring. Two individuals are selected from the current population to breed two new offspring in the

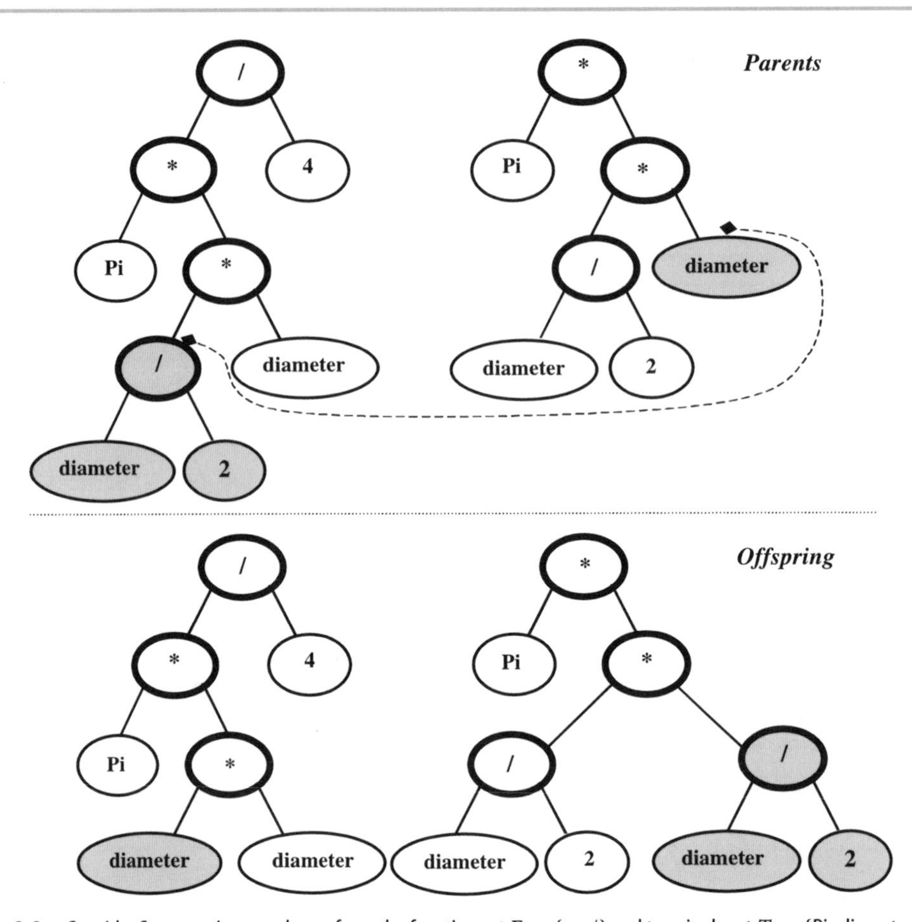

Figure 2.8 Consider S-expressions made out from the function set $F = \{*, /\}$ and terminals set $T = \{\text{Pi, diameter, 2, 4}\}$. The two parent expressions at the top recombine to produce two new offspring depicted at the bottom of the figure. The operation of crossover is defined as selecting a node (usually at random) that marks a sub-tree for each parent individual and then swapping the sub-trees emanating from the selected nodes to create the offspring. The left parent S-expression swaps the sub-tree (*/diameter*2) with the terminal (*diameter*) from the right parent to form two offspring, each of which is an expression that calculates the area of a circle. Since most nodes of a tree are leaf nodes, selecting nodes is usually biased so that a greater proportion of internal nodes is selected during crossover.

case of recombination. The new offspring is/are then added to the new population. The process repeats until the new population contains n new S-expressions.

The genetic operation of reproduction is defined as replication without modification of an individual from the current population. Corresponding to crossover and mutation in the SGA, the genetic operations of crossover (see Figure 2.8) and mutation (see Figure 2.9) are modified in GP to work with S-expressions. The GP crossover operation, as in GA, is claimed to provide the creative/innovative transformations that lead to adaptation with mutation used as a *secondary* operation in order to introduce variation within the population of S-expressions. This claim is based on extending the schemata theorem and building block hypothesis from GA to GP, where schemata are made out of S-expression templates (Langdon and Poli 2001). However, the notion of building blocks in GP is

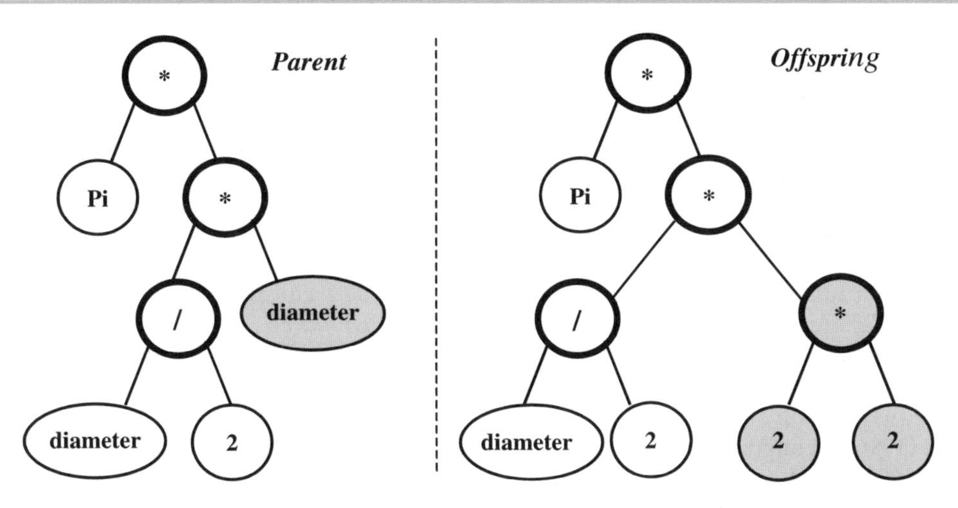

Figure 2.9 The mutation operation is specified as selecting a node at random from the parent parse-tree and then replacing the sub-tree emanating from that node with a new randomly composed S-expression. In the above the terminal node (*diameter*) is selected from the parent (on the left) and is replaced with the expression (∗2 2) in the offspring on the right.

problematic (O'Reilly and Oppacher 1995). This is because GP crossover incurs large changes in the structure of programs. These changes can be substantial enough to disturb building blocks frequently. Empirical observations indicate that GP crossover often behaves as a macro-mutation rather than a structured information exchange akin to the GA one-point crossover (Banzhaf et al. 1998, pp. 148–156). As opposed to GA one-point crossover, which leads to lexical convergence (since crossing identical individuals produces identical offspring), the lexical structure of S-expressions does not converge via the use of GP crossover even though the behavior of S-expressions may converge. GP crossover maintains diversity within a population of individuals since crossing identical individuals is likely to produce different offspring (Koza 1992). Several new crossover operations have thus been proposed that allow a more structured information exchange between parse trees such as: context sensitive crossover (D'haeseleer 1994), one-point GP crossover (Poli and Langdon 1997) and the use of crossover templates (Jacob 1996).

In Koza's original description of GP (Koza 1992) a number of secondary genetic operations are also described. These include (apart from mutation), permutation, editing, encapsulation and decimation. Permutation is a generalization of the inversion operator, described by Holland (see the section on GA). Editing provides the means of simplifying symbolic expressions. Encapsulation allows automatic identification and reuse of potentially useful code segments in S-expressions.

Decimation is used to destroy very low fitness and expensive (in terms of the use of computational resources) individuals in a specified generation of a run, typically the first generation. However, only reproduction and crossover are used in the majority of the work by Koza (1992), as the initial population is deemed to be large enough to contain enough variation for crossover alone to build working programs. Hence only a few experiments described therein use mutation or other secondary operations.

(4) *Replace the current population with the new population.*

Similar to the SGA algorithm, the standard GP algorithm is generational, although variations of GP exist where the bounds between generations are not distinct, such as *steady-state* GP implementations (Reynolds 1994e; Banzhaf et al. 1998, pp. 134–135).

(5) *If the termination criterion t is satisfied stop, otherwise go to step 2.*

Typically the algorithm terminates after a given number of generations, or when a solution has been found (i.e. an individual found with zero standardized fitness). At this stage the output of the algorithm is the best individual(s) in the population.

Koza and others have proposed many extensions to the standard GP algorithm. These include, encapsulation mechanisms that allow for hierarchical problem solving using function decomposition such as *automatically defined functions* (ADFs) (Koza 1994). Strongly typed genetic programming (STGP) (Montana 1995) allows the evolution of programs that do not obey closure, whilst syntactically constrained systems that allow extensions of closure have been proposed by Koza (1992, pp. 479–526), Whigham (1996) and Gruau (1996b). Langdon presents a study of GP using data structures (Langdon 1998). Koza et al. (1999) describe a set of genetic operations referred to as *architecture altering operations* that allow the evolution of an S-expression's architecture (i.e. function set, terminal set and ADFs) as well as its topology. Hybrid algorithms have also been proposed such as GP–ES hybrids (Keller et al. 1999; Sarafopoulos 2001).

Koza's original genetic programming system was written in LISP (Koza 1992, pp. 735–755). However, the relative simplicity and generality of the paradigm, allowed genetic programming systems to be implemented in many high-level (imperative, object-oriented and logic) programming languages (Langdon 1998, p. 38), as well as in low-level machine code (Banzhaf et al. 1998). An in-house GP system in C++ was used to carry out the experiments described in this chapter.

2.3.4 Evolution Strategies (ES)

Rechenberg and Schwefel jointly developed evolution strategies as a method for function optimization inspired by the Darwinian theory of evolution and natural selection. The technique was originally applied to discrete hydrodynamical problems, such as drag minimization of a joint plate (Rechenberg 1965), and structure optimization of a two-phase flashing nozzle (Schwefel 1968; Klockgether and Schwefel 1970; Herdy 2001). The application of the technique was initially based on experimental set-up. In the case of the flashing nozzle, for example, joints of the nozzle segments were physically adjusted and the performance of the adjusted nozzle was evaluated. When current adjustment of the construction performed better than previous ones it was used as the basis for further trials. Each adjustment was viewed as a mutation of the previous construction, and, founded on the observation that in nature large mutations appear more rarely than small ones, the discrete mutations were sampled by means of a binomial distribution with constant variance. Where traditional optimization methods failed to provide sufficiently good results this apparently, simple technique was surprisingly successful.

Schwefel was the first to simulate such ES using computers by testing different versions of the strategy for the optimization of continuous functions of real variables (Schwefel 1995). Different types of ES are known as $(1+1)$-ES, $(\mu+\lambda)$-ES and (μ,λ)-ES, in a notation that describes the selection method applied during a generation of individuals where μ parent individuals selected from the current population produce λ offspring. An early incarnation of ES (Schwefel 1995) is the $(1+1)$-ES or *two-membered ES*, where each generation consists of one n-dimensional real-valued vector and variation is introduced by normally distributed

Outline of the (μ, λ)-ES algorithm

(1) *Set generation count to zero, i.e.* $t = 0$, *and create an initial population of* μ *parent individuals* $Pop(0)$. *Where each member of the population is a vector* $\vec{a} \in I = (\vec{x}, \vec{\sigma}) = R^n \times R_+^{n_\sigma}$, *where* $n_\sigma \in \{1, \ldots, n\}$. *The component* $\vec{x}$ *is a real-valued vector that holds n* object variables *encoding the problem at hand.* $\vec{\sigma}$ *is also a real-valued vector of* strategy parameters *storing the standard deviations that specify the mutation step lengths for components of* $\vec{x}$.

(2) *Create an intermediate population* $Pop(t')$ *made out of* λ *new offspring, where each new offspring* $\vec{a}_i''$, $i \in \{1, \ldots, \lambda\}$ *is generated by: first using the recombination operation r on* $Pop(t)$ *to generate* $\vec{a}_i'$, *i.e.* $\vec{a}_i' = (\vec{x}_i', \vec{\sigma}_i') = r(Pop(t))$, *and subsequently mutating offspring* $\vec{a}_i'$, *i.e.* $\vec{a}_i'' = (\vec{x}_i'', \vec{\sigma}_i'') = m(\vec{a}_i')$.

(3) *Select the most promising* μ *individuals from* $Pop(t')$ *and store them in the new population of parent individuals* $Pop(t + 1)$.

(4) *Increment generation count,* $t = t + 1$.

(5) *If the termination criterion is met stop, otherwise go to step 2.*

Figure 2.10 Pseudo-code for (μ, λ)-ES algorithm outline.

mutations with expectation zero and a given variance. The parent vector is modified using normally distributed mutations with the same standard deviation for each component to generate one offspring. The offspring replaces its predecessor in the event that it improves performance.

Here we concentrate on the $(\mu + \lambda)$-ES and (μ, λ)-ES that are the most commonly used and which, as opposed to the $(1 + 1)$-ES, are population-based. In the $(\mu + \lambda)$-ES case, μ parent individuals produce λ offspring that subsequently compete against themselves and the parents (i.e. $\mu + \lambda$ individuals compete) to form the new generation of μ parent individuals. In the case of (μ, λ)-ES only the λ offspring compete to form the new generation of parent individuals and each individual therefore has a life span of one generation only. The outline of the (μ, λ)-ES algorithm is shown in Figure 2.10. In the case of the $(\mu+\lambda)$-ES, step 2 of the algorithm outline (in Figure 2.10) has to be modified so that the intermediate population, $Pop(t')$, is generated from parents and offspring, i.e. $Pop(t')$ comprises $\mu + \lambda$ individuals, where parents are copied verbatim into $Pop(t')$ from the current population $Pop(t)$.

ES focuses on mutation where recombination is deemed to be a secondary operation (Beyer 1995). Mutation is usually introduced as a normally distributed mutation of the components of the vector to be optimized, where each component is mutated separately. The step lengths (i.e. standard deviations) can be constant or can be adapted during a run. Step length control of mutations is very important. If the step lengths are too long, mutations can produce long random jumps across the fitness landscape that may step over nearby minima of the landscape. If, on the other hand, step lengths are too small, the system might never get close to a distant solution in a reasonable amount of time. If we constantly reduce the step length during the run, we might improve performance towards the end of the run, when arriving close to a possible solution. However, in doing so we run the risk of damaging performance at the beginning of the run, and may need more steps in order to come sufficiently close to a solution than could have been required with a fixed step length.

Clearly there is an exploration–exploitation trade-off; when step control is small we can efficiently search for potential solutions that exist in the vicinity of the current solution, but we lose out on solutions that exist in other parts of the search space. On the other hand, when the step length is long, we explore the fitness potential of the whole landscape more efficiently but we cannot efficiently "fine-tune" the search in order to exploit promising areas. Ideally adaptive mutations learn how big the step length has to be at each stage of the run, so we can maximize performance accordingly.

An example of an adaptive method for controlling the step length was introduced by Rechenberg (1973) and is known as the 1/5 success rule (Back 1996, pp. 83–87) which states: "The ratio of successful mutations to all mutations should be 1/5. If it is greater than 1/5, increase the standard deviation, if it is smaller, decrease the standard deviation" (Back 1996). Rechenberg derived the 1/5 rule from convergence rate results by analysis of the behavior of two model objective functions:

$$f_1(\vec{x}) = F(x_1) = c_0 + c_1 \cdot x_1, \quad \forall_i \in \{2, \ldots, n\} : -b/2 \leq x_i \leq b/2, \quad \text{and}$$
$$f_2(\vec{x}) = c_0 + c_1 \sum_{i=1}^{n} (x_i - x_i^*)^2 = c_0 + c_1 \cdot r^2,$$

where f_1 is the so-called *corridor model*, where progress is achieved by moving only along the direction of the x-axis in a corridor of width b; f_2 is the *sphere model*, which represents a hypersphere or radius r, where the minimum $\vec{x}^*$ is located at the center of the hypersphere. The convergence rate φ is defined as the expectation of the distance k' covered by mutation; that is,

$$\varphi = \int p(k')k' \, dk',$$

where $p(k')$ is the probability of a mutation moving an individual structure distance k' closer to the optimum (Back 1996, pp. 83–87). According to the 1/5 success rule, the mutation operation on individual $\vec{a}' = (\vec{x}', \vec{\sigma}')$ is specified as

$$x_j'' = x_j' + N(0, \sigma_j'), \quad j \in \{1, \ldots, n\}, \quad \text{and}$$
$$\sigma_j'' = \left\{ \begin{array}{ll} c_d \sigma_j' & \text{if } r < \frac{1}{5} \\ c_b \sigma_j' & \text{if } r > \frac{1}{5} \\ \sigma_j' & \text{if } r = \frac{1}{5} \end{array} \right\},$$

where $N(0, \sigma)$ is a normal distribution with mean zero and standard deviation σ. r is the relative frequency of successful mutations (given by the ratio of successful mutations over all mutations) sampled over an interval of a small number of past generations, e.g. 10. Schwefel suggests (Schwefel 1995, pp. 110–113) that the constants c_d and c_b should take values within the interval [0.817, 1].

In self-adaptive (μ, λ)-ES a vector of standard deviations (that represents mutation step lengths) is attached to object variables (Schwefel 1995, pp. 142–145), as described in Figure 2.10. It acts as a set of strategy parameters. An individual's object variables as well as step lengths are then subject to mutations. The expected result of mutations of the step lengths without selection is neutral; an increase of step length is as likely as a decrease, and, if we sample using a log-normal distribution, small changes occur more often than large ones. Good mutations of step lengths therefore gain an adaptive advantage through selection pay-off, and poor ones are filtered out. Using this plan an individual $\vec{a}' = (\vec{x}', \vec{\sigma}')$

is mutated by

$$\sigma_i'' = \sigma_i' \cdot e^{\tau N(0,1)} \qquad \text{and} \qquad x_i'' = x_i' + N(0, \sigma_i''), \quad i \in \{1, \ldots, n\},$$

where τ is a constant that represents an overall learning rate, Schwefel suggests (Schwefel 1995, p. 144) that τ should be inversely proportional to the square root of the number of object variables:

$$\tau \propto \frac{1}{\sqrt{n}}.$$

This equation was also studied by Beyer (1996, 2001), who provides an improved formula (based on an analysis of spherical models) to calculate τ for (μ, λ)-ES, given by

$$\tau = \frac{c_{\mu,\lambda}}{\sqrt{n}},$$

where $c_{\mu,\lambda}$ is the so-called "progress coefficient", which can be computed numerically for different μ and λ. A table of $c_{\mu,\lambda}$ values, for selected (μ, λ)-coefficients, is provided in Beyer (1996).

Schwefel proposed a similar method for self-adapting not only the size of mutation step lengths but also potential linear correlation of mutations of object variables, by introducing another vector of rotation angles as part of the representation (Schwefel 1995, pp. 240–242). There are $n(n-1)/2$ such rotation angles that correspond to linear correlation of mutations of n object variables. Rotation angles are self-adapted using additive, normally distributed mutations (Back 1996, pp. 68–73).

Hansen et al. (1995) propose *generating set adaptation* (GSA) as a method that is independent of the coordinate system of the standard deviations vector $\vec{\sigma}$. This is carried out by linear modification of the basis vectors of $\vec{\sigma}$. A history of successful mutation steps is stored and the weighted sum of all mutation steps of the individual's history is used to alter the coordinate basis of $\vec{\sigma}$. This strategy was extended by Hansen and Ostermeir (1996) with the introduction of *covariance matrix adaptation* (CMA). The use of a covariance matrix allows the technique to be invariant to linear transformations of the search space such as scaling and rotation.

Recombination is often used in ES when $\mu > 1$. Recombination can be applied to object variables as well as strategy parameters. There are several recombination techniques in ES. They are subdivided into sexual and panmictic. In both cases, partners are selected randomly from the parent population. In sexual recombination two parents are randomly selected to generate a new offspring, whereas in panmictic recombination one individual is randomly selected which remains fixed and, for each component (of its vectors), a new partner is selected randomly from the population. Recombination methods are also divided into discrete and intermediate. Discrete recombination of the object variables of individuals in the parent population is defined as

$$x'_{\mu+1,j} = \begin{cases} x_{\alpha,j} & \text{if } u_j \leq 0.5 \\ x_{\beta,j} & \text{if } u_j > 0.5 \end{cases} \qquad \forall j \in \{1, \ldots, n\},$$

where u_j are uniform random variables, sampled for each pair of corresponding components (of parent vectors). The parents are selected once in the case of sexual recombination, whilst the second parent is selected anew for each component in the case of panmictic recombination. Strategy parameters can also be recombined in the same fashion. Intermediate recombination of the object variables of individuals in the parent population is defined as

$$x'_{\mu+1,j} = x_{\alpha,j} + \chi \cdot (x_{\beta,j} - x_{\alpha,j}),$$

where χ takes a value within the range [0, 1]. Typically $\chi = 0.5$, in which case the result obtained is the intermediate value between the two parents. In general, intermediate recombination performs a linear interpolation of the two parents. Strategy parameters can also be interpolated in the same fashion. Again, the parents are selected once in the case of sexual recombination, and the second parent is selected anew for each component in the case of panmictic recombination. However, object and strategy parameters don't have to be recombined using the same method. Back suggested that empirically better results are obtained when discrete recombination is performed on object variables, and intermediate recombination is performed on strategy parameters (Back 1996, pp. 73–76).

Although the representation and genetic operations of GA and ES appear very different, i.e. GA operate on binary strings as opposed to ES operating on real-valued vectors, the emphasis of both techniques is on an adaptive search viewed as an exploration–exploitation trade-off. This trade-off in ES is based on the selection of mutation step lengths. In the case of GAs the trade off is "built-in" within the crossover operation. This may be seen, for example, by looking at Holland's analysis for the optimal allocation of trials for the two-armed bandit problem using GA (Goldberg 1989, pp. 36–39; Holland 1992, pp. 75–88). Alternatively, one could view GA crossover as an adaptive mutation. To do so, we start with the observation that an initial population of randomly generated binary strings will contain individuals that are lexically very different. This implies that crossover operations will produce very different individuals, i.e. mutation step lengths will be very high. However, via selection pressure schemata (i.e. bit-string templates) that perform well will appear more often in the population. The population is thus likely to start converging, and crossover will produce individuals that are likely to be lexically more similar, i.e. mutation step lengths will progressively reduce.

2.4 Case Studies

We use two case studies to demonstrate the applicability of EAs to computer animation and computer graphics in general. The first case study is based on the evolution of procedural textures the second on the evolution of iterated function systems. The emphasis in both case studies is on the use of evolutionary algorithms as a tool for the artist and the application of such as tool (using techniques described in the above sections) to the generation of aesthetically pleasing and challenging visual content.

2.4.1 Interactive Evolution of 2D Procedural Textures

Here we present an application of evolutionary algorithms to the automatic generation and evolution of procedural textures. This is a popular application of EAs to computer graphics (and animation) and leads to interesting visual results. The images and work presented here are based on the animation sequence "textures" generated by one of the authors (Sarafopoulos 1995). In order to use an evolutionary algorithm to generate procedural textures we need to decide on an encoding, fitness function and genetic operations. In this case we choose genetic programming as the EA of choice and therefore the encoding and genetic operations are based on S-expressions (see the section on genetic programming). Using a "conventional" fitness function as in standard GP, however, can prove problematic. To avoid this we will use "interactive selection" as explained later in Section 4.1.2.

2.4.1.1 Encoding of Procedural Textures

Procedural textures (Perlin 1985; Ebert et al. 1998) in computer graphics are simply functions whose domain is $\Re^2$ (in the case of two-dimensional textures) and the range is a vector of intensity or color values. Typically we are looking for a function that maps coordinates to intensity or color values of the form:

$$f : \Re^2 \rightarrow I(R, G, B),$$

where the color components R, G, B belong to the interval $[0, 1]$ and correspond to red, green and blue intensities of a pixel at a given location of an image. A procedural texture can readily be coded as an S-expression. Thus, genetic programming could be used to explore the space of such functions. Karl Sims originally proposed this method of generating procedural textures using genetic programming (Sims 1991). There are several examples of systems that use a similar approach to generating textures, including some commercial applications (Ibrahim 1998; Wiens and Ross 2000, 2001).

2.4.1.2 Interactive Selection

In order to drive the evolutionary process we need a way of assigning fitness to individual textures. In our case a small number of textures are originally generated randomly and displayed on the computer screen and the user is asked interactively to assign fitness to the texture(s) they consider to be the most pleasing; for example, using mouse-driven interaction. Interactive selection avoids the problem of providing a conventional fitness function. Such a fitness function would probably have to quantify what is seen as visually pleasing (under some context), which is itself a very difficult task.

2.4.1.3 GP Architecture

Here, for simplicity, we are looking at expressions that generate gray-scale images. In order to define the components of symbolic expressions using genetic programming we need to decide on function and terminals sets. In our case we chose the following function set F:

$$F = \{\cos, \sin, +, -, *, /, \text{sqrt}\}.$$

That is, the function set consists of simple arithmetic expressions and few simple functions. The terminal set T consists of

$$T = \{x, y, \mathit{ephemeral}\},$$

where x and y are variables that denote pixel coordinates of an image, and *ephemeral* is a terminal that is initialized to contain a random floating-point constant.

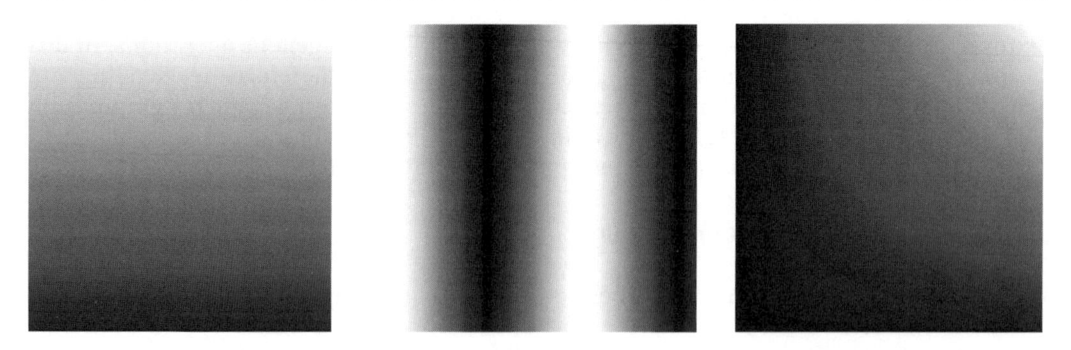

Figure 2.11 The texture on the left is generated by the S-expression (y), the texture in the middle of the figure is made out of the S-expression $(\sin x)$ and the one on the right is generated by $(*xy)$.

Resulting expressions can be used to generate a wide variety of textures. However, textures produced by simple S-expressions (constructed out of one or two function nodes) are also simple (see Figure 2.11). The challenge is to discover combinations of functions and terminals that produce pleasing elaborate patterns and textures. Generation of a texture that corresponds to an individual expression proceeds by evaluating that expression for every pixel of the image holding the texture.

2.4.1.4 Results

After 40 or 50 generations, textures that are visually complex start to emerge. A liquid-like texture is presented in Figure 2.13 and the S-expression that generated this texture is given in Figure 2.12. Figure 2.14 is a screenshot of the interactive system written by one of the authors in order to generate such 2D textures using genetic programming.

```
(* (cos (* (cos -38.81) (+ -75.55 (sqrt (+ (+ (sqrt (+ (cos-
38.81) x)) -38.81) (* (cos (cos (cos (* (cos 38.81) (+-75.55 (+
(sqrt (+ (+ (sqrt (+ (cos -38.81) x)) x) (* (cos(cos (cos (sqrt
(+ (+ (sqrt (+ (cos 38.81) x)) x) (* (cos(cos (cos (* (cos -
38.81) (+ -75.55 (+ (sqrt (+ (+ (sqrt (+(cos -38.81) x)) x) (*
(cos(cos (cos (* (cos -15.13) (+-75.55 (+ -75.55 (sqrt (+ -75.55
(* (cos -38.81) y))))))))))y)))(cos (*(cos-38.81) (+ -75.55 (sqrt
(+ (+ (sqrt (+ (cos-38.81) x)) x) (* (cos (cos (cos (* (cos -
15.13) (+ -75.55(+ -75.55 (sqrt (+ -75.55 (* (cos -38.81)
y))))))))))y))))))))))))) y))))))) y))) (cos (* (cos -38.81)
(+75.55(sqrt (+ (+ (sqrt (+ (cos -38.81) x)) x) (* (cos (cos
(cos(* (cos -15.13) (+ -75.55 (+ -75.55 (sqrt(+ -75.55 (* (cos-
38.81) y)))))))))) y))))))))))))) y))))))) 12.40)
```

Figure 2.12 This complex combination of function nodes and terminals results in the liquid-like texture of Figure 2.13.

Figure 2.13 A liquid-like texture generated using the S-expression given in Figure 2.12.

2.4.2 Evolutionary Morphing and Iterated Function Systems

Here we focus on the generation of interesting visual content without the use of interactive selection. The question then arises as to how we can guide the generation of images automatically since very little is really known (in scientific terms) about the reasoning underpinning aesthetic judgments. The approach adopted here is one used throughout the history of art, namely "visual representation". An evolutionary algorithm is asked to evolve/generate a shape or a form that is the same as or resembles a shape in nature (in the same way a painter may be asked to represent a natural scene or to draw a portrait, a figure, etc.). If we choose this approach interesting results may be obtained in two respects. First, we don't have to ask for the representational goal to be achieved exactly, thereby allowing for variations on a given theme. Secondly, the process of reaching the representational goal might itself be interesting, since it will portray a path starting from noise (a random generation of individuals) to a well-defined outcome, i.e. a type of "evolutionary morphing" (between noise and a specific representational goal) could be achieved.

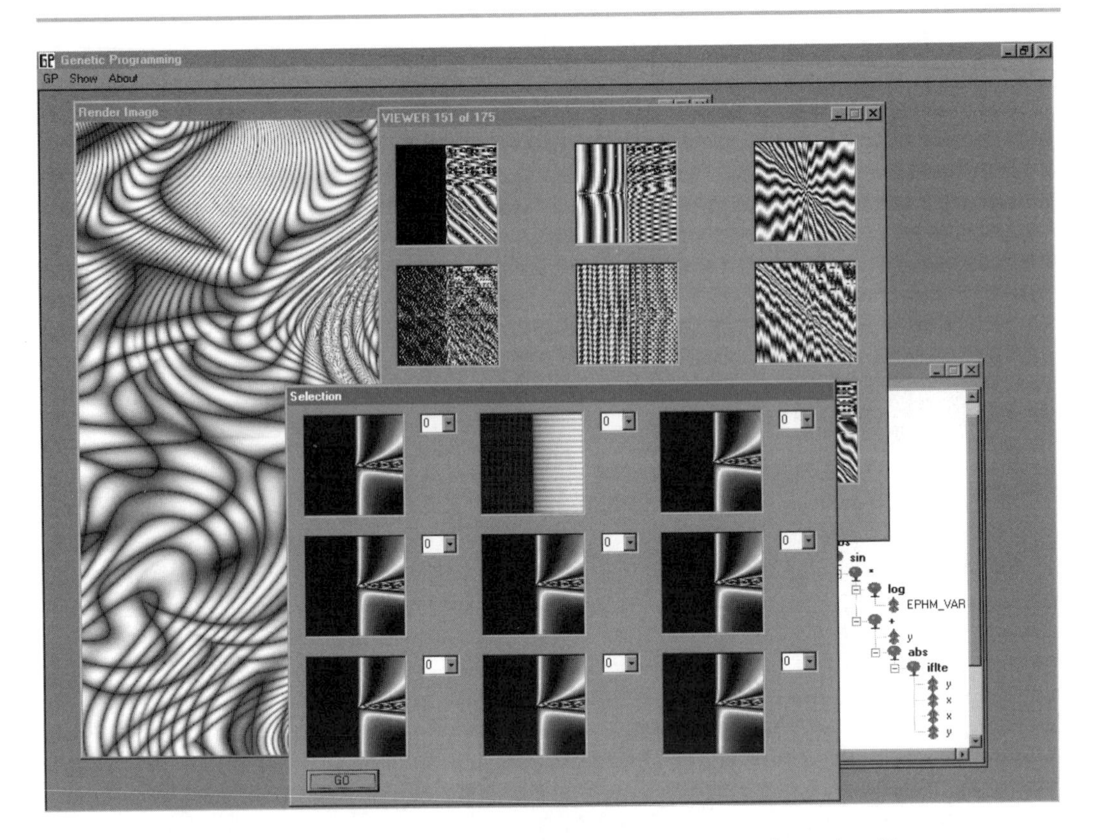

Figure 2.14 Screenshot of an interactive genetic programming system used to produce 2D textures.

The next challenge is to decide on a symbolic technique that will allow us to approximate a shape or form. There are several options. However, for simplicity and because it is a well-understood model we use iterated function systems. An iterated function system (IFS) is a fractal mathematical model that can be used to represent natural shapes and structures. IFSs provide a procedural tool for modeling natural objects in computer graphics applications. 2D and 3D objects can be modeled. However, here we restrict ourselves to the representation of 2D fractal-like shapes that are well studied and the IFS representation of which is known (Barnsley and Hurd 1993, pp. 82–88).

In order to achieve our aim we use an evolutionary algorithm that is a GP–ES hybrid. Because of the nature of IFSs as arrays of floating-point numbers we need an optimization technique well suited to dealing with real-valued vectors (i.e. ES). On the other hand, we require a variable length representation since we don't know before hand the number of codes needed to represent a shape using an IFS (i.e. GP). We call this hybrid approach a "*hierarchical evolution strategy*".

2.4.2.1 Hierarchical Evolution Strategy

Evolution strategies (ES) operate on real-valued vectors. The elements of these real-valued vectors that correspond to the coding of a (optimization) problem are referred to as *object variables* (see the section on ES). Here using notation from Back (1996) an ES individual is defined as a vector a, such as

$$\begin{aligned} &\vec{a} = (\vec{v}, \vec{s}, \vec{r}) \in I = R^n \times R_+^{n_\sigma} \times [-\pi, \pi]^{n_\alpha}, \\ &n_\sigma \in \{1, \ldots, n\} \\ &n_\alpha \in \{0, (2n - n_\sigma)(n_\sigma - 1)/2\}, \end{aligned} \tag{2.1}$$

where the vector v represents *object variables.* Each individual also incorporates the *standard deviation* vector s, as well as the *correlation coefficient* vector r. One of the most important features of ES is variation of *object variables,* which is represented by use of normally distributed mutations (other distributions may also be used for the mutation of object variables). Equally important is the self-adaptation of the *standard deviations* and *correlation coefficients* of the normally distributed mutations (Back 1996, pp. 68–73).

We combine GP and ES using a method that allows us to encode an ES individual as a LISP symbolic expression (S-expression). In the light of this one could implement ES within a GP system by simply allowing ES individuals to be coded as symbolic expressions, composed of GP functions and terminals. The structure of an ES program can be seen as a hierarchical tree as illustrated in Figure 2.15. Such a tree representation allows one to incorporate ES individuals into GP seamlessly. We call this tree an *evolution strategies' individual* (ES individual). We allow ES style mutation as described by Back (1996) to operate on ES individuals. We implemented ES individuals, using *strongly typed genetic programming* (STGP) (Montana 1995) (see the following section). STGP recombination can also be applied to ES individuals.

A *hierarchical evolution strategy* (HES) is defined as an S-expression that contains ES individuals. According to this definition, ES itself is a subset of the hierarchical evolution strategy, defined by a single ES individual. Such a structure (as in ES) will be of fixed size. GP architectures that contain scalar terminals could code these terminals as ES individuals. The potential benefit of this is that each ES individual can be fine-tuned separately using ES-style mutation operations.

The hierarchical evolution strategy has two major potential benefits: (a) it provides a GP architecture that contains a self-adapting mechanism, in terms of mutating scalar terminals and (b) it provides a modified ES that is of variable length. Variable length coding has been reported in early work using ES (Schwefel 1968). Nevertheless, the architecture has to be described in terms of ES individuals, which have to remain of constant length within the evolutionary simulation cycle.

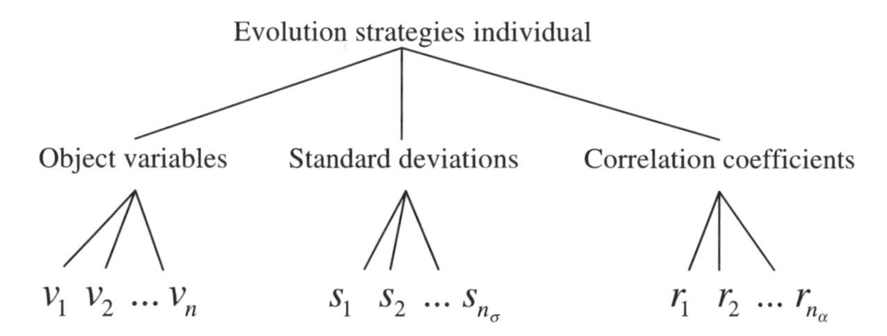

Figure 2.15 A tree that defines an ES individual, corresponding to vector a of Equation (2.1).

2.4.2.2 Strongly Typed Genetic Programming

In GP as described by Koza (1992) individuals are S-expressions composed of two types of functions. Functions that accept arguments are called *non-terminal functions*, and functions that accept no arguments are called *terminal functions*. For brevity we use the term *function* to mean a *non-terminal function*. One important constraint of GP is the property of *closure*. That is, all functions have to accept arguments and return values of the same data type. Koza (1992) has proposed a method to relax *closure* using *constrained syntactical structures*. We use strongly typed GP (Montana 1995) to allow S-expressions that are composed of functions that accept arguments and return values of different data types. The hierarchical evolution strategy is implemented using an STGP system. In STGP we need three sets to describe S-expressions: a set of *data types* D, a set of *functions* F and a set of *terminals* T. In GP as described by Koza only two sets are needed (*functions* and *terminals*). The notation

$$\begin{aligned} &r \leftarrow f(a_1, \ldots, a_n), \quad \text{where} \\ &r \in D, \\ &a_i \in D, \quad i \in \{1, \ldots, n\}, \end{aligned}$$

describes a function f that accepts n arguments and returns a value of data type r, where D is the set of data types. If the number of arguments n is equal to zero, the above notation describes a terminal returning data type r. We use this notation in order to simplify the description of STGP functions and terminals.

2.4.2.3 Iterated Function Systems and the Inverse Problem

IFS offer a method of generating fractal shapes. J. Hutchinson originally developed the idea of IFS (Hutchinson 1981). An affine IFS is a finite set of contractive affine transformations (Barnsley 1993; Barnsley and Hurd 1993). Points in space affected by contractive transformations are always placed closer together. It has been shown that a shape can be represented using IFS by discovering a set of contractive transformations that take that shape onto itself. That is, in order to represent a shape A using an IFS, the union (*collage*) of the shapes of A under the transformations has to generate A. This is known as the *collage theorem*. The notation for an IFS is $\{w_n, n = 1, 2, \ldots, n\}$. Application of an IFS to a finite shape A is defined as a transformation W:

$$W(A) = \bigcup_{i=1}^{n} w_n(A). \tag{2.2}$$

A fractal shape can be constructed by calculating the *attractor* of an IFS. The attractor of an IFS is defined by

$$\lim_{n \to \infty} W^{\circ n}(A), \tag{2.3}$$

where $W^{\circ n}(A)$ is defined by

$$A,\ W^{\circ 1}(A) = W(A),\ W^{\circ 2}(A) = W(W(A)), \ldots,\ W^{\circ n}(A) = W(W^{\circ n-1}(A)).$$

There are two well-known methods of constructing IFS fractals: the *photocopy algorithm* and the *chaos game algorithm* (Lu 1997, pp. 34–41). For a detailed description of IFS see Barnsley (1993). A challenging problem when constructing IFS fractals is the inverse or inference problem. This problem can be seen as a restricted form of the general representation problem we mentioned in the opening section of this case study.

According to the inverse or inference problem for IFS we are asked to find an unspecified number of contractive affine transformations whose attractor is a given image. Here we attempt to solve the inference problem for the Barnsley fern (Barnsley and Hurd 1993, pp. 98–99). The Barnsley fern was chosen because of its complex self-similar shape. The fern shape was also chosen because it has a well-known solution that contains four transformations, each different from the others, one of which is singular (i.e. non-invertible). The shape of the fern leaf is one of many natural forms whose geometry can be represented using IFSs (Barnsley et al. 1988).

2.4.2.4 Architecture Using the Hierarchical Evolution Strategy

An IFS does not contain a predefined number of contractive affine transformations. Encoding is based on S-expressions that represent a list of ES individuals (see Figure 2.15), with each ES individual representing an affine transformation. Therefore we use a coding that allows the representation of a variable length list of affine transformations. The coding is based on S-expressions that contain a *list* of ES individuals.

The set of data types D for this problem consists of

$$D = \{ifs, map, obj, dev, rot, ephObj_i, ephDev_i, ephRot_j\},$$
$$i \in \{1, \dots 6\}, \quad j \in \{1, \dots 15\}.$$

The data type *obj* stores the *object variables* (see Figure 2.15) of an ES individual. The *object variables* in this case correspond to floating-point coefficients $\{a, b, c, d, e, f\}$ of an affine transformation. *dev* is the data type that stores the *standard deviations* of an ES individual. *rot* is the data type that stores rotations that stand for *linearly correlated mutation coefficients* of an ES individual. We have six *object variable* coefficients and six *standard deviation* coefficients. Finally, we have $n(n-1)/2$ *correlation coefficients*, where n is the number of object variables. *ephObj* data types, *ephDev* data types and *ephRot* data types store an ephemeral floating-point variable each, which corresponds to an object variable, a standard deviation and a correlation coefficient, respectively. Each coefficient of an ES individual is given a different data type so it could be recombined only with the same coefficient of a different individual.

ifs stores a list of two or more transformations. *ifs* is the return data type of an S-expression for this problem. The data type *map* stores a list of one or more ES individuals (see Figure 2.15).

The function set F for this problem consists of

$$\begin{aligned} F = \{ifs &\leftarrow f\_ifs(map, map), \\ map &\leftarrow f\_list(map, map), \\ map &\leftarrow f\_map(obj, dev, rot), \\ obj &\leftarrow f\_obj(ephObj_1, ephObj_2, \dots, ephObj_6), \\ dev &\leftarrow f\_dev(ephDev_1, ephDev_2, \dots, ephDev_6), \\ rot &\leftarrow f\_rot(ephRot_1, ephRot_2, \cdots, ephRot_{15})\}. \end{aligned}$$

The function *f_ifs* is used to ensure that IFS have two or more affine transformations. The function *f_list* is used to connect affine transformations in a list.

The terminal set T for this problem consists of

$$T = \{ephObj_i \leftarrow t\_objv_i(),\; ephDev_i \leftarrow t\_dev_i(),\; ephRot_j \leftarrow t\_rot_j()\},$$
$$i \in \{\ 1, \ldots 6\}, \quad j \in \{1, \ldots 15\}.$$

Functions, terminals and data types are designed to construct S-expressions that mirror the graph in Figure 2.15.

2.4.2.5 Fitness Function

Two fitness functions were tested: the *Nettleton* fitness and the *overlap* fitness. The *Nettleton fitness* was also used in Sarafopoulos (1999, 2001), and was originally proposed by Nettleton and Garigliano (1994). The *overlap fitness* combines the Nettleton fitness with a calculation of the pixel overlap of affine transformations of an IFS.

Here we deal with (fractal) *shapes* that are black-on-white drawings. Strictly speaking fractal shapes are defined as compact subsets of 2D Euclidean space (Barnsley 1993). However, in terms of the fitness function the universal set U is the set of pixels of a 128×128 bitmap. A *shape* is defined as a subset of the set of pixels of a 128×128 bitmap. We are searching though the set H of all subsets of U. The *target shape* or *target* for this experiment is the *shape* of the Barnsley fern. Let $N(a)$ be the number of pixels of *shape* a. Let the *collage*, for a given IFS w_n, be the *shape* created by applying the transformation W (see Equation 2.1) to the *target shape*.

We calculate the standardized fitness of individuals, in the Nettleton case by

$$error = N(target) + N(collage \cap target^{-1}) - N(collage \cap target), \tag{2.4}$$

where a^{-1} is the complement of set a. That is, each individual in the population strives to produce a collage that covers pixels of the target shape. The more pixels of the *target* an individual covers and fewer pixels of the rest of the image, the greater the reward. Nettleton fitness can lead to the evolution of "opportunistic" transformations. That is, transformations that cover a large portion of the target will tend to spread quickly among the population. However, such transformations are not necessarily optimal, nor do they necessarily lead to a solution. One way to reduce this effect is to introduce a term in the fitness function which punishes individuals that contain transformations that overlap, thus reducing the spread of a transformation that just happens to cover many pixels of the target and therefore encouraging transformations that cover a small number of pixels.

We calculate the standardized fitness in the *overlap* case as $error + sharing$, where *error* is defined in Equation (2.4), and *sharing* is defined by

$$sharing = \sum_{j=1}^{n} \sum_{\substack{i=1 \\ i \neq j}}^{n} N(w_j(target) \cap w_i(target)), \tag{2.5}$$

where n is the number of affine transformations of a given IFS.

In order to avoid non-contractive IFS individuals, any IFS individual with contractivity greater than 0.85 was not selected for reproduction.

2.4.2.6 Control Parameters

In the case of the HES coding, the selection scheme was a (μ, λ)-ES-type elitist selection with $\mu = 9$ and $\lambda = 16\,000$. Affine transformations for individuals in the initial population were generated by ES mutation of an affine transformation with $\{a, b, c, d, e, f\} = \{0.2, 0.0, 0.0, 0.2, 0.0, 0.0\}$. Standard deviations were initialized to 0.4. The initial values for correlation coefficients were set to 0. ES style mutation operation was allowed to help fine-tune the parameters of transformations within a given IFS. The ES mutation operation was applied to a randomly selected ES individual of an S-expression. The affine transformations were not directly constrained, and therefore were not forced to be contractive after mutation. However, in order to reduce mutations that produce non-contractive maps we repeated the mutation operation until a contractive individual was found or a maximum number of iterations were reached. The population number was set to 16 000, the generation number was set to 30 and the number of runs to 6.

2.4.2.7 Results

The results of a successful run are shown in the following figure and they depict the evolution of the fern shape starting with noise, i.e. a random initial population. Each section/image of Figure 2.16 represents the best individual in a generation. Generations increment from top to bottom, and left to right. The only exception to this rule is the image in the top left-hand corner of Figure 2.16, which depicts the target shape. The attractors of the evolved IFSs have been rendered and then post-processed using a chamfer distance mask (Borgefors 1984), which creates the shading effect for visualizing the distance of points in an image from the IFS attractor.

2.5 Conclusions

We have demonstrated how evolutionary algorithms could be used in the context of computer graphics as an artistic meta-tool. Only still images are presented here, however the techniques can be extended to generate animation sequences (this could be achieved by interpolation between texture representations or IFS codes). Currently there are two major pathways being explored. The first is the use of EA as a tool for interactive exploration of procedural models, such as textures. The second path involves automatic exploration of features important to computer animation using EA as an optimization tool. In this case strict criteria need to be applied to evolve new forms. This is usually considered problematic if these forms are to claim some aesthetic visual quality. We propose a method that avoids this problem; that is "visual representation". The aesthetic judgment is made indirectly before starting the process of evolving new forms by deciding on a subject matter and a method of (symbolic) representation that is aesthetically pleasing. We demonstrated this approach by generating a sequence of images that gradually approximate the shape of a fern. This provides a proof of principle that our method can be applied in practice. Currently we are working on techniques that will allow us to improve the efficiently of the EA so shapes and images of greater complexity can be depicted.

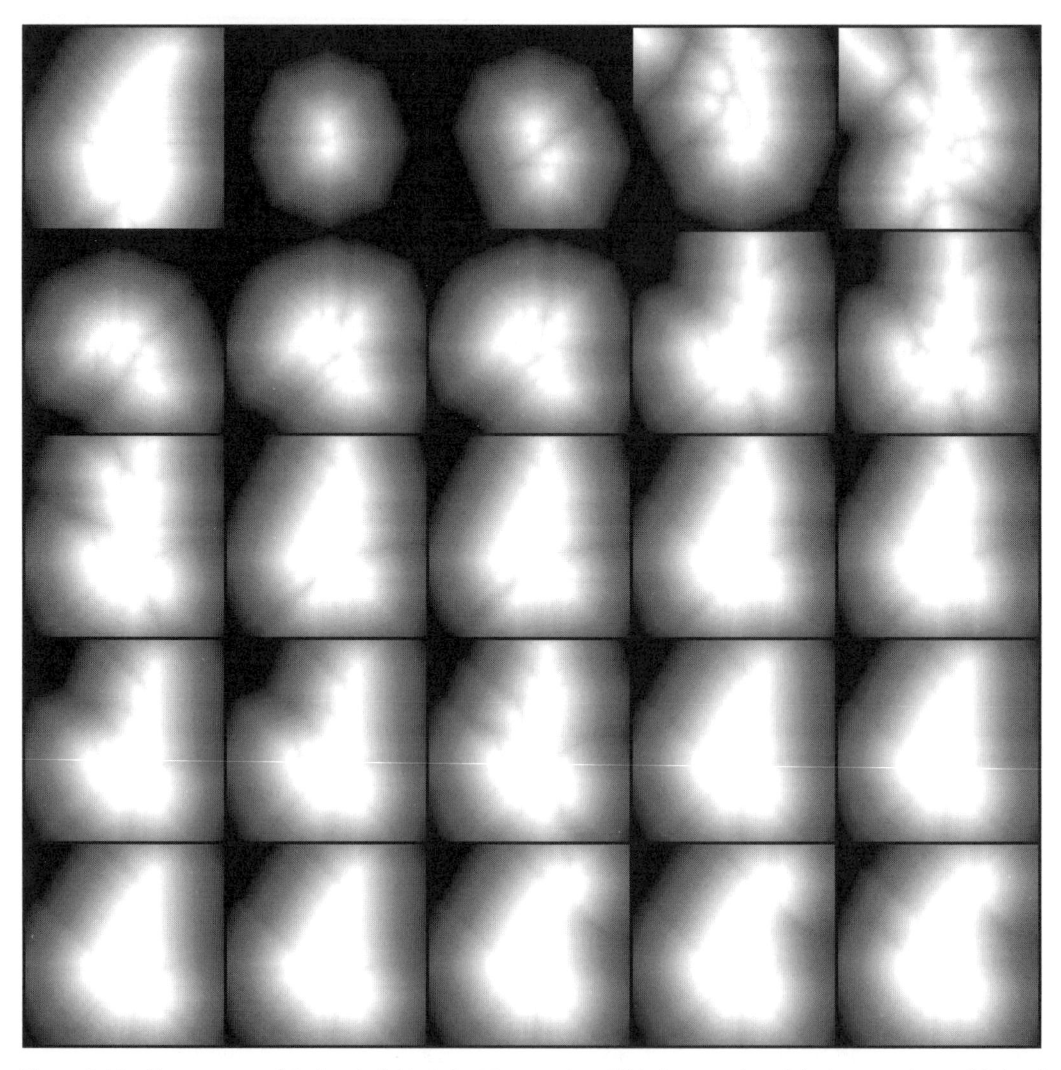

Figure 2.16 The attractors of the best individuals for 24 generations. With the exception of the image at the top left-hand corner, which depicts the target shape, generations increment form top to bottom, and left to right.

Bibliography

Back, T. *Evolutionary Algorithms in Theory and Practice*, 1996, Oxford University Press

Banzhaf, W., Nordin, P., Keller, R. and Francone, F.D. *Genetic Programming an Introduction*, 1998, Morgan Kaufmann

Barnsley, M.F., Jacquin, A., Mallassenet, F., Rueter, L. and Sloan, A.D. Harnessing chaos for image synthesis, *Computer Graphics* 22(4), 131–140, 1988

Barnsley, M.F. *Fractals Everywhere*, 2nd edn, 1993, Academic Press

Barnsley, M.F. and Hurd, L.P. *Fractal Image Compression*, 1993, AK Peters

Bentley, P.J. (ed.) *Evolutionary Design by Computers*, 1999, Morgan Kaufmann

Bentley, P.J. and Corne, D.W. (eds.) *Creative Evolutionary Systems*, 2001, Morgan Kaufmann

Bethke, A.D. Genetic algorithms as function optimizers, *Ph.D. Dissertation*, University of Michigan, Ann Arbor, 1981
Beyer, H.-G. Toward a theory of evolution strategies: on the benefit of sex – the $(\mu/\mu, \lambda)$-theory, *Evolutionary Computation*, **3**(1), 81–111, 1995
Beyer, H.-G. Toward a theory of evolution strategies: self-adaptation, *Evolutionary Computation*, **3**(3), 311–347, 1996
Beyer, H.-G. *The Theory of Evolution Strategies, Natural Computing Series*, 2001, Springer-Verlag
Borgefors, G. Distance transformation in arbitrary dimension, *Computer Vision, Graphics, and Image Processing*, **27**, 231–345, 1984
Caruana, R.A. and Schaffer, J.D. Representation and hidden bias: Gray vs. binary coding for genetic algorithms, *Proc. 5th International Conference on Machine Learning*, 1988, Morgan Kaufmann
Collins, R. and Jefferson, D. Ant farm: toward simulated evolution, *Artificial Life*, Vol. II, Santa Fe Institute studies in the Sciences of the Complexity, Langton C. et al. (eds.), 1991, Addison Wesley
Cramer, N.L. A representation for the adaptive generation of simple sequential programs, *Proc. International Conference on Genetic Algorithms and the Applications*, pp. 183–187, 1985, Lawrence Erlbaum
Darwin, C. *The Origin of Species*, New American Library, 1859, Mentor paperback
Dawkins, R. *The Blind Watchmaker*, 1986, Harlow Logman
Dawkins, R. The evolution of evolvability, *Artificial Life Proceedings*, pp. 201–220, 1989, Addison-Wesley
De Jong, K.A. An analysis of the behavior of a class of genetic adaptive systems, *Ph.D. Thesis*, University of Michigan, Ann Arbor, 1975
D'haeseleer, P. Context preserving crossover in genetic programming, *Proc. 1994 IEEE World Congress on Computational Intelligence*, pp. 256–261, Vol. 1, 1994, IEEE Press
Ebert, D.S., Musgrave, F.K., Peachey, D., Perlin, K., Worley, S. *Texturing and Modeling: a Procedural Approach*, 2nd edn, 1998, AP Professional
Fogel, L.J. *Intelligence through Simulated Evolution*, 1999, John Wiley & Sons
Furuta, H., Maeda, K. and Watanabe, W. Application of genetic algorithm to aesthetic design of bridge structures, in *Microcomputers in Civil Engineering*, pp. 415–421, 1995, Blackwell Publishers
Gen, M. and Cheng, R. *Genetic Algorithms and Engineering Optimization*, 1999, John Wiley & Sons
Goldberg, D.E. *Genetic Algorithms in Search, Optimization, & Machine Learning*, 1989, Addison Wesley
Griffiths, D. and Sarafopoulos, A. Evolving behavioural animation systems, *Artificial Evolution, LNCS*, Vol. 1829, pp. 217–230, 1999, Springer-Verlag
Gritz, L. Evolutionary controller synthesis for 3-D character animation, *Ph.D. Thesis*, The George Washington University, 1999
Gritz, L. and Hahn, J.K. Genetic programming for articulated figure motion, *Journal of Visualization and Computer Animation*, **6**(3), 129–142, 1995
Gritz, L. and Hahn, J.K. Genetic programming evolution of controllers for 3-D character animation, *Genetic Programming 1997: Proc. 2nd Annual Conference*, pp. 139–146, 1997, Morgan Kaufmann
Gruau, F. Modular genetic neural networks for six-legged locomotion, *Artificial Evolution*, Alliot, J.-M., Lutton, E., Ronald, E., Schoenauer, M., Snyers, D. (eds.), *LNCS*, Vol. 1063, pp. 201–219, 1996a, Springer-Verlag
Gruau, F. On using syntactic constraints with genetic programming, *Advances in Genetic Programming*, Vol. 2, pp. 377–394, 1996b, MIT Press
Hansen, N., Ostermeir, A. and Gawelczyk, A. On the adaptation of arbitrary normal mutation distributions in evolution strategies: the generating set adaptation, in Eshelman, L.J. (ed.) *Proc. 6th International Conference on Genetic algorithms*, pp. 57–64, 1995
Hansen, N. and Ostermeir, A. Adapting arbitrary normal mutation distributions in evolution strategies: the covariance matrix adaptation, *Proc. IEEE 1996 International Conference on Evolutionary Computation*, pp. 312–317, 1996, IEEE Press
Herdy, M. Optimization of a two-phase nozzle with an ES, EvoNet Flying Circus Demo, http://www.wi.leidenuniv.nl/~gusz/Flying_Circus/3.Demos/Movies/Duese/index.html, 2001
Holland, J. Genetic algorithms and the optimal allocation of trials, *SIAM Journal on Computation*, **2**, 88–105, 1973
Holland, J.H. *Adaptation in Natural and Artificial Systems*, 2nd edn, 1992, MIT Press
Hollstien, R.B. Artificial genetic adaptation in computer control systems, *Ph.D. Thesis*, University of

Michigan, Ann Arbor, 1971
Hutchinson, J.E. Fractals and self-similarity, *Indiana University Journal*, **35**(5), 1981
Ibrahim, A.E., Genshade: an evolutionary approach to automatic and interactive procedural texture generation, *Doctoral Thesis*, Office of Graduate Studies of Texas A&M University, 1998
Jacob, C. Evolving evolution programs: genetic programming and L-systems, *Genetic Programming 1996: Proc. 1st Annual Conference*, pp. 107–115, 1996, MIT Press
Kang, Y.-M., Cho, H.-G. and Lee, E.-T. An efficient control over human running animation with extension of planar hopper model, *Journal of Visualization and Computer Animation*, **10**(4), 215–224, 1999
Keller, R.E., Banzhaf, W., Mehnen, J. and Weinert, K. CAD surface reconstruction from digitized 3D point data with a genetic programming/evolution strategy hybrid, *Advances in Genetic Programming*, Vol. 3, pp. 41–65, 1999, MIT Press
Kimura, M. *The Neutral Theory of Molecular Evolution*, 1983, Cambridge Univ. Press
Kirpatrick, S., Gelatt, C.D., Vecchi, M.P. Optimization by simulated annealing, *Science*, **220**, 671–680, 1983
Klockgether, J. and Schwefel, H.-P. Two-phase nozzle and hollow core jet experiments, In D.G. Elliott, (ed.), *Proc. 11th Symp. Engineering Aspects of Magnetohydrodynamics*, California Inst. of Technology, Pasadena, CA, pp. 141–148, March 1970
Koza, J.R. *Genetic Programming: on the Programming of Computers by Means of Natural Selection*, 1992, MIT Press
Koza, J.R. *Genetic Programming II: Automatic Discovery of Reusable Programs*, 1994, MIT Press
Koza, J.R., Bennett III, F.H., Andre, D. and Keane, M.A. *Genetic Programming III: Darwinian Invention and Problem Solving*, 1999, MIT Press
Langdon, W.B. *Genetic Programming and Data Structures*, 1998, Kluwer Academic Publishers
Langdon, W.B. and Poli, R. *Foundations of Genetic Programming*, 2001, Springer-Verlag
Lewis, M. Aesthetic evolutionary design with data flow networks, *4th International Conference and Exhibition on Generative Art*, 2000, http://www.accad.ohio-state.edu/~mlewis/
Lewis, M. Visual aesthetic evolutionary design links, http://www.accad.ohio-state.edu/~mlewis/aed.html, 2001
Lim, I.S. and Thalmann, D. How not to be a black-box: evolution and genetic engineering of high-level behaviours, *Proc. Genetic and Evolutionary Computation Conference*, Vol. 2, pp. 1329–1335, 1999, Morgan Kaufmann
Lu, N. *Fractal Imaging*, 1997, Academic Press
Lund, H., Pagliarini, L. and Miglino, O. Artistic design with GA and NN, *Proc. 1st Nordic Workshop on Genetic Algorithms and Their Applications (1NWGA)*, Univ. Vaasa, Finland, xiii+417 pp. 97–105, 1995
Magurran, A.E. *Ecological Diversity and its Measurement*, 1988, Princeton University Press
Man, K.F., Tang, K.S. and Kwong, S. *Genetic Algorithms*, 1999, Springer-Verlag
Martin, E. and Hine, R.S. (eds.), *A Dictionary of Biology*, 2000, Oxford University Press Market House Books
Mitchell, M. *An Introduction to Genetic Algorithms*, 1996, MIT Press
Montana, J.D. Strongly typed genetic programming, *Evolutionary Computation*, **3**(2), 199–230, 1995
Nettleton, D.J. and Garigliano, R. Evolutionary algorithms and the construction of fractals: solution of the inverse problem, *Biosystems*, **33**, 221–231, 1994
O'Reilly, U-.M. and Oppacher, F. The Troubling aspects of a building block hypothesis for genetic programming, In L.D. Whitley and M.D. Vose (eds.), *Foundations of Genetic Algorithms*, Vol. 3, pp. 73–88, 1995, Morgan Kaufmann
Perlin, K. An image synthesizer, *Computer Graphics*, **19**(3), 287–296, 1985
Poli, R. and Langdon, W.B. Genetic programming with one-point crossover, in Chawdhry, P.K., Roy, R. and Pan, R.K. (eds), *Soft Computing in Engineering Design and Manufacturing*, pp. 180–189, 1997, Springer-Verlag
Provine, W.B. *Sewall Wright and Evolutionary Biology*, 1986, The University of Chicago Press
Rechenberg, I. Cybernetic solution path of an experimental problem, Royal Aircraft Establishment, Library translation No 1122, Farnborough, Hants., UK, August 1965
Rechenberg, I., *Evolutionsstrategie: Optimierung technischer Systeme nach Prinzipien der biologischen Evolution*, 1973, Frommann-Holzboog, Stuttgart
Reynolds, C.W. An evolved, vision-based behavioral model of coordinated group motion, *From Animals to Animats (Proc. Simulation of Adaptive Behaviour)*, 1992, MIT Press

Reynolds, C.W. An evolved vision-based behavioral model of obstacle avoidance behaviour, *Artificial Life III, SFI Studies in the Sciences of Complexity*, Vol. XVII, pp. 327–346, 1994a, Addison-Wesley

Reynolds, C.W. Evolution of obstacle avoidance behaviour: using noise to promote robust solutions, *Advances in Genetic Programming*, pp. 221–241, 1994b, MIT Press

Reynolds, C.W. The difficulty of roving eyes, *Proc. 1994 IEEE World Congress on Computational Intelligence*, pp. 262–267, 1994c, IEEE Press

Reynolds, C.W. Competition, coevolution and the game of tag, *Proc. 4th International Workshop on the Synthesis and Simulation of Living Systems*, pp. 59–69, 1994d, MIT Press

Reynolds, C.W. Evolution of corridor following behavior in a noisy world simulation of adaptive behaviour, *Proc. Simulation of Adaptive Behavior '94*, Cliff, D., Husbands, P., Meyer, J.A. and Wilson, S.A. (eds), Bradford, 1994e

Ridley, M. *Evolution*, 2nd edn, 1996, Blackwell Science

Russell, S., Norvig, P. *Artificial Intelligence a Modern Approach*, 1995, Prentice Hall International

Sarafopoulos, A. Textures, Animation sequence shown at the *"Cabaret électronique" 6th International Symposium on Electronic Art*, Montreal, Canada, 1995

Sarafopoulos, A. Automatic generation of affine IFS and strongly typed genetic programming, *Genetic Programming, Proc. EuroGP1999, LNCS*, Vol. 1598, pp. 149–160, 1999, Springer-Verlag

Sarafopoulos, A. Evolution of affine transformations and iterated function systems using hierarchical evolution strategy, *Genetic Programming Proc. EuroGP2001, LNCS*, Vol. 2038, pp. 176–191, 2001, Springer-Verlag

Schwefel, H.-P. Experimentelle Optimierung einer Zweiphasendüse Teil I, AEG Research Institute, Berlin, *Technical Report* No. 35 of the Project MHD-Staustrahlrohr, No 11.034/68, 1968

Schwefel, P.H. *Evolution and Optimum Seeking*, 1995, John Wiley & Sons

Sedivy, J.M. and Joyner, L.A. *Gene Targeting*, 1992, Oxford University Press

Sims, K. Artificial evolution for computer graphics, *Computer Graphics Siggraph '91 Proceedings*, pp. 319–328, 1991

Sims, K. Interactive evolution of dynamical systems, *Towards a Practice of Autonomous Systems: Proc. 1st European Conference on Artificial Life*, pp. 171–178, 1992, MIT Press

Sims, K. Interactive evolution of equations for procedural models, *The Visual Computer*, pp. 466–476, 1993a, Springer-Verlag

Sims, K. *Genetic Images*, Media installation allowing the interactive evolution of abstract still images, Exhibited at the Centre Georges Pompidou in Paris, Ars Electronica in Linz, Austria, and the Interactive Media Festvial in Los Angeles, 1993b

Sims, K. Evolving virtual creatures, *Computer Graphics Siggraph '94 Proceedings*, pp. 15–22, 1994a

Sims, K. Evolving 3D morphology and behavior by competition, *Artificial Life IV Proceedings*, Brooks, R. and Maes, P. (eds.), pp. 28–39, 1994b, MIT Press

Sims, K. *Galápagos*, Media installation allowing museum visitors to interactively evolve 3D animated forms, Exhibited at the ICC in Tokyo and the DeCordova Museum in Lincoln Mass, 1997

Syswerda, G. A study of reproduction in generational and steady-state genetic algorithms, in *Foundations of Genetic Algorithms*, Rawlins G.J.E. (ed.), pp. 94–101, 1991, Morgan Kaufmann

Teller, A. Turing completeness in the language of genetic programming with indexed memory, *Proc. 1994 IEEE World Congress on Computational Intelligence*, Vol. 1, pp. 136–141, 1994, IEEE Press

Todd, S. and Latham, W. Mutator, a Subjective human interface for evolution of computer sculptures, IBM UK, *Scientific Centre Report* 248, 1991

Todd, S. and Latham, W. *Evolutionary Art and Computers*, 1992, Academic Press

Whigham, P.A., Grammatical bias for evolutionary learning, *Ph.D. Thesis*, School of Computer Science, University College University of New South Wales, Australian Defence Force Academy, 1996

Wiens, A.L. and Ross, B., J. Gentropy: evolutionary 2D texture generation, late breaking papers at the *2000 Genetic and Evolutionary Computation Conference*, pp. 418–424, 2000

Wiens, A.L. and Ross, B., J. Gentropy: evolutionary 2D texture generation, *Computers and Graphics Journal*, **26**, 75–88, 2002

Wright, S. Evolution in Mendelian populations, *Genetics*, **16**, 97–159, 1931

Zhao, K. and Wang, J. Path planning in computer animation employing chromosome-protein scheme, *Genetic Programming 1998, Proc. 3rd Annual Conference*, pp. 439–447, 1998, Morgan Kaufmann

3 Shooting Live Action for Combination with Computer Animation

Mitch Mitchell

3.1 Introduction

Computer animation is often combined with live-action material, and is a process, which consists of three separate stages: creating the computer animation, photographing the live action and combining the two into a seemingly single integral image. Such shots may be divided into two types. On the one hand are the computer graphic shots, which are almost complete in themselves, apart from requiring the addition of one or more live-action elements. An example of this type would be where a CGI space station has been created, and needs the addition of some actors seen within, to give it the final ring of authenticity. The alternative scenario, is where computer-animated elements are to be added to an otherwise live-action scene. An example of this would be when we are now within the space station – a physically built set – but need to see parts of its exterior through the window. These would obviously be CGI elements which, in this case, have to be added in to the live action.

All live-action–CGI combinations will fit into one of the above two categories. The final step in their creation is combining the CGI and live-action elements created in the first two stages. This process is generally referred to as *compositing* and can take many forms, both analogue and digital (although the former is becoming less popular with every day that passes). In general, although there are exceptions, the compositing is done by a third party separate from either the live-action crew or the CGI team. Usually the CGI elements will be output to tape or a data receptacle in the form of a master (the actual desired image or components thereof such as front light, highlights, shadows, etc.) and perhaps, a matte which enables the master to be automatically delineated, for combining with other elements. For example, a satellite model might also have a matte, which corresponds on a frame-by-frame basis to its external bounding edges. This matte could then be used to composite the satellite into any other pictorial element it might need to be combined with. Similarly, live-action material will be digitized and converted into a common format with the CGI elements, so that they can be manipulated together, in a common environment.

The success (i.e. believability) of a composite image is dependent on two things: the technical quality of the combination itself and the much more difficult to analyze match between the elements that have been combined. In the case of live-action elements the former involves mainly, although not entirely, technical attributes such as the alignment of matte edges, the elimination of keying colors from the foreground and the optimum adjustment of keying threshold, color slice, opacity and so on. This subject is dealt with in detail in Chapter 4.

The other process on which the success of a combined image depends, is that of matching the live action and CGI elements. This can often be a very difficult thing to do, with some of the mismatches being very subtle and difficult to observe. This is one of the main reasons why matte shots are so often technically perfect, but for some unknown reason, just do not "gel". The requirements of shooting live-action material specifically for combination with CGI is therefore almost entirely the art of "matching" the two types of image together, and feeding back information to the CGI team, so that they can match the "real"-world shooting situation. It is with this particular part of the equation that this chapter deals.

3.2 Who Matches Whom?

Although at first sight it should be very simple for the imagers working on whichever element is created second, to match to the already shot material, this is not necessarily the case. In the CGI world there are few limitations to the way in which a subject is imaged. Ultra-wide angles with no distortion are easy, as are lighting set-ups with invisible luminaires shining from positions directly between the observation point and the subject. If the live-action shoot were to mimic this, it could well be that lighting would have to come from a slightly different direction (thus causing different shadows and differently positioned highlights). Additionally, a matching lens would be impossible, so a less wide angle would have to be used, thus adding mismatching perspective lines and impossible to match distortions to the side of frame. In this example as with all CGI–live-action matching scenarios, the only solution is to pre-plan the shot and preferably storyboard it, so that what is required of both teams is something actually attainable. Having got both sides into the same ballpark, it is then a much simpler matter to match one to the other.

If the funds are available then a pre-visualization stage might be used. In this, low-resolution CGI is created which represents the intended end product of the entire process, including the live action. This may incorporate either CGI-generated approximations of the live action or "scratch" or "sketchpad" video of the live action, perhaps done in an office or other stand-in for the final setting. Some directors like to cut these *pre-viz* scenes into the rough cut, and then gradually replace them shot-by-shot, and even layer-by-layer, as the final composites are completed – a fantastic bonus for the producers, who have a viewable cut at any time they need to assess progress.

3.3 Information Required for a Shoot

Ideally it is better for the CGI to be matched to the live action rather than the other way around, because computer-generated elements can be imbued with an almost infinite variation in parameters. For example, lens angle and lighting position can be almost anything in the virtual world, whereas in the real world, both may be very limited indeed. However, in many circumstances it may be necessary for the CG elements to be generated first and then matched by the live-action team. Two examples might be, where the live action is a small insert into an otherwise wholly computer-generated setting, or, where the schedule requires that the live action be shot in the last weeks of a project, on which the CGI has taken many months to create.

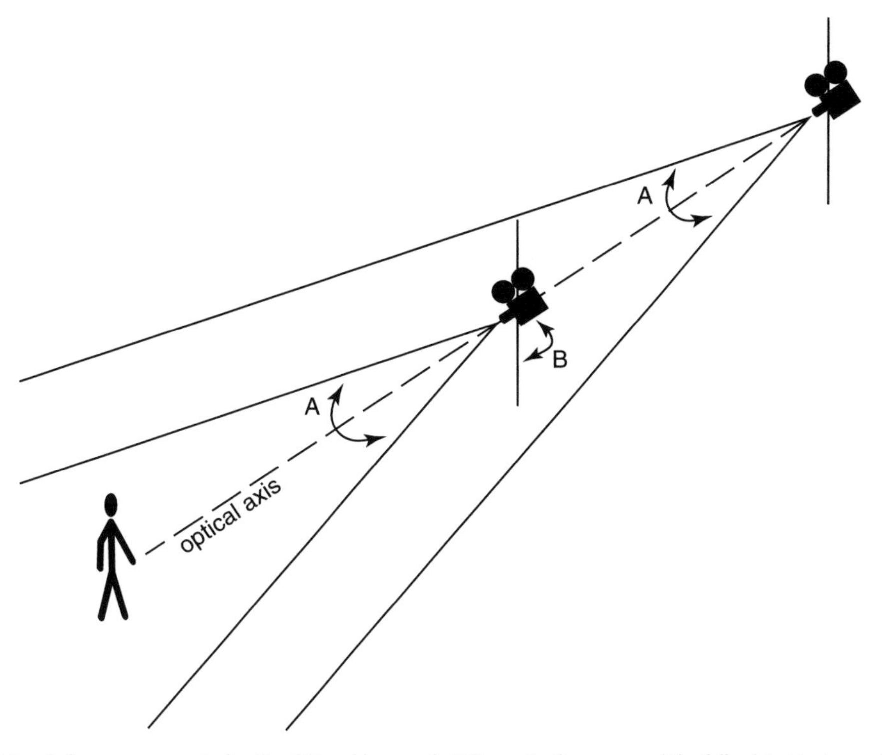

Figure 3.1 As long as camera inclination "B" and lens angle "A" remain the same and the full subject is shown, then the camera can be moved to any point along the optical axis and resized in post-production.

In these circumstances the CGI team must be aware of real-world shooting limitations. Most important of these is the *taking lens*. For example, what range of focal lengths does the cameraman have at his disposal? It is absolutely essential that the angle of view and distortion of the virtual lens and the real lens are identical. The focal length is an easy one, the CGI department must simply be provided with a list of the lenses used on the shoot. Matching the distortion is something that can be achieved at the compositing stage, and will be discussed below.

The next consideration is the effective f/stop, which also dictates to depth of field. If the CGI and live action both cover a number of planes in the image, it is important that the depth of field characteristics of both match, otherwise focus will be inconsistent across the depth of the image. It should be remembered that unless shooting outside in bright sunlight, it is very unlikely that the cameraman will be able to make a stop of f/16. Indeed, most film cameramen like to work with a shallow depth of focus, so as to delineate their subject matter more effectively.

The next issue is where to place the camera. Whether the real camera is in a studio or on a location, the virtual camera must be placed in a position, which it can match. It cannot be positioned beyond the walls of a four-walled studio for instance, with the one exception being where the live-action element is going to be placed at a reduced size within the CGI master. If the correct focal length is used, and the camera can be set at the same angle, then it can indeed be closer, but the camera must be accurate in those two particular specifics: the camera to subject angle and the lens angle (see Figure 3.1). This latter example is particularly relevant to shooting a blue screen composite where the actor against blue is going to be very small in the frame. As long as the perspective lines and angle to the subject match, and assuming that the action does not break the frame, then the subject may be much bigger in frame, than would be required if it were not going to be resized in post-production.

Camera movement must be within the boundaries of possibility if it is to be matched by a corresponding live-action move. A virtual camera is effectively floating in space and is not bounded by the laws of gravity or friction – but the one in the studio is! Thus, care must be taken to provide movement that can be matched both in speed and in space. Of course, if the shoot involves motion control and non-human subject matter, then this rule can be relaxed slightly, because using stop-motion techniques the motion-control camera can also perform movements that are impossible in the real world. With regard to this particular parameter, it is essential to discuss any movement before committing to it. Some software does now exist to check virtual camera moves against those of a motion-control (MoCo) rig, so as to ensure that the CGI department is not moving the virtual camera in a manner that a MoCo rig cannot emulate. If such software can be applied to the rig and CGI systems in use, then it is worth employing.

Mismatches in lighting are one of the most common "give-aways" in composite shots. Even quite subtle differences seem to be picked up quickly by the human eye and so should be avoided. The CGI team should be aware of the limitations of lighting in the actual live-action shot. Lights must be put in positions that are physically possible in the location or set, and should be types of lamp practical in the shooting circumstances. For example, an infinite soft light stretched across the scene, where live action is to be shot at night in a forest, is not really very helpful.

Image format parameters should be checked to make sure that the same format is being used as will be expected by the live-action crew. Aspect ratio, frame rate and whether the image will be originated on film or video should be ascertained and matched. If the live action is to be shot on film, then 24, 25 fps or whatever frame rate should be used in progressive mode – 50 or 59.94 fps field interlace should not be used. Video has two fields per frame. In film images on video both of these relate to the same moment in time, whereas original video images would have both fields representing different moments in time. Thus, when compositing is carried out, an interlaced CGI image will have to advance by two pictures for every one of the film images. For example, consider a live-action character (shot on film) who is to be shown running along the top of a moving CGI model (rendered in fields). In the first field the live-action element will take a step and the model will also move on field one. On field two the live-action character will not move, but, the model will move again. On the third field both will move again, and on the fourth only the model will move. Thus the actor will appear to *jitter* along with the model, and by this separate and different motion will reveal himself to be a separate element from the model, and the illusion of coincidence will be lost (see Figure 3.2). Alternatively, where the CGI is being added to video-originated material, then it must be interlaced just as the electronic images are or the same situation will occur in reverse.

If the CGI and live action are shot in different aspect ratios, then one or the other will have to be cropped for combination into a single image (see Figure 3.3). If anamorphic wide screen were being used for one component, but not the other, then this would make things even more complex. The result would be a definite loss of quality in one of the elements and an obvious visual mismatch in grain, optical quality and probably resolution

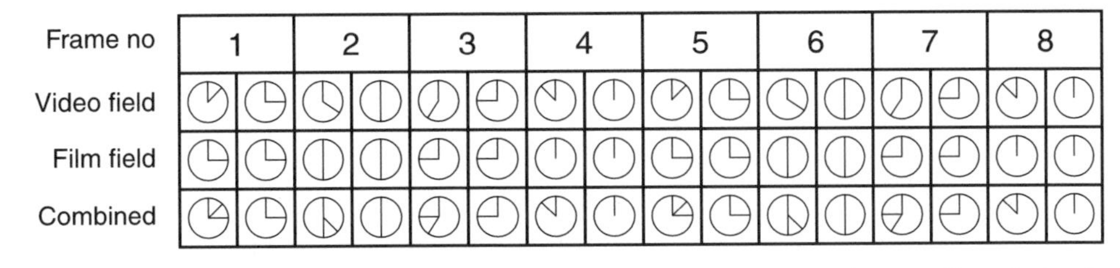

Figure 3.2 The two fields of a film frame are identical because in telecine they are scanned twice and in projection they are exposed twice. Note: video fields and film fields individually produce smooth and consistent temporal flow, but the combination results in a juddery effect with a double image every other field.

too. This would be because only a small portion of the anamorphic image could be used and enlarging its particular optical look would make it appear visibly different from the "flat" material.

Where elements originating from television systems are to be mixed with those coming from other sources, extra care should be taken. 576 (Euro) or 480 (US) by 720 pixels are the sizes of standard digital TV frames, but they do not have square pixels as is standard in all other computer systems. European TV pixels are 1.067:1 (i.e. wider than square) and US pixels are 1:1.125 (i.e. taller than square). Obviously film or any other originating medium (including CGI) will be converted into this format when transferred into the PAL or NTSC domain. However, care must be taken in going the other way, since if the pixels are assumed to be square, then the resulting frame will be squeezed in one direction or another. Most professional software supplies tools to deal with this peculiarity, but it requires somebody to decide to use it!

Having endeavored to create images that can be matched by the live-action unit, it is necessary to supply them with data, which will enable them to quickly and painlessly

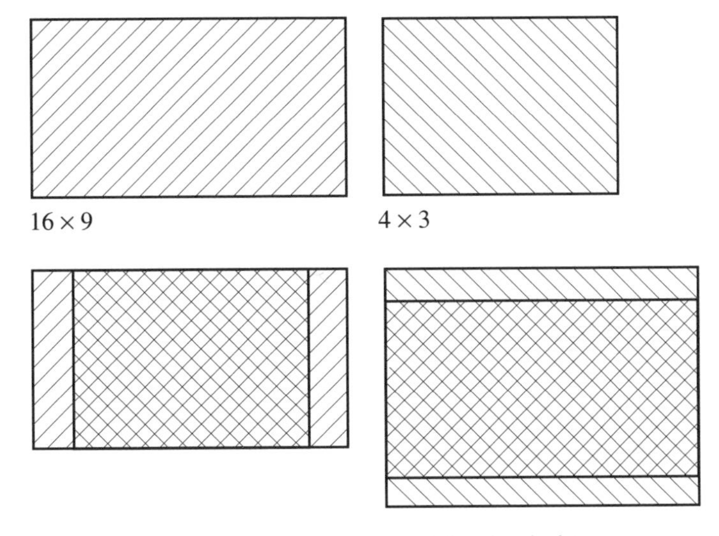

Figure 3.3 Neither way of combining fully utilizes both images.

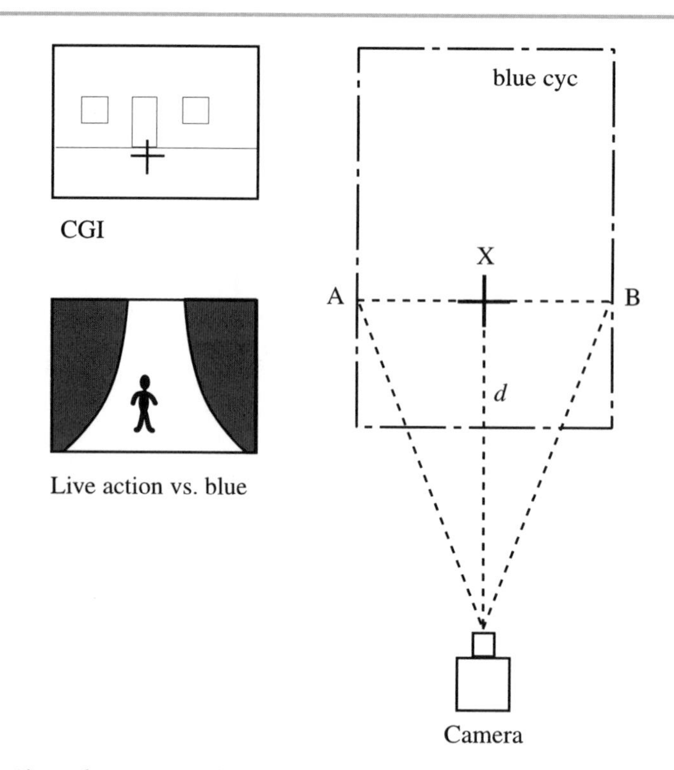

Figure 3.4 CGI provides a reference cross and distances *d* (camera to X), AX and BX, camera to A and camera to B, which can be replicated on the live-action set. If the lens angle and height are known then the subject can be accurately placed.

position their camera in the same relative position as the virtual camera in the CGI element. For this they will need to know the lens, the subject distance (or focus setting if possible) and other spatial information so as to physically position the camera, and pan and tilt it in the correct direction. Small adjustments are possible and usually necessary, but at the very least, a ballpark position should be easily found, from which those necessary adjustments are made. Useful measurements might be the distance from the lens to the subject, the height of the lens above the floor, the distance along the floor immediately below the camera to a subject, and the inclination of the camera.

If it is possible to provide a "cross" in the CGI image in the position where the subject of the live action is to be set, and measurements made to and from this, then the job is even easier (see Figure 3.4). In addition to camera information, a lighting plot clearly delineating heights and plan positions of all the lights used, and their respective beam angles (if hard) and relative diffusion if soft. This should be a scaled diagram, which can be accurately utilized on the set.

If the shoot is using a computer-controlled camera-movement system such as motion control, then it is important to liaise with the operator and/or software developer so as to feed them motion data that can be used to drive the system. It may seem obvious to say it, but the data must be compatible and correctly formatted to work on the MoCo system, and extensive tests should be made well in advance to prove the systems can be made to understand one another.

It is most important that commonalities be established between the shooting and CGI crews. Where measurements are being used, for example, they should be in a common unit and nomenclature. The shoot crew should never be expected to do conversions on the studio floor during shooting, so the idea of CGI working in metric and the camera crew working in imperial is beyond belief (but happens often even in NASA where the Mariner crashed for precisely this reason!). Anything that can help the shooting crew should be provided, particularly in the form of visualizations. Photographic printouts of existing elements annotated where appropriate, and test renders of the composite are particularly helpful. It is important to remember that shooting conditions will often make it impossible to view tapes, so "pieces of paper" that can be carried in the pocket are often the most useful materials possible!

3.4 Shooting Live-action for CGI Use

Material being shot for CGI use involves three distinct processes. First, it must fulfill the script and director's requirements – the basic mandate of any shoot. Secondly, it must be within the specifications of the CGI usage it will be put to. Thirdly, it must provide as much information as possible to the post-production team who shall be compositing it.

Providing adequate data about the shots is probably the most important special requirement when making shots that are going to be combined with others at a later stage, and yet it is the most often neglected aspect of this work. Without adequate information, huge amounts of time will be wasted in post-production, trying to fathom out what specific shots are for, and where other apparently necessary fragments may be located. Even once all the components of a particular scene have been found, without usable data describing how they should fit, it can be a nightmarish and time-consuming affair trying to align them to the necessary standard. The result is that effects shots are often discarded or used in much simpler forms than were intended, or indeed, without using many of the elements on which precious time and money were expended.

Fulfilling the technical requirements is probably the easiest aspect of the job. This has hard-and-fast rules and basically demands that the shots should fulfill certain technical specifications and supply the stipulated components. There is no particular latitude here – it is simply a matter of providing what has been asked for. The only incredulous aspect of this is where information channels are not flowing freely, and therefore, it is unclear exactly what is being asked for!

Finally, interpreting the script and the director's requirements can prove a major problem on visual effects and computer animation shoots. The "creative" team may not care about the difficulties that their requirements may cause at a later stage, and often the shooting crew will go along with these aspirations. What is easy for compositing or mimicking in CGI, may not be the production team's idea of how something should be shot. For low-cost compositing, static shots for example, are the most effective fiscal "super-saver", but many directors and cameramen are totally allergic to static cameras, and basically won't have it. Where this is the case it would be wrong to have a static camera anyway. Because the style of the rest of the material is constantly in motion, then a suddenly static CGI shot would be an instant give away that something is afoot.

This emphasizes a very important aspect of making visual effects shots of any kind, but particularly those which it is hoped will be invisible to the audience. CGI and indeed most VFX (visual effects) must match the style of the rest of the piece. This is a different sort of matching from the technical matching of image quality, but is every bit as important, and is often the reason why shooting crews may seem unsympathetic to the requirements of the post-production team.

A cameraman is dedicated to matching all of the shots in a scene so that the cuts between them are apparently seamless. If every shot has been on a moving crane, or shot through a heavy fog filter, or illuminated with colored light, then the cinematographer is not going to take kindly to being asked to alter that method. On the other hand, it will sometimes be necessary to recreate a stylistic effect by some different methodology, such as adding it once the compositing is complete. Thus it may be that the shots are to be hand-held or even "wobbly-cam", or just generally very rough looking – because that is how the surrounding sequences have been shot.

It is essential to understand the crew's reasons for doing things. Sometimes these may seem very strange and even perhaps idiotic, but there is usually a reason for the way of doing things, and more often than not, it is to do with time and thence schedule. For this reason, along with many others, a CGI supervisor on set needs to be discretion itself if he/she is not going to be the focus of all pent-up frustration. It can take substantial political acumen to avoid becoming persona non gratia on the set, and comments about the inappropriateness of certain preferred techniques is not going to help. Schedules are becoming ever tighter, and it may be difficult just to shoot the required material, let alone "mess about" with some effects shots. Thus unthought-out/undiscussed fixes may be required on shots, in the "fix it in post" vein. This is almost always because of problems with the schedule (or lack of it). For example, a blue screen may be unevenly lit because the lighting company sent the wrong type of luminaires in error. To get them replaced and hung might take four hours and with 50 various technicians and artistes standing about could prove very expensive. Even a day extra in post, fighting with the unevenness would still be an order of magnitude cheaper. Actor availability can even instigate inappropriate blue screens where, for instance, due to scheduling difficulties an actor has to be shot against a blue screen half way up a mountain, because that is the only time they are available – and yet this is hardly the ideal location!

Expense and time may also influence the way in which the shooting crew wishes to photograph a set-up. Where blue or green screen may be the compositor's preference, the cameraman may fight for black or difference matteing. In the case of shooting a large area against black rather than blue, it may be a matter of time and number of lights (which in turn equals time and money). Flagging lights off bright spots and hanging black drapes is a far easier way of dealing with a large area than trying to evenly light it blue. Black and difference backings also have the advantage of not causing color spill on to the back of the subject, and it is for reasons of lighting continuity that a difference matte is preferred by many cameramen, directors and designers. With a difference matte the artiste can be surrounded by approximately the color of the background into which he or she is going to be put. A hole in the matte will probably go unnoticed compared with a large dark or blue edge. Additionally it can be lit to precisely match the background's lighting conditions, whereas a blue screen shot must always be backed by a totally even flat tone.

The message is therefore that the post-process departments should try to find out the reasons why particular paths are being followed by the main shoot and try to help them as much as humanly possible.

3.5 Matching Specific Photographic Parameters

It should be evident that one of the most difficult tasks in compositing is matching between the pure digital world of computers and the impure analogue world of cinematography. A number of considerations, both concerning this impurity, and the chosen specifications of the photographic elements, require a more detailed look.

3.5.1 The Lens

Seen by most cameramen as the principal tool in their arsenal, the characteristics of the lens must be closely, if not identically, matched in all the elements of a composite image. If not, then elements within the image will in all probability be impossible to match precisely, either positionally or dimensionally.

The first priority is to match exactly the angle of view of both the virtual camera and the real-world camera. This is essential if perspective lines within the image are going to match and therefore the composite is a believable integrated whole. If we were matching the shooting elements of two real cameras to be combined, then maintaining a common focal length lens and image size would do this. It must be remembered that the focal length alone does not determine the subject size within the frame. It is the combination of the focal length and the "gate" or actual area behind the lens which together decide what will be included in the image. Thus, for instance, if two camera bodies are being used, then they must both have identically aligned and sized gates if the focal length is going to guarantee matching the subject magnification. This is common practice on visual effects' shoots and where possible is simply resolved by using the same camera body with the same gate, viewfinder graticule and lens (not a similar lens but the same lens!).

To make sure that alignment is correct a chart will often be shot with a center cross and lines or marks in the corners aligned to the graticule markings in the viewfinder. This procedure is repeated on both of the set-ups so that in post-production they can be aligned for a precise fit. Where a video tap (or assist) is being used then the tape of the shoot being compared should include the chart at its head (exactly sized to match all other material which has been telecine'd) and the graticule markings matched to those coming off tape. This will involve enlarging/reducing and repositioning one of the two elements while they are superimposed on the monitor. It is essential to provide a photographed chart representing what was seen through the viewfinder and on the video tap. Either/both of these may be misaligned and a filmed chart is the only way of conveying what was seen in the viewfinder, to the post-production team, where the indicated section of frame can then be made full screen to match the intended image.

Matching a CG image with a real-world one immediately begs a number of questions. If the software being used allows a focal length to be specified, it is also important to set the virtual image size to which it relates. In other words, if it allows you to set a focal length of say 50 mm, does it also specify that this relates to 35 mm motion picture camera image size. If the options are 35 mm still, 35 mm movie and 16 mm movie, then it will not be accurate apart from the still mode. 16 mm has two generally used sizes of "standard" and "super", but more importantly 35 mm has a huge number of variations in image size depending on, amongst other things, the intended use and the aspect ratio. If the software has an "other" or "user-definable" choice, then this should be used, and the actual camera gate dimensions keyed in.

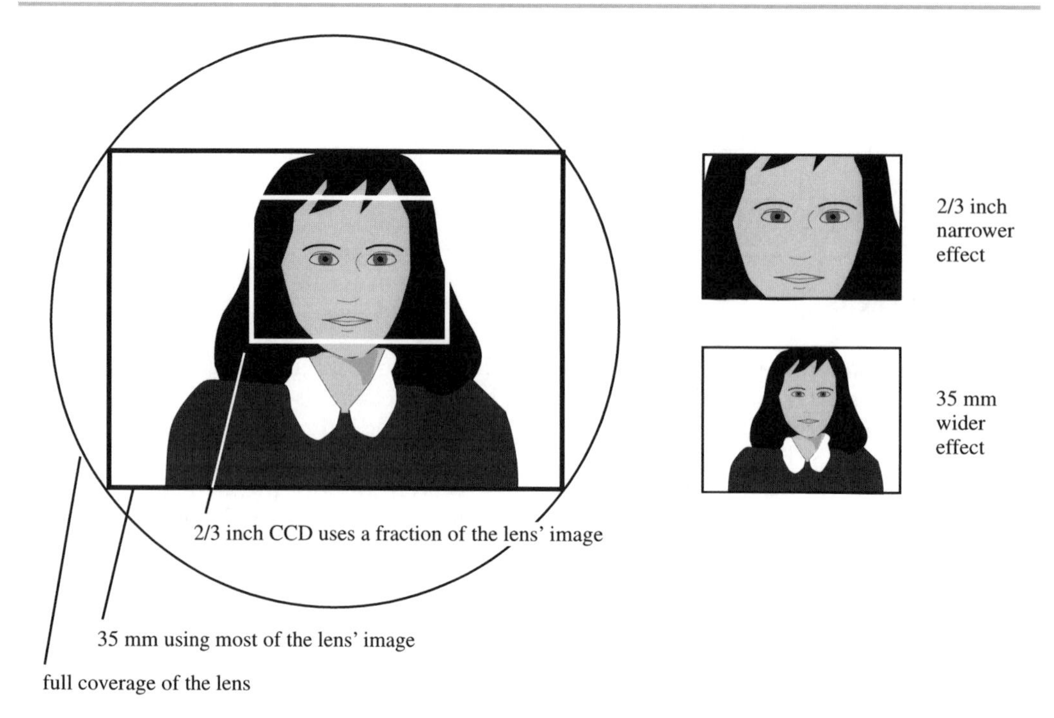

Figure 3.5 The same lens produces different coverage depending on the size of the imaging medium.

If the shooting medium is video (or digital video) then the target area or actual image size will be that of a $\frac{2}{3}$, $\frac{1}{2}$ or $\frac{1}{3}$ inch chip. With these it is only the vertical figure that alters according to the aspect ratio, so horizontal angles will always be consistent for a particular chip size of a particular type (e.g. standard resolution or high definition). It is important to note that if a film lens, say one designed for use with 16 mm, is used on a video camera, then the image size will be different, and the image cast on say a $\frac{1}{3}$ inch chip will be that of an apparently narrower or more telephoto lens. This is because the solid-state imager in the video camera is smaller and therefore only "sees" part of the whole image cast by the lens (Figure 3.5, lens coverage).

If it is not possible to find the exact focal length and/or image dimensions, then it may be necessary to use the "angle of view", which is an alternative, system-independent and very accurate parameter. A combination of aspect ratio and angle of view should be as accurate as using the focal length and image size. Unfortunately, angle of view is not one of the standard parameters used by motion picture crews and therefore it will have to be worked out either from lens data supplied by the manufacturer in respect of the image size being used or it will have to be measured. The American Cinematographer manual will give certain angles of view for particular image sizes (both film and video) but does not offer them all.

If all else fails then it is necessary to calculate the angle of view. This may be achieved by aligning two points (lighting stands are good for this) to the two edge lines of the viewfinder graticule, and measuring the distance from the focal plane to the objects thus aligned. During this measurement the two markers should be adjusted to be an equal distance from the lens and this distance should be more than 6 ft. Next, the distance between the markers

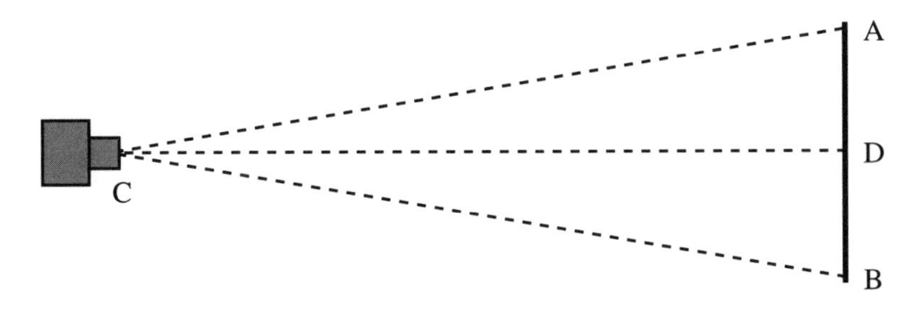

Figure 3.6 Scale drawing of the setup used to calculate the lens angle. AB is the object width, AC = CB and CD is the distance (measured from the nodal point).

(object width) should be measured. Armed with these three distances (two of which are equal) the horizontal angle of view can be calculated according to the formula:

$$\text{lens angle} = 2\tan^{-1}(\text{object width}/2)/\text{distance}.$$

Alternatively a scale drawing of the setup can be made and the angle measured with a protractor. Where the lens angle has been measured using the graticule side markings, it is important that framing should be within them, and that in post-production these will be judged to be the edge of the frame and the CGI elements aligned accordingly (Figure 3.6).

Zoom lenses are much more subject to distortion than "primes" and for effects work should be avoided. Pincushion and barrel distortion are particularly prevalent in this type of lens and cause increasing geometric distortion as the image nears the edges of frame. Additionally, the symmetry of the image can change with the focal length adjustment, and so the same lens can produce different results according to its settings. If two live-action elements were to be shot it can be quite difficult to set the zoom to precisely the same focal length on two occasions, and hence, this type of equipment really should be avoided where possible. Where it must be used, a distortion map (see below) is absolutely essential, if anything is ever going to fit precisely.

Finally, if anamorphic lenses are being used, or a widescreen video system which fits a squeezed picture into the conventional 4×3 image dimensions, then this too will have to be precisely replicated in the creation of the CGI elements. Additionally, any curving of the picture, particularly at the edges, will have to be matched during compositing.

Having equated the lens angle in both environments it is then necessary to make sure that other lens parameters are also the same, since these will adjust the depth of field and the focus position. Again, if in a composite the two images do not share a common depth of field and focus point, then the shot will appear to be wrong. Elements that one part of the image would suggest should be out of focus, may be in focus – an immediate give away to the visually sophisticated modern audience. To control the depth of field, the lens aperture (f/stop) should be set to be the same in both environments, and the focus should also be matched. If this is not possible in the computer software, then it will be necessary to set an "everything in focus" scenario, and then soften the image by eye during the compositing stage, to match the depth of field of the real-world photography.

If the live action is to be scaled with respect to the CGI image then the focus setting should be scaled accordingly. The camera must be placed on the focal axis and at the same inclination as it would have in the correct position. For example, let us say that the CGI scene requires an artiste to be placed 60 ft. from the camera, but the studio being used is too small. As long as the same focal length lens and camera tilt are used, then the camera can be placed at a distance of 15 ft. from the artiste and focused at this distance. In post-production the shot may be reduced by $\frac{1}{4}$ scale (assuming that all of the artiste is held within the frame and not cut off) (see Figure 3.1).

It should be noted that in all calculations involving lens distances, the distance should be measured from the focal plane, which is marked on the majority of cameras, and if it is not, can be measured by removing the lens.

Having managed to accurately match the lens size and its adjustments, the CGI and real camera images may still not align exactly! This is because the virtual lens is perfect, whereas the physical one is subject to the imperfections of the real analogue world. Most lenses suffer from a number distortions or inaccuracies such as barrel distortions, chromatic aberrations and shading at the edges, or a "hot spot" in the center.

To compensate for these it will be necessary to either "fix" the real image by applying corresponding and exactly complementary effects so as to iron out these inaccuracies, or alternatively, to add equal distortions to the CGI to degrade it to match. Both of these methods are applied during the post-production phase, and for either to work, it is necessary to establish information describing the faults with the taking lens. To do this it is normal to shoot a grid with each of the lenses being used for composite shots. These grids should be clearly marked with lens information and photographed exactly parallel to the focal plane. Consisting of parallel vertical and horizontal lines preferably white on black, they will reveal any geometric distortion in the image produced by the lens, which can then be matched in the compositing. Special software is also available for automatically calculating lens distortion and compensating for it. Requirements vary with programs and facilities, so check what your colleagues prefer.

Photography of a gray scale will provide accurate information about the color reproduction and a gray card filling frame will show any shading, color or tonal. In all of these cases accuracy depends on the image being lit completely evenly. It is usual to have these test shots made at the same time as the lens tests – usually made by the camera assistant prior to the start of a project.

Finally, another lens or optical system characteristic that may need to be considered is "flare". This is basically stray light, which bounces around within the glass surfaces of a lens system and contaminates (lightens up) the darker areas. A typical example is where white titles on a black background have an area of lightness surrounding them, a halation, but more importantly, it may also cause the entire picture to be lifted in brightness which results in two unwanted results. On the one hand, it can unevenly alter the image making it difficult to match, and on the other hand it can desaturate color, and thus make work such as blue screen much more difficult to execute. When adding synthetic flare to compensate, be careful not to overdo it, as is all too common!

3.5.2 Other Optical Systems

A variety of optical enhancements are commonly used during principal photography. These include lens adapters, filters and various lens attachments. The use of these can effect the optical quality of the image, and make integration with CGI more difficult.

The most common element which cinematographers add to their lens is a color correction (compensation) filter. Some of these are merely to correct the color temperature of the light source and are designed to cause the absolute minimum of optical impairment to the final image. These types of filters will have no appreciable effect on the combination of images and indeed are a necessity if the color is going to be correctly reproduced.

However, there are additional and often stronger filters that are used for pictorial effect and these can have a disastrous impact on later compositing processes. Although the visual effect of filters is to add light, a red filter for instance makes everything appear red, they actually subtract or filter out light, and so a red filter, will actually filter out cyan (blue–green) light. If a dense high-absorption filter is used, then it can remove so much of a particular color (or colors) that it is difficult to replace them if this is required at a later stage. Particularly in compositing it may be necessary to have a full gamut of colors, where for instance, a blue or green screen process is being used. In these circumstances it would be best not to filter the live-action photography. Of course this results in the particular shot, not displaying an effect which possibly all of the shots surrounding it in the sequence will have used. Obviously this effect must be present for the VFX shot not to stand out, and so it must somehow be applied.

The best solution is to make the final combined shot first, with clean elements from both the live-action crew and the CGI, and then, once it has become an integral shot to be cut in to the rest of the piece, it can have the effect applied overall. This will have the additional benefit of helping to hide the effect by crossing any compositing boundaries.

For the shooting crew the most important thing is to define precisely what filtration they have been using. A verbal description is not enough, and it really is necessary to provide a good reference to show exactly what the effect is. The best solution would be to run at least a couple of the shots where the effect is used, with and without it, so that the post-production team can see a *before and after* of the effect. It is also essential that the element destined to be composited (which does not have the effect on it) should also provide a pass with the effect, so that the post crew can easily see the visual qualities of the desired final image. There is the additional benefit that part of this version can be cut and pasted into the final, if practical. For the reference version, it is not necessary to have the precise action as used in the main takes – a rough walk through by the main artistes would usually be quite sufficient.

Filters designed to alter the optical quality of the shot, are not usually acceptable under any circumstances. It is very difficult to merge the join line between composited elements which have the same optical qualities as edges within the separately filtered individual elements. Devices which fall into this category include diffusion, star, fog and starburst filters. Any such devices will have the same general effects, so nets behind the lens and wire mesh filters, are all unacceptable from a post-production point of view.

The situation is made even worse if an automated compositing system such as blue screen is in use. The softening or diffusing effect will make the edges against blue less defined and thus much more difficult to adequately key. Even if the keying is satisfactorily executed, there will still be the problem of the diffusion on the two halves of the composited shot being made to act in the same way along the matte edge that they would have if they had been shot simultaneously in situ. As with color filters the same solution is preferred – to apply a digitally matched effect after the combination has been made and over the whole of it, thus crossing any compositing boundaries and therefore helping to obscure them. Again the use of a before and after reference shot will help enormously in doing this. Where specular effects are involved, it is very worthwhile to shoot a black card with some holes punched in it and lit from behind. This arrangement will define very precisely the exact amount and quality of the spread in light produced by the filtration. For example, with a star filter the rear lit hole would be surrounded by a star clearly delineating the dispersal characteristics of that filter.

Where any "distorting" lenses are a main component of the pictorial design, then these will have to be precisely delineated so that they can be matched during the post processes. An example of this would be a fly's-eye prism lens that produces multiple images, or a cylinder lens that deliberately curves the image as if by excessive barrel distortions. Bug's-eye or fisheye lenses also produce massive distortion, and this needs to be analyzed by shooting suitable grids as described earlier. It must be said that if matching together is required for a number of such elements it should be carefully considered whether a digitally produced effect might not be more suitable. Most of these optical effects can be generated in post-production and have the advantage of being more easily matched with one another than if done at the time of principal photography. As with all of these types of discussions it is also important to remember that there is no going back if the effect is added during the shoot, whereas post effects can be remade if the initial effect is rejected by the producers.

3.5.3 Camera Speed Parameters

Exposure time and temporal effects are controlled by the shutter in a film camera, or scan timing and the electronic shutter in a video camera. It is in this particular area that the principal differences between video and film exist. Standard video uses a system called interlace to scan two fields for each frame of the picture. This means that it scans down every other line on the first field and then fills in the remaining alternate fields on a second scan. The result is that 25 frames per second video actually has 50 distinctly different images per second, rather than the 25 one might at first imagine. For most uses in television this has little effect apart from giving video a more realistic, transparent and often flatter appearance than film. However, in the area of effects it has major ramifications. (For US television this becomes 29.97 frames per second and 59.94 fields per second, respectively.)

Film for television (European) is exposed at 25 frames per second and these are genuine single frames. When scanned by a telecine the two fields of each TV frame are made from these single film frames and are thus both snapshots of the same moment in time. In video-shot material, there are two moments of time per frame. Thus, if you were to play back a videotape of a rapidly moving car shot on video, the image would jitter as the still frame cycled between the two fields during which the car was recorded in two different positions. The same shot made on film and played back off videotape would not shimmer, since the two fields of the frame would be temporally coincident. Like vodka and gin these systems are fine on their own, but they do not mix! If one shot had fields which were all different, and the other, with which it was being combined did not, then the result would be one element juddering, behind the other apparently floating alongside it. Thus it is essential that all elements being combined must be shot either in fields or frames, the two simply cannot be mixed for a believable composite.

Recent video cameras such as the 24P from Sony (HDW-F900) produce a picture with the same temporally coincident fields as film and can be happily mixed with it. The "P" stands for progressive, wherein all the lines are scanned in a single pass. Most 3D computer graphic systems will allow rendering either in film- or video-type frames and this setting should be applied according to the intended live-action shooting method. If for some production reason interlaced fields are the preferred system and film must be used for the shooting, then the film should be run at the field rate of the video. In this example the film should therefore be run at 50 fps, and then in post, the frames can be integrated into fields. In all

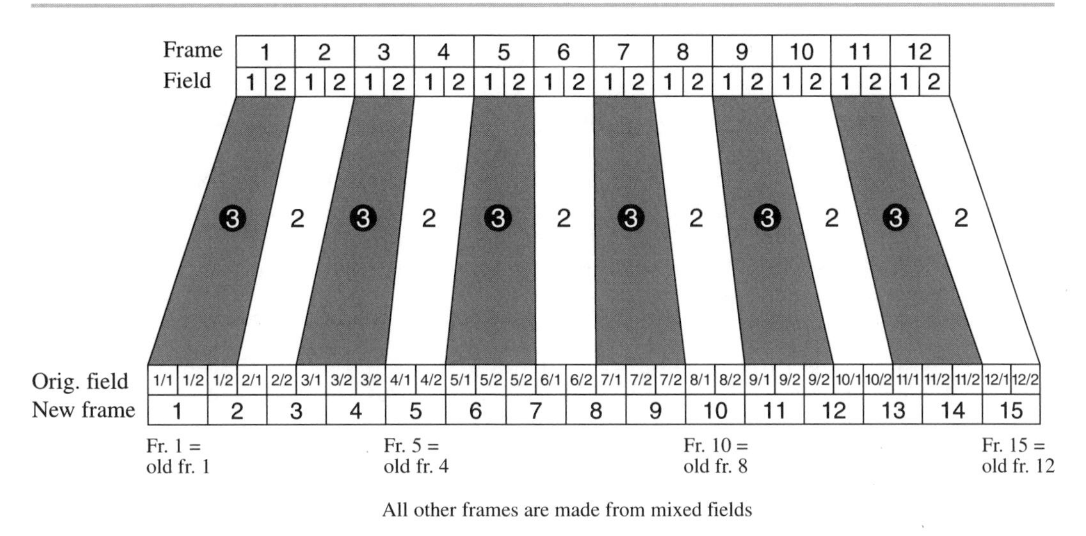

Figure 3.7 The 3:2 pulldown method as used in US television.

of the above we have used the 25/50 fps European system, but the same rules apply to the 29.97/59.94 fps frame rate of the US television system.

In US television it is common practice to shoot film at 24 fps and then to telecine it with so-called 3:2 pulldown. This basically alternates between transferring three and two fields for each successive frame. Film frame 1 will have two fields extracted from it to make up frame 1 of the video. Frame 2 of the video will then have field 2 of film frame 1 repeated as its first field and field 2 of video frame 2 will be field 1 of film frame 2. Video frame 3 will start with film frame 2 field 2 and have film frame 3 field 1 as its second field. Finally, video frame 4 will consist of film frame 3 field 2 repeated for both its fields bringing us back to the start of the sequence with video frame 5 and film frame 4 displayed as normal (see Figure 3.7).

This extraordinary arrangement makes 24 frames fill all the frames of 30 in a reasonably smooth fashion. Obviously some frames of the resulting video will be made up from two fields from separate moments in time and therefore will "jitter" when shown as a frozen frame and would be impossible to draw a matte for. To draw a matte for every field would result in some fields being drawn twice since some fields are repeated – and this would be very inefficient.

The best strategy to use when working with material which has been through the 3:2 pulldown mill, is to either obtain a copy which has been telecine'd frame for frame (and therefore appears the wrong speed when run), or to take the 3:2 pulled down version and "unstitch" it to recreate the original 24 fps sequence of frames. In both cases this means normal film frames where both fields are temporally coincident are available to matte in the normal way and only after the compositing has been done, should they then have the 3:2 pulldown applied.

Currently video can only run at one dedicated speed (24/25/29.97/30 fps) unless using very expensive specialist equipment such as "supermotion", making this aspect of it, at least, simple to deal with. The exposure time is nominally half of the shooting speed, so 25 fps gives an exposure time of 1/50 second. Where fast movement is involved this should be borne in mind when matching motion blur, and settings in CGI systems should be adjusted accordingly.

There is nothing the cameraman can do to alter the frame rate, although some modern cameras do allow an adjustment of shutter speed so that exposures of faster than $\frac{1}{50}$ s permit a fast moving subject to be sharper in each frame, where say still frames are going to be analyzed. Coordination of this feature is obviously necessary if motion blur of elements is going to be invisibly matched. It should be noted, however, that, if high shutter speeds are used it usually results in "stuttery" unrealistic looking motion similar to that in early examples of stop-frame animation, and most visual effects specialists would recommend having as much motion blur as tolerable for realistic looking results.

Adjustment of shooting speed allows modification of temporal flow and so things can be sped up or slowed down. This is used a lot in visual effects work to help scale subject matter. Water, for instance, when used with miniatures gives itself away because it always behaves the same way irrespective of scale. Thus miniatures in water are usually shot at higher speeds to slow down the water and give the impression that it has greater mass. Pyrotechnics and fire effects also can be lost when shot at normal speeds, so are often slowed down too. It will be necessary for the compositors to know at what speeds things have been shot, if they are to match them with other elements, particularly if these have been shot at other speeds.

In film the variations available are far greater, since the shooting speed and the shutter angle can both be adjusted on most cameras. This means that exposure time and frames per second can both be altered with a resultant greater flexibility in the creative possibilities of the scene. This is one of the principal reasons for the continued use of film in, particularly the commercial's and visual effect's arenas. It is important to coordinate the frame rates of all the elements being combined, otherwise the join will be revealed by different motion attributes. If, for instance, a CGI car is being put into a live-action background, then irrespective of the frame rate employed in the live action, the CG material should start and stop in the appropriate number of frames. With computer control of aperture, shutter angle and speed it is common to alter speed nowadays, and exact knowledge of such speed variations must be fed to the post team, if they are going to have any success whatsoever in combining shots.

On a film camera the shutter angle can be adjusted. This allows the relative exposure time to be altered independent of the shooting speed. Exposure in a film camera is best imagined as a wheel with a segment removed. This wheel spins between the lens and film and obviously the size of the open segment defines the exposure time (see Figure 3.8). The normal setting is 180° which at 24 fps gives an exposure time of $\frac{1}{48}$ second, but, it is possible to close down to say a 45° shutter which gives a $\frac{1}{192}$ second exposure. These shorter exposures will obviously result in less motion blur and so the same considerations as mentioned above come into play. As previously stated motion blur must match between elements and should be at a believable level. Computer-generated motion blur can successfully be added to shots lacking it or with too little, so it should be possible to match shots, but the keyword here is "add". It is very easy to add motion blur in post, but almost impossible to remove it, so once again, if in doubt keep it to the minimum so that it can be added in the correct quantity during post.

3.5.4 Shooting Processes

The principal shooting technologies are film and video. Although film is a major acquisition format for television, standard video is not generally considered suitable for theatrical film

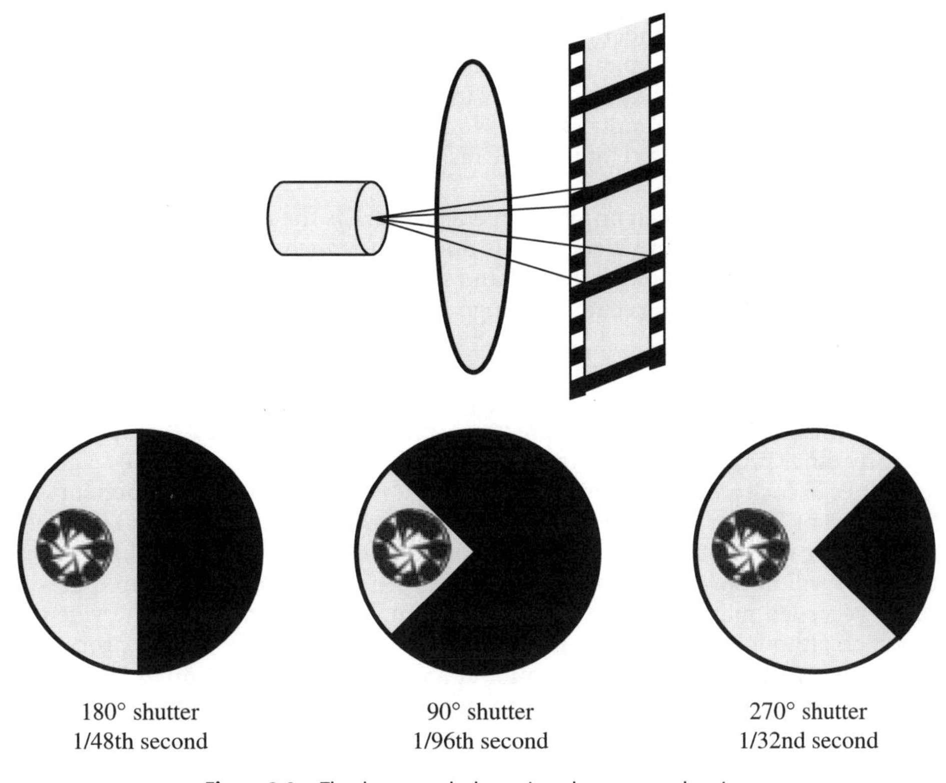

Figure 3.8 The shutter angle determines the exposure duration.

(cinema) presentation. This is because of the field issue discussed above, which does not translate well to the situation in film where each frame is projected twice to give two fields (which are therefore identical). The two fields of video (which are different) can never be combined to provide a perfect single frame, and the much lower resolution of current standard video formats compared to film (576 (Europe) and 480 (USA) versus over 4000 lines on 35 mm) all conspire to make it a poor source for cinema projection. On the other hand, the new 24P cameras which Sony and Panavision (using the same basic head) have developed, offer a vertical resolution of 1080 lines. Although this falls well short of 35 mm's inherent resolution, it does catch up with it, at least partially, during the duplicating and distribution stages, because its quality never decreases from the originating stage, due to the no-loss nature of digital copying, whereas film suffers badly from each duplication. But areas such as reframing and color grading are more limited, as discussed below, in relation to all types of video.

For television use there is a choice between film and video and although this will usually be made for production preference reasons, the pros and cons of each are worth considering. Video has low noise (the equivalent to film grain), has a completely steady frame with no movement in it and has a very realistic looking appearance because of the 50 or 60 field system. These factors could all be advantageous when combining CGI images that similarly exhibit all of these same characteristics – given that they are both systems with common electronic origins. These are of course the very same aspects of video that most film-image enthusiasts dislike, but, dependent on the type of programming the images will be used for, could provide a very good match and consistency of look. Finally, being an electronic medium it is possible to feed the video signal directly into other electronic equipment. Thus, for example, an editing or CGI suite could have a direct feed from the studio floor, and thus make combinations "live" so as to guarantee results, or experiment with a finished product on the monitor.

One other difference that can impinge on effects work is the much greater depth of field available for a given f/stop in video. This is due to the much smaller light sensitive area used by video systems compared with film, and could make shooting certain sorts of model effects somewhat simpler to execute, although film's ability to use long exposures largely compensates for this.

Against video is its look (artistically) and the fact that the 50 (or 60) individual pictures to matte makes for double the work if rotoscoping or other frame-by-frame procedures are going to be done. Video recording for effects use must never be made in composite (i.e. D2 or D3 or any other PAL or NTSC system), because these all combine the color signals with the luminance in such a way that they cannot be successfully recreated. Component systems such as D1, D5 or Digital Betacam must be used. However it should be noted that these all record the color signal at half the resolution of the luminance (so-called 4:2:2 sampling) and therefore will not produce a perfect blue or green screen matte compared to where a separately recorded matte is made. Digital Betacam further compresses the video signal by 2:1, which although it works for most uses can cause problems in CGI processes. DV formats go even further and record the color at only a quarter resolution (4:1:1) and so are even less suitable for chroma key type techniques. Film, on the other hand, may be output from a telecine machine at full bandwidth (4:4:4).

Video is unable to run at different camera speeds and so is generally unsuitable for shooting pyrotechnics, miniatures or any other subjects requiring temporal adjustment. Finally, since the video camera has the same specification as the system it is part of, there is a limited amount of adjustment possible in post-production to the color, contrast or gamma of the image and very limited amounts of enlargement possible if reframing is required. For accurate matching of the many elements going into a complex effects shot such adjustments are usually essential.

Film has greater resolution than a TV system and can therefore be enlarged a substantial amount. When shooting film for effects it is usually recommended that the shots be framed allowing some space around the edges so that, if it needs to be reframed, it can be adjusted without having to crop the intended frame area (see Figure 3.9). The color gamut and contrast range of film far exceeds that of video (including the 24P camera currently on offer) and therefore very large adjustments can be made to it in post, allowing matching of almost any extreme. As discussed film may also be run at a variety of speeds, permitting miniature and other effects photography. It is also perfectly suited to use in systems that require very low frame rates such as once every half hour (time lapse) or where long exposures are required (motion control of models and stop-frame animation). All of the above and particularly the non-interlaced nature of film combine to create the famous "look" of film much loved by most production and camera personnel.

On the negative side, film exhibits some artifacts that are not so good for visual effects work. It weaves mechanically and although this is usually acceptable it can become excessive with worn or badly maintained equipment – if this happens then it must be stabilized in post or its frame by frame movement added to any stable elements (such as CGI) with which it is being combined. Film also has grain that continuously moves over the frame. This texture

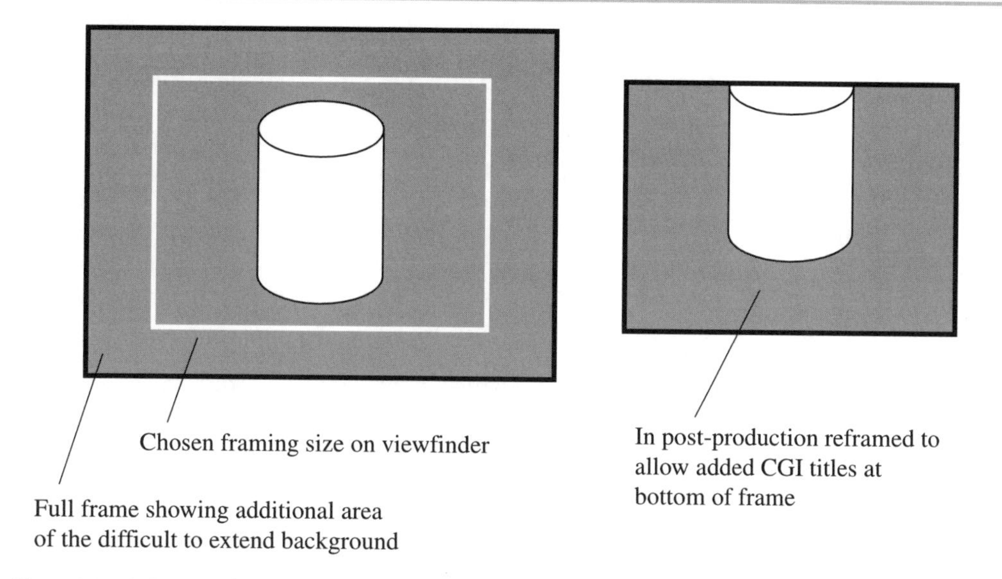

Figure 3.9 Reframing of shots in post-production using extra space left around the originally intended framing size.

is not present in CGI-originated, or any other electronically generated images, and so must be very carefully added to the non-film images to match it. Most compositing software has tools to achieve this, but it must be done with care if the look is to be maintained. Finally, film stock and processing are expensive compared to videotape, and have a limited maximum duration shootable in one film load, depending on the format and camera in use.

There are many film formats offering greater and lesser resolution and other benefits compared to the standard 35 mm. 8 mm, although used for "effect", is not suitable for effects work due to its inherent instability and small frame area. 16 mm although the mainstay of most television dramatic output, is not really suitable for effects either. This is not because of the resolution it offers, but because of stability problems. 35 mm is a full photographic system designed for professional applications. A 35 mm negative has so-called B&H (Bell and Howell) sprocket holes which are "tight" fitting, meaning that they have squared off corners and are designed to fit the pull-down claw of a camera very precisely, thus sitting absolutely steadily in the camera gate. For projection, where such perforations would be destroyed by continuous use in ill-maintained projectors the so-called KS (Kodak standard) sprocket holes are used. These have rounded corners to help guide the projector claw into position and are "loose fitting" so as to compensate for imperfect claws.

16 mm film, on the other hand, only has KS sprocket holes and in the case of super 16 only has one hole per frame on one side of the frame – the other sprocket hole having been sacrificed in order to accommodate the extra space needed for the wide screen aspect ratio. Although it is possible to stabilize 16 mm in post this is never totally successful where the newer and more common super 16 is used, because due to it only having a sprocket hole on one side, the whole frame moves inconsistently across its area, and if one side is stabilized the other may still be moving – basically super 16 is rather like a flag blowing in the wind!

Other higher-resolution film formats also exist such as vistaVision where 35 mm film is run through the camera sideways using a frame of eight sprocket holes length as opposed to the normal four of standard 35 mm. This process has the advantage of giving a much bigger frame area and therefore higher resolution and smaller grain. It also has the advantage of being able to be processed in any standard 35 mm lab and of having available all 35 mm film stocks currently in production. Other large formats such as 65 mm (which prints to 70 mm) suffer from a reduced range of stocks being available and only being serviced by a small number of laboratories, camera and accessory manufacturers. On the other hand, these are excellent systems where high resolution is required to allow the final image to be moved around within a larger frame. For the vast majority of high-end television and most cinema production 35 mm is the standard shooting method and is likely to continue in this role for some years to come due to its variable shooting speeds, high resolution and inherent stability.

Having chosen to shoot on 35 mm (or another film format) there are still other standards to be chosen between. First, the aspect ratio which can be anything from the original silent gate (also called full aperture or open gate), which fills the gap between the sprocket holes on either side, top and bottom of the next frames, respectively, and has an aspect of 1.33^{r}. Academy is a smaller area displaced to one side to allow for a soundtrack running up the side of the film with an aspect ratio of 1.37:1. This has been the standard for television use since its inception up until the last few years and was the standard in cinema until the development of widescreen in the early 1950s.

Most wide-screen systems are academy with a crop added top and bottom to make, for example, 1:1.66, 1:1.85 or 16 × 9, the latter being the world widescreen television standard. An additional format in cinema use is the 1:2.35 anamorphic system, often called CinemaScope, in which a 2:1 squeeze lens is used on the taking camera and then an inverse de-squeeze is applied during projection. In television this system is often used where a matte shot requires an in-vision pan. Here the background element is shot "scope" and then the normal TV frame can be panned across it (see Figure 3.10). Whichever of these formats is used, an upfront agreement must be made between the live-action crew and the CGI team so that framing will be in a compatible format.

Figure 3.10 Panning of a subject over a "scoped" background.

The quality of film is also dependent on another variable – the film stock used. Unlike a video camera which has a built in sensor, the film camera must run film through it and this can be any one of many current film stocks. Indeed, it is interesting to note that a film camera made in the early 1920s can still run through it the latest high-speed, t-grain ultra-modern technology film stocks and thus produce an up to date result. For effects work, it is best to choose a finer grain film, since matching the large grain of high-speed stocks can be difficult, and the moving texture of the grain can also break up keying edges in, for example, a blue screen shot, thus making it very difficult to derive a clean matte edge. Where blue or green screen work is involved Kodak currently offers a special emulsion called SFX200 and this is the best compromise in stock for this type of work, although it is fearfully expensive. Next to this the Kodak EXR100T (5248) and slower Fuji stocks seem to suffer least from the blue halo problem around the subject's edges. The finer grain slower stocks are also best for combination with CGI due to their much finer texture, which is easier to match when adding in pseudo-grain to blend the elements together.

There are various other processes that can be applied to film to control its general look. To achieve particular photographic effects cameramen will choose to expose and otherwise manipulate it in a variety of unorthodox ways. Over- and under-exposure with or without subsequent pull or push processing are often used to alter the exposure curve and contrast of the negative. The fact that the photographic process utilizes a negative and a positive also means that an under- or over-exposed negative can be printed lighter or darker, respectively. For TV or digital use this may be done at the telecine or scanning stage, respectively.

A more radical and very popular technique at the time of writing is to use the skip bleach or bleach bypass process where the removal of silver salts from the negative is only partially carried out or omitted entirely. This is most usually carried out on the print and results in an image, which has much deeper blacks and less saturated color. Another technique, in some respects the opposite, is that of flashing the negative by either mildly exposing the raw film to an even overall illumination, or achieving the same end by reflecting light into the lens at the time of principal photography. In either case the idea is that the exposure threshold at which photons will trip the light sensitive crystals is lifted and so the sensitrometric curve is altered.

All of these techniques can result in interesting looking pictures, but since they alter the emulsion's response to light they can adversely affect special processes such as blue screen or difference matteing. Bleach bypass for instance reduces the color saturation and so a negative that had been so treated would not be helpful to the compositor trying to derive keying information from that negative. Basically one should avoid doing anything which will affect the color separation in the negative. Where these affects must be used then the same advice offered in respect to physical filtration also applies – provide before and after examples of some shots so that the effect can be mimicked in post-production and added to cleanly processed and combined elements.

In conclusion, it has to be said that careful discussion up front will avoid most problems. Any process that deteriorates the image in a technical sense should best be shunted into post-production, and the negative or videotape offered to post should be as clean and "down the middle" as possible. Information passed on should be as complete as humanly possible so that decisions can be rapidly made and matching can be as automatic as practical.

An important point resulting from all of the above parameters is that of resolution. The CGI work should always be done at a resolution that matches the other elements in use. Thus, if it is for television use and the live action is shot on digital betacam then the CGI work need only be carried out at 576×720 (for European video) or 480×720 (for US). If money is no object or the ultimate quality is required, then the CGI may be carried out at a higher resolution and then subsampled down to the target resolution. In certain circumstances this can result in a superior looking result, just as a 35 mm negative transferred to video will often look better than 16 mm, even though both are theoretically of higher resolution, than the television system

For film a variety of resolutions can often be used depending on the size of an element within the final frame and how sharp it is. Most CGI work is, in fact, carried out at 2K resolution even though film is theoretically of a much higher resolution – it is only the most difficult and tricky blue screens, for instance, that are executed at the full 4K. As with everything, the financial imperative is one of many factors that must be balanced.

3.6 Requirements for Specific Compositing Methods

By the very fact of integrating CGI and live action some form of compositing must occur. This could be simply laying the computer-generated image over the live action (as background) using a self-generated matte, but it could also involve complex multilayered composites in which many elements coming from different operations are combined. Basically where the live action (in compositing terms acting as a foreground) is to be added to the CGI there are two scenarios. Either the material is specially shot to allow automated compositing (e.g. against a blue or black background), or, it is shot with the intention of being manually rotoscoped (i.e. somebody laboriously drawing around the subject matter to create a silhouette mask, which can then be used by compositing software, to combine it with the background).

3.6.1 Live-action Foregrounds

Looking at the hand-drawn, labor-intensive situation first, there are a variety of ways in which the live-action material can help or hinder the matte artiste. Most importantly he or she must be able to see what they are cutting out, so it is really much better if there is some illumination all around the edge which is to be cut out. If it must be dark, then of course it can be made darker in post after the matte has been drawn, but trying to draw around a black object on a black background is difficult and frustrating work, if possible at all.

Another helpful aid is to record a short sequence of image without the matted subject in place. This permits a difference matte to be created, at least in part, by subtracting the empty frame pixel by pixel from the action frame – the difference is the subject matter. For this to work nothing in the frame should be altered between the two passes as this would create an unwanted difference. The best procedure where live action is concerned, is to wait until there is a preferred take (a "print it") and then with the cameras still running, ask the actors to come out of frame – this way things are as close as they will ever be.

Obviously if the camera is moving then on extra passes the repeat move should be the same, and this would require a repeating (mimic or memory) head or motion-control system, so that the camera travels identically for each pass (with the subject and without it). In certain cases tracking software can be used instead but with its own set of requirements (see later).

Where the *roto* work is to remove wires or other rigs that are supporting moving objects within the setting, then the same rule applies – always try and get a clean pass which, in this case, can be used to replace (restore over) the area being fixed. It is also important on the floor to keep an eye on reflections and/or shadows of the wires or rigs since these are not very easy to remove.

Automated systems basically use a tonal, or a colored background, to specify the outline of a subject placed in front of it. Most common are black, blue and green, but other colors and tones are possible, and difference matteing using separate passes with and without the subject may also be used (see above).

Where a tonal difference is being used (most commonly black) it is important to make sure that the subject is well defined against it. A man in a black dinner jacket, is not what one would shoot against black, for instance. Black is often the preferred alternative for a shooting crew because it is so easy to provide a black background and there is no spill light on to the backs of the subjects. It can, however, be quite difficult to pull a decent matte from it, if the lighting is "contrasty" and with dark tones.

In model shoots or motion-control situations such as stop-frame animation, a variation often used is the so-called "front-light back-light" method, where two frames are recorded for each position of the subject and camera. For the first, the normal lighting is used (front light) and then for the second, a white card is inserted behind the subject (if it is against a light backing this is unnecessary), and the backing is evenly lit with the front lights turned off. The result is a silhouette of the subject and this may be used as a matte in post-production. The result is equivalent to the alpha channel output of many CG programs but comes on the original negative in the form of checkerboard of alternating production and matte frames.

People wearing blue are best shot against green, just as Robin Hood and his merry men are best kept against blue. The arguments about blue vs. green in digital compositing are many and often heated. Basically, blue has always been the favored color for compositing systems used to shoot people, because there is almost no blue in the human form but lots of all the other colors including green. However, since digital compositing came along, and is a lot more forgiving and thus makes other colors more practical, it has been realized that the blue layers of film emulsions are more noisy than the others, and that, in fact green is the quietest. Thus there is an argument that the key from green is much cleaner (smaller grained) than that from blue and therefore gives a better result. Of course against this you could argue that there is less latitude because of the higher green content in the actor standing in front of it. In the end effects people are split over the issue. This writer for example tends to favor blue, partly because it usually emits less light and therefore decreases the illumination on the back of the subject, but obviously if using blue grain or noise should be minimized by, for example, using slow fine-grained film stock.

Light reflecting off the screen is a major problem and results in a blue (or green) halo (spill) surrounding the subject. Modern compositing can get rid of the color, but not of the lighter tone left behind, which can make the artiste look cut out. For this reason the subject should always be as far from the color screen as possible with a minimum of around 12 ft. A further aid is to illuminate the screen at about one stop below the key level on the artiste – in other words, if an incident reading of the key is f/4 then the backing should be f/2.8. A bit higher is acceptable, but basically this is a pretty successful compromise between the CGI requirement for minimum noise and the photographic requirement to minimize spill. It is a good practice to reduce the size of the backing, preferably to just slightly wider than the subject – with a large screen this can be done by draping black cloth over the sections which are not immediately adjacent to the subject. Finally, backlight can help sell the effect, but of course it should not oppose what is happening in the backplate lighting. In general, as much backlight as you can get away with will be a big help!

With all of these techniques it is always well worth shooting a pass without the subject. This will give post information on the density of the screen to more easily balance it, but of course if there is a moving camera then a repeat head will be necessary so as to exactly align the passes. If a move is being made on the blue screen, which is to be matchedmatching with movements created in the computer, then tracking markers should be placed in the blue. Where possible these should be in the approximate position of whatever they will be replaced by, so that the precise movement of the camera can be determined. If 3D models are involved, then measurements of the distances between the markers should also be provided (e.g. they should be placed in a grid with known distances between them). The markers should be well defined, but not too big, since they will obviously have to be "lost" during the post process! One very good method is to use white markers if the blue screen is being illuminated with blue light (or a green screen with green light). The marker will appear on film or tape as light blue and will, with careful alignment, be blue enough to key, but light enough to track.

The color screen may be created by lighting a white backing with blue light, by lighting a blue backing with white light, or a blue backing with blue light. The latter is by far the best solution, since the blue fabric or paint used will always reflect back some non-blue light. By illuminating the backing with pre-filtered light it will have a very limited spectrum from which to reflect. The best blue source is a specially made blue fluorescent tube (e.g. by Kino-Flo), but conventional lamps with blue filtration also work well – HMI lamps which have a higher color temperature than tungsten also help in achieving a better result.

An alternative to a front-lit backing is a rear-illuminated screen which can be of special blue material or illuminated with blue light. This is the best method of providing a color screen, but on occasion, it will be necessary for the artiste to be seen standing on the floor, and in this case blue light cannot be used because of its potential to infect the subject. In this case the lighting of the backing should be very even indeed and with shadows cast by the subject of a similar density and traveling in the same direction as those in the image to which the matte will be added. With even blue, apart from the subject and his or her shadow, it should be possible to carry shadows across with the matte, and thus have very realistic moving shadows within the composite.

If the subject is reacting to something such as a scary robot or monster, then any interactive elements that are needed should be used. For example, a flash from the robot's lights or its shadow as it approaches, should be made on the blue stage so that they really do appear to interact with the actor. To put a flash of light over the actor in the studio is relatively easy, whereas to do it in post would be quite difficult.

Finally, in all such situations where characters are being shot without the element with which they are reacting, it is very worthwhile to have a stand-in for the missing element. Thus with an actor in the blue stage reacting to a non-existent gigantic robot, it would be good to have a stand-in acting out the CGI monster's part. If necessary, a large pole representing the true height of the monster could be used to give the actor something to react to – placed in the relatively correct position.

As with all elements for compositing exact measurements are very useful for establishing where and how the various elements should all fit together.

3.6.2 Live-action Backgrounds

The converse of the blue screen element to be put into a CGI environment is the background plate into which the CGI element will be inserted. This is similar to the backgrounds which might otherwise be shot for the conventional blue screen of an actor, but with a few peculiarities special to working with computer-generated elements.

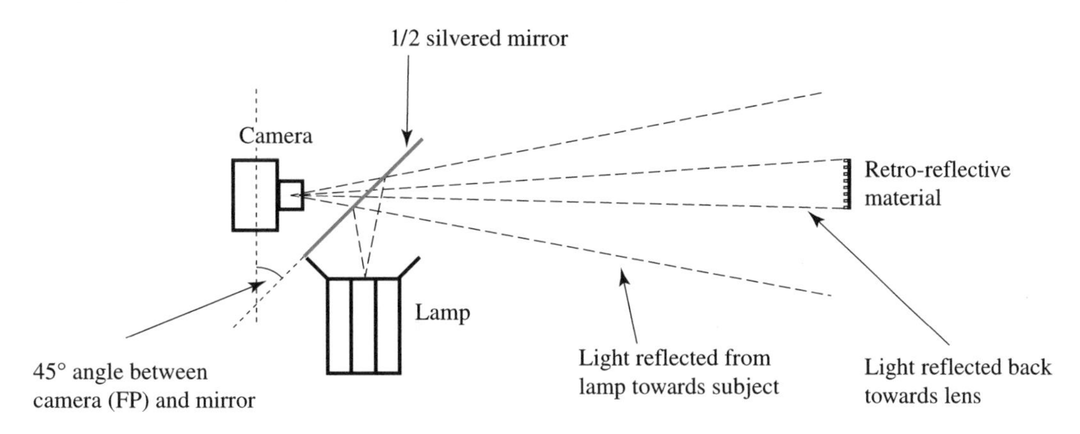

Figure 3.11 Lighting of tracking markers in dim lighting conditions.

Shooting background "plates" for the addition of CGI images is the most common instance of live action and computer imaging combination. Since creating large and believable 3D environments can be time consuming, processor intensive and very expensive, whilst shooting empty locations or sets is relatively cheap, it is normal to shoot real backgrounds into which the dinosaurs, alien craft or robot models created in the computer world are inserted. Creating CGI models is, on the other hand, usually easier and more flexible than building physical miniatures. As always the first imperative is to have upfront discussions about the shots, and if possible, to create storyboards. All the requirements of carefully recording relevant data about the shot and how it was made are also particularly important in this instance, most specifically if 3D extensions are to be added to the setting (e.g. adding the futuristic tops to large buildings of which only the bottom floor have been physically built).

To exactly scale the CGI and real setting together, very carefully made measurements are necessary, so that the computer model can be built to fit exactly, and it should be remembered to provide information that relates the physical universe in three-dimensional space. A good way of providing this information, in a manner that is easy to assimilate, is to get a frame enlargement of the image, or of a specially taken still photograph and annotate it – otherwise a precisely drawn diagram with the measurements clearly delineated is the minimum useful information.

Where shots are moving it will be necessary for the CGI team to be provided with specific tracking points (or markers). Where well-defined architectural or decorative features exist these can often be used, and indeed are preferable because they do not have to be removed from the shot after use. But, if no such clearly discernible points exist (i.e. which can be seen on the final film/tape), then tracking markers must be put into the scene so that any movement in 3D space can be clearly analyzed in post. The form of these markers depends very much on the size and nature of the setting, but a number of convenient and cheap materials are readily available, starting with white or reflective yellow ping-pong balls, then reflective tennis balls and up to various types of football. The advantage of all of these are that they are full 3D spheres and therefore are the same when viewed from all angles. It is of course important to ensure that these are adequately illuminated and not obscured as the camera makes its move. One solution to lighting these, if they are distant from the camera in dim surroundings, is to coat them with 3M retro-reflective paint and then beam a light via a half-silvered mirror at 45° to the camera (see Figure 3.11). Another popular light source for markers is LEDs. Red shows up particularly well but care should be taken to keep them away from demarcation edges as their color is very saturated.

For many 3D tracking programs the requirement is to see six markers at all times with four in one plane and the other two in a different plane. When working quickly it is sometimes beneficial to have a cube frame built with markers in the eight outer corners. This may then be set into the shot at short notice and can be quickly adjusted for best visibility. In dark settings or night shoots it is sometimes necessary to use small light bulbs as markers so that they can be seen in the gloom – so-called pea-lights or small round golf-ball-sized opalescent bulbs tend to be best.

The problem with special tracking markers, which have been placed in the set, is that they then have to be removed afterwards. This can be a very time consuming and a thankless job, not to mention very difficult, particularly in the case of the small light bulbs used on night shoots. To make the compositor's job half-way tolerable it is imperative that these do not have translucent materials passing in front of them. The worst example is smoke and flame, since it is then very difficult to fix, particularly, if there is a long and slow camera move passing in front of it. Also, although seemingly obvious, but apparently not, try to keep the markers to a minimum – removing hundreds of markers from a shot where they have not been used is nothing short of insanity! Finally, precise measurements of the markers relative to one another and the setting are essential, if the job of analyzing the shot and being able to accurately track it, are to be straightforward.

To match the environment precisely it is necessary for a comprehensive lighting plot to be provided. This includes day exteriors lit by sunlight, where a diagram describing precisely the lighting conditions, overcast/clear and the position of the sun relative to the camera and subject, are a great aid. To help in doing this a useful piece of equipment is a white or 18% gray globe, 8–10 inch is a good size, which can be held in front of the camera at the end of a take and in the position of the subject. This will automatically reveal the direction and quality of light falling on the subject area. Additionally, a reflective globe of the same size, or slightly bigger, will also show the actual highlights of the main sources of illumination and their relative directions. All that needs to be done is to hold these two globes in the relevant position for a few seconds at the end of a take and to ensure that they are passed on to the post department along with the rushes.

If the CGI model that is being inserted into the scene is highly reflective, then it will also be necessary to provide the imagery that should go into the reflective surfaces. This will be the view from the position where the 3D model is going to be placed, looking towards the camera – in other words, it will usually be approximately what the camera would see if it were swung through 180°. This can be done using the reflective globe mentioned above but held in the position where the CGI would be if it were real. This will give a fisheye-like image but unfortunately the camera will appear quite large in the middle of it. The positive point of using a globe is that it can be very quickly stuck in front of the camera at the end of a take, and so will not use up valuable production time as well as accurately showing what was where during the take. An enlargement of this will be quite grainy and blurred, but usually will provide sufficient information for just the reflective elements that will not be strongly visible in any case.

Two alternatives are to stand in the position of the CGI model and take a series of stills as tiles (i.e. stills that can be joined together when printed) covering the whole of the area, again looking towards the movie camera shooting the scene. The alternative is to do this but with a fisheye lens. In either of these cases it is preferable to have the camera equipment

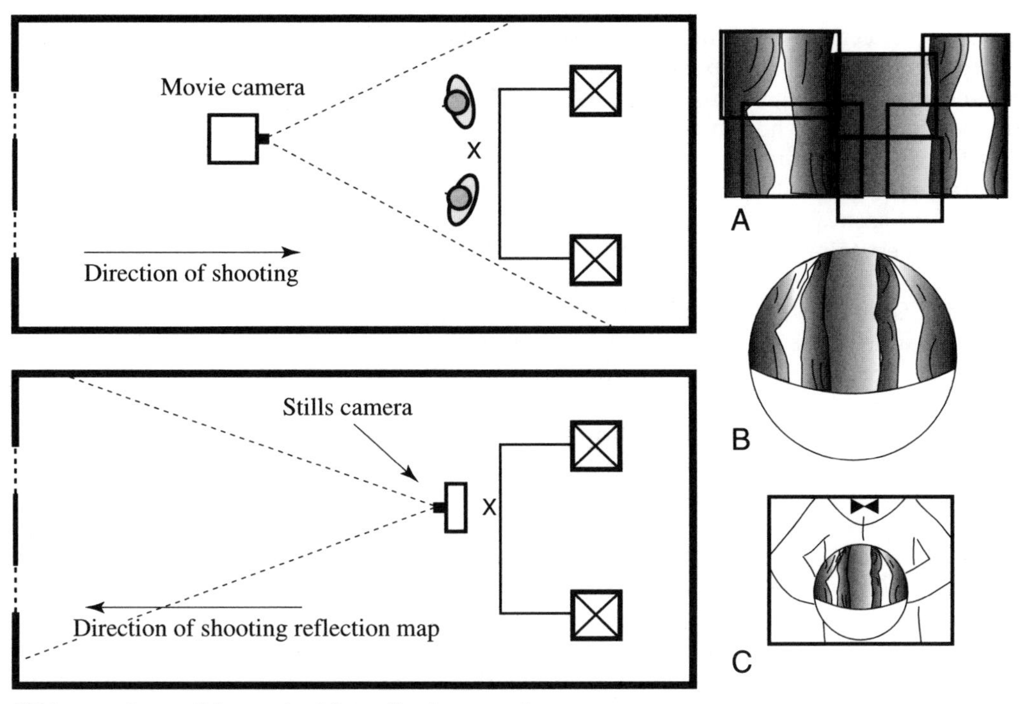

Widest angle possible required for reflection mapping
A: using still camera as shown and tilting for max. coverage
B: using still camera and fisheye lens
C: reflective globe (position X) shot from original movie camera position

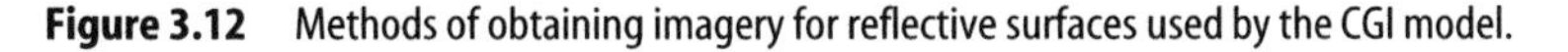

Figure 3.12 Methods of obtaining imagery for reflective surfaces used by the CGI model.

removed so that a clear view is obtained without necessitating substantial retouching. This may sound horrendous, but in most cases can easily be done when the camera is moved for the next setup – the only necessary help is in arranging to have the lighting left as it was during the scene. It should be noted that in most cases this work need only be approximate, since the reflections will not be that clearly defined or avidly viewed by the audience (Figure 3.12).

In addition to a pass showing a reflection globe and a lighting reference globe there are a number of other extra passes that can be helpful to the post-production team. First and foremost is a reference pass showing approximately what the CGI element will be doing (if it is doing something) and approximately where it will be. Obviously it is only in certain types of shot that this can be done, but where for instance the CGI represents some sort of creature or being playing against live actors, this would be very useful, particularly for the actors. Thus a stand-in, possibly even dressed to vaguely represent the CGI model, can act out its part amongst the actors, thus giving them eye lines and causing shadows and other interactive effects within the scene. This is very important to help the believability of the whole, and should be done wherever possible. It may even be that where a large-scale addition is being made, a mechanical mock up of some sort can be used – again so as to produce shadows, reflections and other effects which would be difficult to mimic successfully at the compositing stage. If at all possible the main take should have this reference figure playing the scene, and indeed, as long as the CGI model is going to be larger than the stand-in, then this will be simple to administer.

If the reference must be removed – in part at least – then it will be necessary to make a separate pass without it so that the empty setting can be used to paint out the unwanted parts of the scene. If, on the other hand, the reference cannot be used in the shot at all, then it will be best to make a separate pass with it in situ so that the compositors can at least gauge size and lighting effects from its interaction with the surroundings.

If the CGI model is required to interact with its surroundings in other ways such as, for instance, if it flashes with a bright light, then this effect should be applied in the set with the actors present. It may be that the reference entity could have lamps attached to do this, or it might be that conventional luminaires can be used from off camera, either way it is very easy to establish such lighting effects in the studio and difficult for them to be done convincingly in post. Other effects such as smoke or pyrotechnics should similarly be executed on the floor if that is at all possible.

If aerial or atmospheric effects are to be used in the scene, then reference for these should also be provided, preferably in the form of takes with and without the effect (just as we would do with optical filters). The action take can be either way depending on the actual work that is being done in post, and this must be discussed at the time. In some cases it is possible to have the smoke effects in the setting during the action take. With a "clean" reference pass the effect can then be added on top of the CGI to match. In other instances it is better to shoot clean and add the entire effect in post by matching the reference pass. Fire and rain, snow and other such effects all operate in the same way. If possible it is always best to have the effect in the setting because of their aerial effect – that is to say, features of the set become more indistinct with distance, whereas if the effect is added over the whole frame in post, it will all be of a single density and therefore less believable.

If the CGI model is to pass behind a foreground object then a matte can be created for this by placing a white or black sheet of poly (foamcore) or a flat behind it, and making a special pass. This will then provide a silhouette mask of the object. Obviously this only works for static features in the frame and a static shot, but if possible can save a lot of time for the compositing team.

If complex moves are being made and a number of different elements as described above are needed, then it is best to employ a motion control or repeating (field or mimic) head so that the various passes can be combined seamlessly. Where such a system is used, the motion data should be saved and passed on to the CGI team so that they can automatically extract the data describing the moves.

If the camera is only panning or tilting, but CGI elements are being added, say to extend the setting, then a nodal point head should be employed. This is basically a head, which is larger than usual and capable of being balanced and rotated around points other than the center of gravity of the camera itself. With such a camera it is possible to set the pivot points for both the pan and tilt such that they coincide with the nodal point of the lens. The nodal point is where the rays of light cross within the optical system and so if the camera is moved around this point there will be no parallax effects in the shot. Thus close and distant objects will not slide against one another as the camera moves.

You can mimic this effect simply by looking through one eye and moving your head too and fro whilst observing some close object (say your outstretched hand) against a more distant one. As you move your head the close object will appear to move with respect to the background. Now move your eye about by panning around your eyeball but not your head – the two objects will no longer move with respect to one another – the situation with a nodal point head. When shooting a pan with such a head, CGI elements will be able to track precisely without having to compensate for any slippage and hence be executed much more quickly and economically.

Obviously there will be circumstances where markers and exact motion information are not possible, such as shots taken from moving vehicles such as helicopters or tracking cars. In these cases, as much detailed information as possible should be gathered such as camera height and orientation on the tracking vehicle.

If more than one view is required, then multiple cameras running simultaneously should be used, as in the case of a shot requiring panning to be added in post-production. Here it might be best to shoot either an anamorphic negative for a standard format or with three cameras giving overlapping (or tiled) images, which can be stitched together in post to give a giant cinerama-like frame around which the standard frame can be moved.

When shooting any of the above effects for either background or foreground, it is always worth shooting slightly wider than normal. Having done this, adjustments may be made during post-production if the elements do not quite line up. It also allows for stabilizing of the image if camera moves are slightly uneven or there is a fault on the camera that permits the film to weave more than is acceptable for effect work.

3.7 Essential Equipment

There are a small number of items of equipment that are required in addition to those always carried by camera crews and which could help in setting up shots that are to be combined with CGI elements.

A hand-held computer such as those made by Psion, Palm Computing or Handspring running a cinematography software package will permit almost instantaneous calculation of time lapses, angle of view and other such esoteric data required for specialized photography. These programs can be obtained from either: www.davideubank.com or www.zebra-film.com

A sensitive *spot meter*, which can be used to check the evenness of blue or green screens. Most cameramen carry these but some prefer to use only incident meters and with a large screen it can be difficult to check the entire surface with these. Also it is best to measure the reflected light, because some screen materials absorb so much of the light falling on them.

Not everyone's cup of tea, but the old fashioned way of checking a blue or green screen is to use a colored *viewing glass*. This is like a small looking glass and Tiffen make both a blue and green screen one. Looking through these you can, once used to it, instantly see any irregularities in the evenness of the screen – it is very quick and if you are a visitor to the set will be the most inconspicuous way of checking the screen is OK!

An *inclinometer* is not often carried by normal shooting crews and will be an essential tool for working out the camera angle relative to the floor. Working out the tilt of the camera is necessary if the shot is to be exactly matched either on a further live-action shoot or in the computer environment.

A *laser measuring device*. High-quality measuring systems of this type are manufactured under the "disto" name by Leica (the camera manufacturer) and are essential for rapidly measuring up the distance from camera to various features of the set, and the set itself. With a copy of the set designer's (art director's) floor plans, precise measurements can be made very quickly using this method. Basically these devices project a small laser dot and whenever activated display the distance to that dot. This means that the distance to even quite thin objects can be measured and distances to say the camera on top of a high crane can be measured instantly from the ground. It is advisable to carry a small reflective white card so that if the feature being measured is very dark then the card can be stuck to it for measuring purposes.

A good long *tape measure* (best are the type sold for building sites or cricket) as an alternative to the laser measure and as a backup in case it fails and a small steel measuring tape for short distances such as the placement of objects on a table.

Reflective and white *globes* for giving a reference for lighting and reflections.

A set of *camera sheets* with a list of all the measurements and data that should be gathered for each set up. The use of a form to fill in will avoid things being overlooked in the heat of the moment (see the Appendix to this chapter).

A *gray card* for recording a lens' shading and general color defects and a *white on black grid* to record lens aberrations particularly barrel distortion.

A *still camera* with a wide-angle lens to take lots of reference stills which can be knitted together to show the whole setting or annotated with measurements. This camera should have a fast lens so that images can be taken in studio lighting conditions without the use of a flash. This is essential if the actual set lighting is to be shown for reference purposes. The best practice is to use film of the same ASA/ISO rating as that being used for the shoot, and then exposing with the stop set by the cameraman – this should get as close as possible to the look that will end up on film or tape.

An alternative or addition to a "film"-based still camera is a *digital camera*, which has many advantages such as instant gratification, the ability to load into an on-set computer and do tests, and even e-mail the images back to base. Care should be taken here though, since the resolution and color may not match the system being used for the "real" job. On the other hand, digital stills taken on set (and video too) have often been integrated into the final composite!

A roll of 1 inch white, blue and green camera tape for quick fix tracking markers and other such use when using colored screens or for anything else for that matter. Although crews normally carry all three the one you need may be "out" just when you need it!

Finally, although not part of the VFX supervisor's personal equipment, it is worth remembering that the film stocks of particular relevance to blue/green screen work are SFX200 and EXR100T (5248).

3.8 Summary of Variables that Should Be Analyzed

Notes should be clear and where appropriate should be made on floor plans where possible supplied by the art or location department. All of the parameters listed below should be considered and noted for later use. They should also be clearly identified as to when, where, by whom they were shot, where they can be located (by listing for instance roll number, timecode or slate and by scene and take) and by whom the notes have been made.

3.8.1 Parameters Relating to the "Look" or Quality of the Image

A description of the shot as taken and the final composite it is to be part of. For example, this could be a two-shot against blue screen to be combined with a CGI background.

Serial numbers should be logged to determine which camera body, lens and other equipment have been used. This could be for later matching or determining the causes of problems.

- Format (e.g. 35 mm film or $\frac{1}{2}$ inch CCD recorded on digital betacam); stock used (e.g. Kodak SFX200); grain (film) noise (video) a product of film speed (ASA/ISO) or video gain in dB (a DoP may not use the manufacturer's recommended speed); rez (resolution): number of pixels and whether square, PAL or NTSC; aspect ratio.
- Color temperature; focal length of the lens being used (and/or lens angle) + lens make and type; f/stop (on film lenses often quoted as T/stop); focus setting; depth of field (if relevant include hyperfocal distance).
- Camera speed (frames per second); shutter angle (degrees) and/or speed (fractions of a second); field-frame rate (in video frames per second and whether progressive or interlace).
- Filters (color, compensation and optical).

Additional information in the form of specially shot charts should be provided to describe distortions or misalignments inherent in the photographic path:

- color registration
- lens distortion (e.g. barrel or pin-cushion distortion)
- flare in system-lens-optics-actual scene.

If the image has been telecine'd then information about the grade (preferably including a copy of the colorists disk for the session) should be noted.

If the above information does not adequately describe any special adjustments to contrast and gamma (e.g. skip bleach process) or perspective (e.g. scale of miniatures), then these factors should also be noted.

3.8.2 Position in Space

Details of the actual location (e.g. inside or outside):

- lens angle
- camera inclination (degrees)
- coordinates (real or relative) defining where the lens is in space and relative to the subject and setting – height (elevation) and distance from the subject
- camera tilt (degrees from the horizontal and either up or down) and of pan cant/dutch angle if the camera is not level.

As a help to placing the camera in 3D space, lens and perspective lines could be drawn in scale on a floorplan and GPS data might be recorded to give precise geographical placement if the camera needs to be relocated in the same spot at a later date.

3.8.3 Movement of Subject or Viewpoint

- Details of any change of position of the camera should be taken and relative data recorded; for instance, by shooting tracking markers or recording motion-control data.
- On tracking vehicles exact camera positions should be noted (e.g. pointing back at 60°, height from the ground and tilt down).
- Movement of the subject should be described and recorded if necessary.

3.8.4 Environment/Subject

Full descriptions of the pictorial qualities of the subject should be noted including:

- lighting (direction and quality of lighting)
- reflections (including pictorial representation using a globe or stills)
- natural phenomena (smoke, rain, etc. should be defined preferably by shooting passes with and without the phenomenon).

Details of any special processes in use (e.g. blue screen or process projection and information such as brightness of screen and its distance from camera and subject).

Appendix Shoot Data Sheet

Episode		VFX shot		Date:		Sample
Scene		Element no		Time:		
Slate		Preferred take		Notes by:		
Take		Roll no:		Storyboard:		VFX data

Shot information	Setting:						Interior	Exterior
	Time of day	Day	Dark	DFN	Dawn	Dusk	Locked	Moving
	Director						Sound	MOS

Camera	Camera body no		Camerman	
	Speeds (fps):		Stock:	
	Shutter speed:		Shutter angle	
	Format:		Aspect ratio	
Lens	Lens no:		Type	
	Focal length:		f/stop:	
	Focus setting:		Filter used:	
Position	Elevation:		Subject dist.:	
	Inclination:		Screen dist.:	

Visual effects	Method (e.g. blue screen)	
	Element decription (e.g. fgd)	
	Shot description (diagram over page)	
	Characters in shot:	Shot length (e.g. 2 s)

4 Elements of Digital Effects

Steven Hubbard

4.1 Introduction

The purpose of this chapter is to examine the role of digital effects in the film-making process, to show how elements feed into the digital effects pipeline and to show the processes that take place within that pipeline.

4.2 What Is Digital Effects?

With any examination of the elements of digital effects, we must first answer the question "what is digital effects?". At first glance this may seem to be easy to answer, but for this very reason it is open to misinterpretation. There is both a narrow definition and a broad definition. A narrow definition is "particle systems and compositing". One of the reasons for this is that within digital effects studios, there is the job of "effects animator" and an effects animator creates particle systems, simulations and environmental elements such as fog, rain, sunbeams, etc.

A broader definition such as "a digitally created component of visual media, that cannot be created with conventional cinematography or editing techniques" is a more useful one. Therefore, any images produced by a digital effects studio are arguably digital effects. The CGI (computer-generated imagery) cityscape or digital creature is as much a digital effect as the particle explosion or the water simulation.

Another aspect of digital effects is that it is not really a discipline in itself. It is more of a functional description, or an umbrella term, which encompasses a number of other disciplines. The two most obvious of these are CGI and compositing, with all the activities that they encompass. However, it also includes such activities as digital matte painting, photogrammetry and 3D tracking (see Figure 4.1).

4.3 A Short History of Analogue and Digital Effects

We can not go any further into the realm of digital effects, without first considering the analogue effects that preceded it.

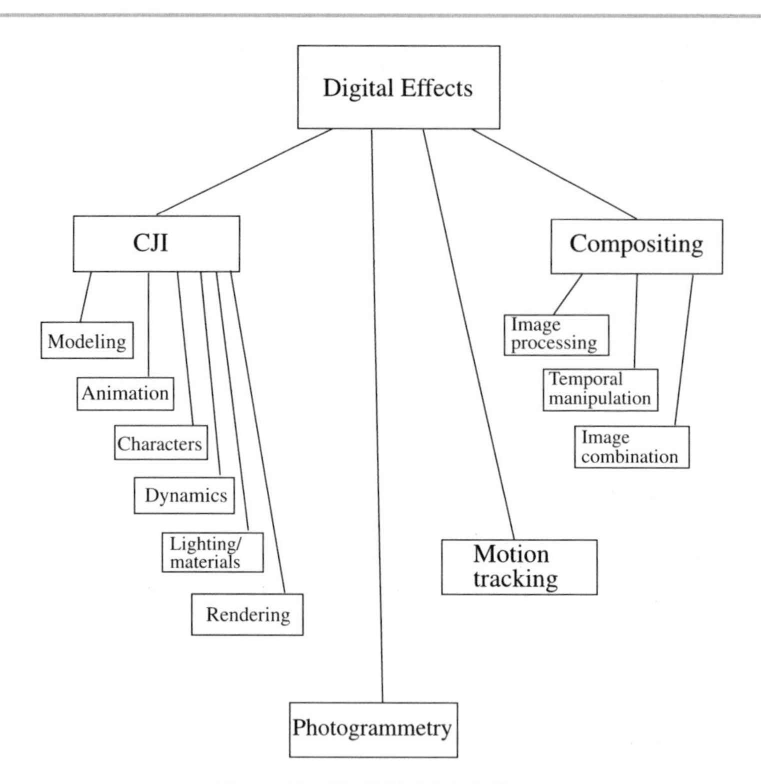

Figure 4.1 The field of digital effects.

A smattering of digital effects emerged in feature films in the mid-1970s, and then developed in sophistication during the 1980s. It was only in the early 1990s that digital effects rose to dominance. Prior to that, even as late as 1993, analogue optical and miniature effects reigned supreme.

A detailed history or analysis of optical and miniature effects is beyond the scope of this chapter. See Perisic (2000) and Fielding (1985) for a definitive treatment. However, as much of the terminology and methodologies of traditional effects have been carried into the digital domain we should touch upon them.

The essential core of digital effects is combining images. Prior to the computer becoming involved this was done photographically, using a process called "traveling mattes". Disparate film elements were rephotographed on special filmstocks, with special laboratory processing, to create masks, or *mattes*. The simplest way of visualizing the function of a matte is to imagine two photographic transparencies, one of an airplane and one of a cityscape. If you sandwich the two together and look through them, you will find that the city is visible through the plane, and the sky around the plane is mixed with the cityscape. The matte is a separate element, which prevents the city showing through the plane, and the sky mixing with the city. When rephotographed the result is a "persuasive" composite image in which the plane looks as if it is traveling over the city.

The device used for this rephotography of elements was an *optical printer*. The simplest version of this would be a single projector, a lens and a camera all fixed in line on a heavy base. The projector and camera could be synchronized to rephotograph every frame, or skip frames between them. The more advanced optical printers could have four projectors, each with as many as three strips of film passing through them. One of the real problems with this rephotography process was that the quality of the image would degrade for each rephotography or *generation* it went through. Also, the process for extracting the mattes from the original elements allowed little room for error. The original element had to be close to perfect.

The genesis of digital effects was as standalone inserts in otherwise analogue films (see Masson 1999). This was invariably in the form of 3D animation, as 2D compositing only really came into existence in the mid 1980s. The first, rather abortive, breakthrough for digital effects were the feature films *Tron* and *The Last Starfighter*. The former largely took place in a fantasy world inside a computer, and was therefore the obvious choice to do the major effects sequences with 3D computer animation. As it was an extremely stylized design, it was very sympathetic to the limited rendering capabilities of the time. *The Last Starfighter* was a slightly different case, as it sought to replace conventional optical and miniature effects with photorealistically rendered 3D computer animation. Some of the sequences betrayed their synthetic origins, while others still hold their own against digital effects of today.

Unfortunately, these two films did not engage the film studios' enthusiasm for digital effects. The real breakthrough was a trio of films for which Industrial Light and Magic provided the effects. These were *The Abyss*, *Terminator 2* and *Jurassic Park*.

With *The Abyss*, the digital effects were back to a supporting role, providing a featured sequence in an otherwise analogue film. This is the famous water *pseudopod*. Its importance is not so much as it being a character sitting in a real scene, interacting with actors, as in the fact that the effect could not have been achieved in any other way but with digital techniques. It simply would not have been in the film otherwise.

Terminator 2 could have been done without digital effects, but it would have been significantly different. Its *T1000* antagonist was designed around the capabilities of digital effects. 3D character animation, 3D metamorphosis and 2D morphing were all heavily featured. Of significant importance, but invisible to the audience were *digital compositing* and *rig removal*. As we shall see, the former is the foundation of modern digital effects.

It was the CG dinosaurs of *Jurassic Park* that really put digital effects on the map. Up until that point, there was still a lot of resistance from other effects studios to "going digital". After *Jurassic Park*, everyone went digital and many new effects studios were created (many of which subsequently closed). The CG dinosaurs in *Jurassic Park* were only on screen for about seven minutes, however, they were breathtakingly persuasive, giving a convincing impression of life. Not only was the 3D animation awe inspiring but the compositing of these creatures into real settings was virtually flawless. Digital effects had come of age.

4.4 Context of Digital Effects in the Film-making Process

Another issue we should address is the changing role of digital effects in the film-making process. Originally, digital effects was clearly a specialized form of visual effects. To carry our earlier description of digital effects a stage further: visual effects is "a component of visual media that cannot be produced using conventional cinematography or editing techniques, but is instead created in post-production". It is clear that digital effects is a sub-set of visual effects.

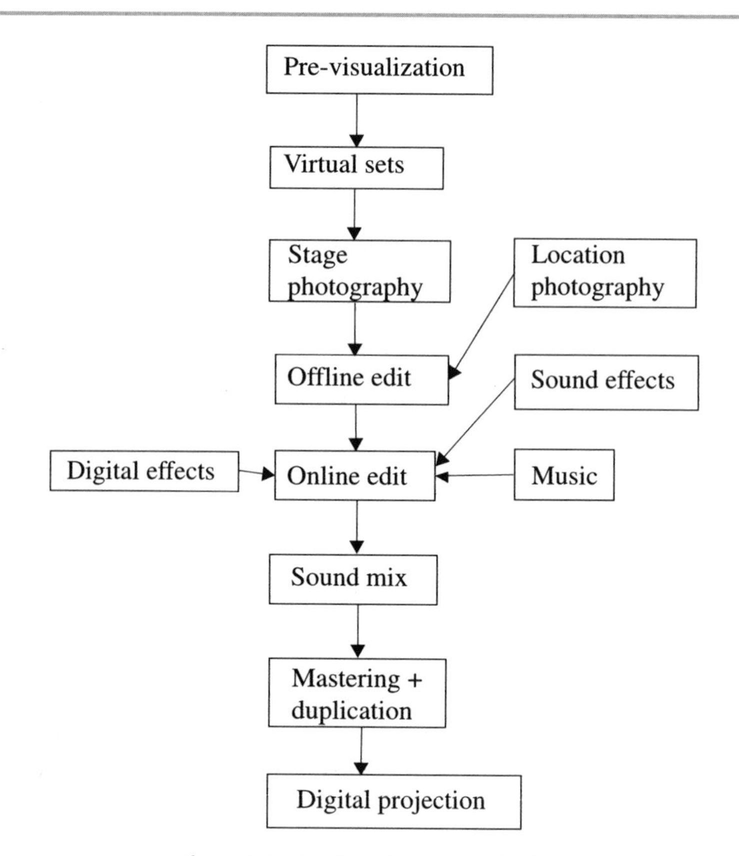

Figure 4.2 The digital film-making pipeline.

However, this definition is going to be increasingly challenged. Currently many elements of visual effects are originated on film. Within the next ten years one can foresee the majority of elements being originated on some form of digital video. When this happens it will be possible to argue that the definition of digital effects has broadened to embrace that of visual effects.

A more far reaching aspect is the increasing digitization of the whole film-making process. At the moment, the vast majority of movies are still mastered on film. i.e. editing is done digitally, but is still merely a pattern for the physical cutting of the film that passed through the camera. Likewise, digital effects are scanned from film, processed and then mastered back to film. Moves are already well under way to shoot movies on digital video. Moves are also afoot to project movies using digital projectors (indeed this has already happened with some digitally originated movies such as *Toy Story 2* and *Shrek*). Once this happens, the film-making process will move wholly into the digital domain (see Figure 4.2).

The computer used by the director for pre-visualization, will be similar to the computer used by the digital effects artist, which will be similar to that used by the editor, the sound editor and so forth.

In pre-visualization, the director will be able to prototype digital effects. On-set elements for effects shots will be processed in real time, to show how the final shot will look. In the edit it will be possible to do some effects, using the same software as effects companies, as the film is mastered. This won't negate the need for specialist digital effects companies. Difficult shots will still take a great deal of time to complete, but digital effects will permeate the whole film-making pipeline.

Terminology

- A "composite" is a combination of two or more images.
- An "element" is an image or sequence of images that is one component of a final composite.
- A "matte" is a monochrome image or sequence of images that controls the amount of mixing or "print through" of composited elements.
- A "background" is a base element over which the other elements are composited.
- A "foreground" element contains a subject that is to be overlaid over a background. For this to happen it requires a matte, or a suitable backing from which a matte can be made.
- A "pass" is a sequence of images of one aspect of a subject, either a motion-controlled miniature or CGI. For instance, a front light pass and a rim light pass which can be composited later.
- A "beauty pass" is a sequence of images of a subject with the final aesthetic lighting.

4.5 Critical Concepts

One critical aspect of digital effects is problem solving. Specifically, creating filmic elements that cannot be created by any other means. Another critical aspect is seamlessly combining disparate elements, often created in radically different circumstances, so that they look as if they were shot at one time using conventional cinematic techniques.

In reality, there are two kinds of effects shots. There are those "hidden" shots, which could have been shot in reality, but were not, due to commercial expediency or safety, or other considerations. Then there are "blatant" effects shots: the digital monsters, starships and exploding planets. The latter are the easier to achieve, as the audience understands that they are effects shots, just as they understand the film they are watching is a work of fiction. They "buy into" the fantasy.

With the former "hidden" effects, the audience is not buying into the fantasy in the same way. They do not know that what they are looking at is an effect, so they are not making any allowance for it. Any flaw in the effect will immediately flag this fact and spoil the illusion. This fragility in the suspension of disbelief is heightened by the fact that the audience will often be familiar with the subject of the "hidden" effects shot, whether it be an historic building or a wasp flying past. Their critical faculties will be highly tuned. So the quality demands on the "hidden" effect are much higher than those on the "blatant" effect.

4.6 Elements Feeding into Digital Effects

Compositing, or the combining of images, is the core process in digital effects. However, it is not absolutely necessary for a shot to be composited to be a digital effect. For instance a CGI shot of a starship against a starry background could be a complete shot. For the most part though, even if they are not combined with another shot, most effects elements

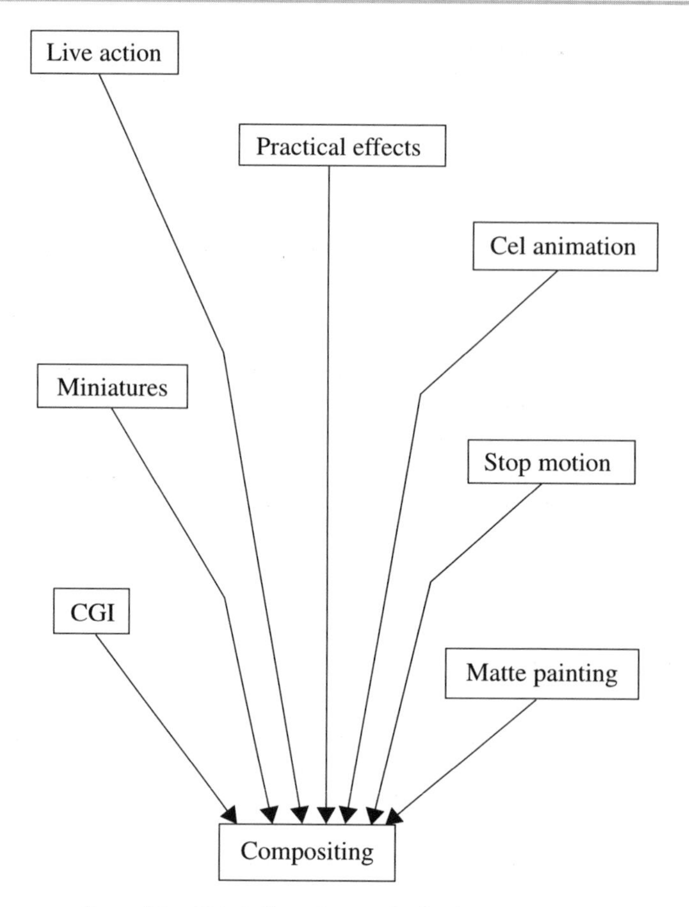

Figure 4.3 Digital effects elements feeding into compositing.

will be digitally processed by a compositing system in some way. A defining feature of a compositing system is that it cannot work alone. Unless images are fed into it, it will sit there doing nothing (apart from wasting money). There are a number of possible sources for these images (see Figure 4.3).

4.6.1 Live Action

Live action can perform a number of functions in digital effects. It can form background plates, over which other elements can be composited. This could be shot with a normal film or video camera, without any special equipment or processes. It could be pyrotechnics, e.g. gunpowder or gasoline being detonated against a neutral background, more often than not shot with a high-speed camera, which will make everything appear in slow-motion when viewed at normal speed. This has the useful effect of making a small explosion look like a larger one.

Probably the aspect of most consequence to digital effects is green screen or blue screen cinematography (in the interests of brevity, hereafter referred to as blue screen). With these, an actor will be shot in front of a colored screen, usually green or blue (see Figure 4.4 – the remaining figures for this chapter are shown in the Plates section), which can be replaced with another background in post-production. This kind of shot is the core of the digital effects process.

Another kind of shot would be an apparently complete shot that needs digital fixing. This could be the removal of wires on stunt people, the removal of incongruous features in a scene, or repair to a damaged negative.

4.6.2 Miniatures

Motion-control cinematography can be used with live action, but it's more commonly associated with models or miniatures. Essentially it's putting the camera on a computer-controlled motorized crane. The clever bit is that the computer can repeat motions exactly. This means that you could shoot a miniature in front of blue screen, then move it, reshoot it with the same motion, and when you composite the "passes" together, it will appear there are two miniatures acting in concert.

A motion-control rig is often used together with a model mover, which can rotate or translate the model at the same time as the camera is moving. This model mover is connected to the same computer as the camera rig. It allows multiple passes of the model. For instance, on a spaceship model, there could be a "beauty" pass (correctly lit), a matte pass, an internal lighting pass and an engine light pass. All of these will have the same movement, and can be mixed together in the compositing process.

Another aspect of miniatures cinematography is changing camera speed. As we have already discussed with pyrotechnics, shooting at high speed, and then projecting at normal speed gives a slow-motion effect. The use of high-speed cameras in regular live action is limited as it's obvious when a person is moving at the wrong speed. Miniatures have no such limitation. No one knows what speed a spaceship moves, so the camera speed can be set low to speed up the action, fast to slow it down or actually vary through the length of the shot.

There are several types of traveling matte that are very specific to miniatures. This is because they require two motion-control passes, one for the beauty pass and one for the matte pass. This rules out their use in live action as actors are not very good at accurately repeating their performance (though in the past, it has been used for live action, but it requires a camera that can run two pieces of film simultaneously). These traveling matte techniques are: infrared, ultraviolet and reverse blue screen, all of which are patented. With infrared, the matte is created with infrared light and infrared sensitive film. It's similar for ultraviolet. Reverse blue screen is a special case. With normal blue screen, the model has to have a matt finish, otherwise holes will appear where it reflects the blue screen. With the reverse blue screen process, a shiny model is shot in front of blue screen as normal. Then another pass is shot against a black background, with the model illuminated with ultraviolet light. The model has been painted with a transparent lacquer that glows blue under ultraviolet light. This allows another matte to be created to prevent print-through of the background image in shiny areas of the model (for more details on infrared, ultraviolet and reverse blue screen, traveling mattes, see Fielding (1985)).

4.6.3 Cel Animation

Conventional cel animation (i.e. cartoon animation in the Disney style) is already a composite. Different elements of a shot are painted onto transparent flexible cels. Once the areas occupied by the element have been painted and opaqued, all the cels can be sandwiched together in a composite and shot by a film camera. Of course, these days all the elements are scanned and composited digitally.

Cel animation has not just been used for animated features, it's been used in the service of visual effects for many years. This was facilitated by a device called a *rotoscope*. What this did was project a film image through the viewfinder of an animation camera onto the animation easel. The animator could then draw to the movement of the live action. Prior to the advent of digital effects, this was how most lightning or energy effects were done.

It was also the technique for creating a *rotomatte*, which is a term still used in digital effects, even though we do not use a rotoscope any more. This is when an animator draws around the silhouette of the foreground subject for every single frame, to create by hand, a traveling matte. This was a principal matte generating technique on *2001: A Space Odyssey*.

Related to this is something called a *garbage matte*, the terminology for which is still in use. This is a very rough matte, used to block out extraneous elements in a blue screen shoot. Namely light stands, edges of the blue screen, model mover base, etc. This does not necessarily have to be made for every frame, only when the subject of the shot approaches the edge of the garbage matte does it have to be revised.

4.6.4 Stop Motion

Stop-motion animation is the manipulation of a puppet or model frame-by-frame, so that when the film is projected, the puppet or model has the illusion of continuous motion. Before the advent of digital effects, it was the principal way of doing full-motion creature effects, of the kind made famous by Ray Harryhausen in films such as *The 7th Voyage of Sinbad*, *One Million Years BC* and *Clash of the Titans*.

In the case of Harryhausen, there was no post-production compositing, all composites were done in camera at the time of stop-motion shooting, using a combination of rear projection and "glass-painted mattes". Nowadays stop motion is rarely used for visual effects, one honorable exception being David Allen's final project *The Primevals*, but it is still used extensively as an animation style in its own right.

As far as compositing stop motion is concerned, one interesting method is Front Light/Back Light. With this the animation puppet is shot, correctly lit, against a black velvet background. Then the background is replaced by a white screen (for instance a lightbox) and the foreground lights are turned off. A frame is then shot of the silhouette of the puppet on a white background. When a complete animation sequence has been shot, this results in a checkerboard film strip of beauty and matte frames. This, in the past, would be put through a skip-framing optical printer to create a self-matting beauty pass and a male matte for bipacking with the background plate.

This technique is also applicable to motion-control miniatures (see Figure 4.5). A beauty pass is shot, then the lights are turned off and the white screen turned on, and the motion-control rig repeats the move to create a matte pass. This allows very high contrast lighting with dense blacks, and was used on *2010*.

4.6.5 CGI

So long as you have the time and resources, you can create most effects these days using CGI. There are certain difficult subjects, namely realistic humans, and environmental effects such as fire, but these will be successfully addressed in the near future.

Subjects that CGI is especially well suited to are spaceships, cityscapes and technological elements. Recently, there have been major advances in the creation of realistic synthetic characters.

The CGI element is one of the easiest to handle in digital effects. Rather than having to create a separate matte, one is generated as the color image is rendered. This matte is called the *alpha channel.* Where it is white it is opaque and where it is black it is fully transparent, and it allows any level of gray (transparency) in between. The color image has three channelss red, green and blue, which mix together to create any color. The alpha channel forms the fourth channel. So whenever CGI is loaded into a compositing system, it is loaded along with its matte.

Another aspect of a CGI image is its lack of film grain. Which is very good, until you try and combine it with an image with film grain, and suddenly it looks far too sharp and clean, which means that you need to add film grain onto it.

CGI images can also be generated at any resolution, or variety of resolutions you want.

4.7 Compositing

Compositing is the core of digital effects, yet the term "compositing system" is something of a misnomer, as it implies that all images are being combined in some way. While it's true that this is a major function, a lot of operations are singular (unary) rather than multiple (binary, or more), i.e. operating on a single image or sequence of images. Compositing system functions broadly fall into the following categories:

- input
- image combination
- matte generation
- matte processing
- filtering
- color correction
- image transformation
- image deformation
- painting/retouching
- time modification
- output.

4.7.1 Compositing Systems – Tiles or Timelines?

Before proceeding with a detailed description of possible processes, we need to consider the architecture of the compositing software. Historically these have fallen into two methodologies, and even though the distinction is now quite blurred, these broadly still hold true. These methodologies are *timeline* and *tile.*

4.7.1.1 Timeline Architecture

This is the oldest architecture as it was the one adopted by Quantel for its video post-production tools in the mid 1980s. The timeline display mimics a strip of film, in that the frames in a sequence are displayed next to one another, and this strip can be dragged up or down to cycle through the frames. Different image strips can be displayed at the same time and dragged onto one another, using a stylus and graphics tablet, to create mixes or composites (see Figure 4.6).

This working method set the trend for digital video post-production tools and subsequent Quantel products have followed this pattern, as have competing products such as Discreet Logic's *Inferno*.

4.7.1.2 Tile Architecture

This approach uses a graphical *worksheet* onto which can be dragged tiles, which represent different functions. These tiles can be connected together into a network, which has a flow from input to output. The tiles can be opened to set parameters or directly manipulate the image at that stage. Once all the functions are connected and set, the network can be rendered to produce the final image sequence. (see Figure 4.7)

This type of compositing grew to dominance in the 1990s, especially as large digital effects studios developed, requiring large numbers of compositing systems running on standard computers. Early examples were *Digital Fusion*, *Eddie* and *Cineon*, which were originally developed in Australia. Later examples are *Shake*, *Chalice* and *Rayz*.

The tile paradigm has also been heavily utilized in the Side Effects 3D animation system *Houdini*, which also incorporates its own tile compositing system.

This distinction between architectures has been blurred in recent years as tile-based software, such as *Cineon*, have incorporated timeline *lineup* panels, and timeline systems have incorporated tile-based effects setup functions.

4.7.2 Input

As previously indicated, compositing, unlike CGI, cannot live in isolation. It has to have imagery fed into it in a format it can understand. Imagery from different sources has different characteristics.

4.7.2.1 CGI

In a way this is the simplest image form to input as it's completely computer generated. Each image is a bitmap, the parameters for which can be selected by the animator at render time. Resolution for video is fixed by the video standard. Resolution for film can be standard resolutions such as 2K (image approximately 2000 pixels across) or 4K (see the following film section). However, as CGI often seems unnaturally sharp a common practice is to render it at 1K then do a filtered scale up to 2K, though this is dependent on the fineness of the detail in the image.

More important than resolution is bit depth. This is the number of binary digits used to represent the color of a pixel (the smallest discrete unit of a bitmap). The larger this bit depth the finer the color graduations, but also the larger the image file.

The standard bit depth is 24 bit, or 8 bits per color channel (red, green and blue). This gives 16.7 million different possible shades, or 256 graduations each for red, green and blue. This is perfectly good for output to video, however it's inadequate for film use, especially if the images which they'll be composited with are in *Cineon* format, which has 10 bits per color channel. The solution is to increase the rendered image to 16 bits per color channel, which gives 64 000 graduations.

Most CGI images will have four 8 bit channels: red, green, blue and alpha, the matte channel. This will give a 32 bit image file.

The actual file format is dependent on what you want to use the frame for. For compositing work, you should not use lossy compression, which drops "unnecessary" detail from the images. This rules out the use of JPEG and MPEG. It should have at least 24-bit color, which rules out GIFs. For video, TGAs or TIFFs fit the bill. For film the best formats are TIFF16 or SGI16.

4.7.2.2 Video

These days the large majority of video is digital, so it has much in common with CGI images. However, it is bounded by very rigid image and file formats dictated by broadcast TV standards and international agreements.

Resolution is fixed to one of either two TV standards, PAL or NTSC. The digital standard for these TV formats is called CCIR 601. Under this standard, PAL video images are 720 × 576 pixels at 25 frames per second, and NTSC images are 720 × 476 pixels at 30 frames per second. In both standards each frame contains two sub-frames or "fields", one of which will occupy the even scanlines, the other the odd scanlines.

This "interlacing" of fields causes difficulties with compositing, so most effects elements intended for television broadcast are originated on film and then telecined to video. In practice this means that images have no visible interlacing.

The image file format for PAL defined by the CCIR601 standard is YUV, where Y is the luminance channel, and U and V are the color difference channels. The color sampling is expressed as a ratio between these channels. So 4:4:4 implies that the luminance and color are sampled at every pixel. Most digital tape formats use 4:2:2 sampling, which samples the color at every other pixel.

Hi-definition video is a completely different matter, with numerous different flavors. Most of these are broadcast formats, in that they are interlaced and related in some way to the dominant TV standard in that area.

The most important hi-definition video format is *Cine Alta* from Sony and Panavision. This has a high-definition image of 1080 lines, at 24 frames per second, i.e. the film frame rate. It uses 4:2:2 sampling and approximately 5:1 compression. The most important aspect of it is *progressive scan*. Instead of scanning two fields made of odd and even scanlines, the camera will scan every line in sequence, removing the interlace, and some of the attending problems that this causes with digital effects.

4.7.2.3 Film

Film is very much an analog format. It's an image recorded on a piece of celluloid using silver halide crystals and color dyes. This means that each frame in a sequence has a different crystal grain structure. So on a macro level there may be similarity between adjacent frames in a sequence, but on a micro level there is none. This gives a much softer appearance to a film sequence than to a video sequence.

To perform digital effects on a film sequence, the film has to be converted from analog to digital, or *scanned.*

Scanners either use a laser or a CRT tube to scan a beam (or beams) of light over a frame of film, usually the original camera negative. The intensity of the light passing through the frame is measured and digitized, and then stored as pixels in a bitmap.

The resolution of the scanned image depends on the resolution of the sensor in the scanner. In the case of Kodak scanners, this is a 4096 element linear CCD. This gave the maximum resolution, across the width of a piece of 35 mm film, of 4096 pixels or 4K. A 4K image file is large and unwieldy (around 40 MBytes, depending on bit depth) so most effects work is done at 2K or 2048 pixels wide.

Unfortunately things are not that simple. A "full aperture" frame covers all the available image width of a piece of 35 mm film. *Academy* and *Cinemascope* frames leave space for an optical soundtrack, and so have a horizontal resolution of 1828 pixels. This still falls under the term 2K.

As the height of a frame is governed by how many samples are taken as the film is pulled past the sensor, there is a lot more scope in the vertical resolution. In Kodak scanners, *full ap* and *Cinemascope* have a vertical resolution of 1556 pixels, and *Academy* 1332 pixels.

There are too many film aspect ratios to go into in great detail, however, *full ap* and *Academy* have aspect ratios of 4:3. Within this frame, images are composed for projection at 1.85:1 or 1.66:1. So if you're working with a full-ap or academy effects plate, it's probable that a large strip at the top and bottom will never be seen by an audience. The exception to this rule is *Cinemascope.* This is shot through an anamorphic lens, which puts a 2:1 horizontal squeeze on the image. The image recorded on the film is almost square and is almost all projected onto the screen.

A standard bit depth of eight bits per channel only gives a tonal range of 256 graduations, which is much less than is recorded on the film. Therefore the output from the scanner needs to be higher. We've already seen that 16 bits per channel can be used. There is another file format that was specifically designed for film, *Cineon.* This was developed by Kodak as the interchange format between its scanners, recorders and *Cineon* compositing systems.

Cineon has 10 bits per channel, giving 1024 graduations. On its own this is enough to provide an accurate representation of the image on the film. However, unlike other formats, *Cineon* has a *logarithmic* mode. Normal formats are in a linear mode, which means simply that each graduation in the tonal range is of equal weighting. In a logarithmic scheme, the graduations differ over the tonal range and are sympathetic to the actual light response characteristics of the film (Figure 4.5).

In software packages such as *Cineon* and *Rayz* image processing can take place largely within the logarithmic domain. In other software the images will have to be converted to 16-bit linear.

4.7.3 Image Combination

There are a number of possible operations for combining images. Some of these use mattes, others don't. Many operators will produce summed results, which can be strange for RGB images, but is very useful for mattes. The simplest combination operations, which don't use mattes, are mathematical operators. Namely *add* and *subtract,* sometimes referred to as *plus* or *minus*. In the following examples we assume normalized pixel values, i.e. values between 0 and 1 rather than 0 and 255.

Add

Adds the luminance of a pixel on *A*, to that of its equivalent on *B*. This is of most use in matte manipulation.

Subtract

Subtracts the pixel luminance of *A* from its equivalent pixel on *B*. Again, of most use for matte manipulation.

Multiply

Multiplies the luminance of a pixel in *A* with that of its equivalent in *B*. For instance, 0.8 (*A*) multiplied by 0.6 (*B*) gives a resulting pixel value of 0.48.

Over

Over is the basic compositing tool. It uses an alpha channel matte on *A* to overlay the subject areas of *A* over those of *B*. The quality of the result is largely dependent on the quality of the original matte (see Figure 4.8a).

Inside

This results in an image containing *A*'s subject, but only within the area of *B*'s matte or vice versa (see Figure 4.8b).

Outside

This results in an image containing *A*'s subject, but only outside the area of *B*'s matte or vice versa (see Figure 4.8c).

Atop

This overlays the subject of *A* over *B*, but only within the matte area of *B* (see Figure 4.8d).

Minimum

Minimum compares two images, most usefully mattes, and records the lowest pixel luminance value into the output image.

Maximum

Maximum compares two images, again most usefully mattes, and records the highest pixel luminance value into the output image.

Mix

An additive mix combines two images by taking a percentage of the pixel value from *A* and adding it to an inverse pixel percentage from *B*. For instance, 80% of the pixel value from *A* added to 20% of the pixel value from *B*. This mix factor can be animated, from 100% on *A*, 0% on *B*, to 0% on *A* and 100% on *B*. This is called a *dissolve*. For an in-depth examination of the combination options, see Brinkmann (1999).

4.7.4 Generating Mattes

The core function of the compositing package is to combine images. To do this without superimposing the images, one of them, the foreground, will have to have a matte. As we have already seen, CGI generates its own matte at render time. Some double-strip or double-pass traveling-matte processes also create an accompanying matte. For the most part, however, live-action effects elements will come without a matte and one will have to be generated in the compositing system. There are several "semi-automatic" methods of matte generation.

4.7.4.1 Luma Key

The simplest way to automatically generate a matte is *luma key*. This operates on the luminance of the source image. Intensities above a certain level will create a white pixel in the matte image and below this level a black pixel (see Figure 4.9). This technique originated in analogue video and has been interpreted into the digital domain.

This may be quite effective for combining abstract or "supernatural" images, but for conventional live action, for instance seamlessly putting an actor into a burning building, it is less useful. To generate an effective matte the foreground will have to be flatly lit so that no areas are in deep shade and there are no deep shadows. The background will need to be uniformly dark. Any deep shaded area of the foreground will appear as holes in the final composite.

Obviously most foreground elements will be shot with a wide range of shading, so will not be instant candidates for a luma key matte. However, it is possible to improve a luma-key matte through the use of contrast controls. By manipulating the contrast, gray pixels can be lifted to white. One problem with this is that the edge of the matte can take on a "chiseled" appearance. In the final comp this will make it appear roughly pasted on.

4.7.4.2 Chroma Key

Like luma key, the concept of *chroma key* originates in analogue video. In an analog chroma-key unit, there will be two inputs: one on a colored background (almost always blue, though it could be other colors) and the other a regular image for the background. As the chroma keyer scans through the foreground video raster, where the background is blue it switches to the background image, when it reaches a non-blue pixel it switches to the foreground image. This switching is the chief weakness of chroma key as it is either on or off, which means you always get a hard edge. Also the switching process takes a small amount of time to perform. This means that fine detail, such as hair, is poorly handled, and fast-moving objects tend to smear.

Luckily the digital chroma keyer in the compositing system is an approximation and avoids many of the shortcomings. The digital chroma keyer obviously doesn't have to switch in real time.

A digital chroma keyer will take in a foreground image with a colored background (see Figure 4.10). This is normally blue as human skin tones contain very little blue. You select a hue color and the amount of falloff either side of it. The obvious limitation is that the chosen hue color should be absent from the foreground image, i.e. for a blue screen the subject should not be wearing any blue clothing. Digital chroma key is much better at differentiating between different shades of blue. More sophisticated chroma keyers also allow ranges of luminance and saturation to be set.

4.7.4.3 Difference Mattes

A difference matte is a radically different way of creating a matte from a piece of live-action footage. No blue screen is required. The way a difference matte works is as follows. A foreground is shot against a background, it could be any background. Then the foreground subject leaves the frame, and the background is shot on its own. The shot with foreground is then compared with the background shot. Where there is a difference between the images the pixel is white, where there's not, it's black, hence a matte is created.

Evidently, anything that is moving will be part of the foreground. Anything that is not will be part of the background. Usage is slightly limited. If you move the camera, everything becomes foreground. To overcome this you'd need to have a motion-control rig to repeat the foreground and background passes. Another problem area is that everything in the background has to be static, so there can be no wind blowing the curtains, for example. Problem areas can also be caused by unwanted reflections or shadows as well. Some of these problems can be negated with the use of garbage mattes. Difference matting is a specialized tool that requires specialized preparation.

4.7.4.4 Primatte

Primatte is a patented digital chroma keyer, found in compositing systems such as *Shake*. It uses a very advanced algorithm that allows multiple chroma values to contribute to the key. Unlike regular chroma keyers which use hue and luminance, *Primatte* also uses the saturation component (see www.primatte.com).

4.7.4.5 Ultimatte

Widely regarded as the finest matte generator available. This is a commercial product, produced by Ultimatte Corporation. It's available as a plug-in for a wide variety of software packages. It is also fully integrated into *Cineon*, *Chalice* and *Rayz* (see Brinkmann 1999; www.ultimatte.com).

History

In the 1950s Petro Vlahos developed an optical effects process called the *color difference traveling matte*, which he patented. This used a combination of optical printing and complex laboratory duplications. Its objective was to handle transparency and liquids better than existing traveling matte processes. Its first feature film was *Ben Hur* in 1959.

In 1976 Petro Vlahos formed the Ultimatte Corporation. *Ultimatte* was a real-time analog video processor, that implemented the *color difference* process electronically. Prior to *Ultimatte* being available all video compositing had been done with keying, with its attendant problems of fringing and poor detail retention. Unlike chroma key with its switching, *Ultimatte* sets the backing of the foreground to black and adds a male matte of the foreground to the background. It then does an additive mix of the two. For real-time live TV compositing, *Ultimatte* units, albeit in digital form, are still the best solution.

In parallel with the advent of digital *Ultimatte* hardware, there came the software, which implements the *color difference* algorithm for compositing systems, originally in the stand-alone Cinefusion product, and later as plug-ins or built in elements of systems.

Process

Ultimatte has two principal processes: One suppresses the backing color (for instance, blue to black). The other creates a matte from the blue backing. To suppress the backing to black, a new blue channel is created, where if a pixel has a larger blue value than green it takes on the green value. In the case of a blue background, there will be no green so it becomes black. If the green value is larger than the blue, then the pixel takes its value from the old blue channel. To create the matte, the maximum of the red and green channels is subtracted from the blue. For a more in-depth examination see Brinkmann (1999) and www.ultimatte.com.

The following *Ultimatte* tools are the *Rayz* and *Chalice* implementation of the technique. These appear as separate tiles which are connected together (see Figure 4.11). In other systems, such as *Cineon*, these may appear as a single tool with multiple facets.

Classic Screen Correction (CSC)

CSC is a form of difference matte. Its purpose is to suppress imperfections in the blue backing. You shoot the original foreground, and then move the subject away and shoot just the background or the "screen correction pass". The resulting difference matte is used to turn the backing to a uniform *Ultimatte* blue. The same limitations and considerations as for a difference matte apply.

Color Correction (CC)

This allows the color and levels of the foreground to be compared and matched to those of the background, so allowing a persuasive composite to be created.

Grain Killer (GK)

One of the problems with a super-fine compositor such as *Ultimatte*, is that it is so sensitive that it can detect film grain. What can happen is that the software can extract the film grain from the blue screen and treat it as part of the foreground. This means that when the foreground is composited over the background, the grain from the blue screen is composited over the background, so doubling the grain. *Grain killer* is used to suppress the grain in the blue screen areas of the foreground element.

Process ForeGround (PFG)

Process foreground is the principal *Ultimatte* tool. It takes as input a prepared blue screen shot, and outputs an image with an added alpha channel containing the matte, and suppresses the blue backing to black. The resulting image can then be used by a standard "over" compositing node to combine it with the background.

The controls for this tool fall into three categories: one to pick the backing color and the appropriate algorithm, the next to control matte density and print through and the final one deals with flare suppression. Flare is also often referred to as "spill", and flare suppression is often referred to as "despill". Spill is when light from the backing "spills" onto the foreground subject. This contaminates the subject with backing light and make it partially transparent. With sophisticated color analysis and control, it is possible to remove the spill light from the foreground. However, in doing so you may modify the foreground colors. For instance, removing blue spill may modify colors containing a lot of blue, such as purple. A despill tool will also be able to color correct these aberrations in the foreground.

AdvantEdge (AE)

Ultimatte AdvantEdge is actually a post-processing tool. If you have a good quality comp you shouldn't need it. If, however, you've pulled a matte with PFG and comped it with an "over", and the foreground and background elements do not sit together very well, you can use *AdvantEdge* instead. What it does is take the foreground matte and use it to generate an edge mask. Through this edge mask the interface between the foreground and background elements is slightly blurred and color corrected.

4.7.4.6 Manual Mattes

If we have semi-automatic mattes, it stands to reason that we can have manual mattes. We've already come across the basic terminology in our discussion of cel animation: garbage mattes and roto mattes. These have again been adapted to the digital domain.

While it is possible to use a digital paint package to create a roto or garbage matte, simply by superimposing the image you want to rotoscope and then painting a white on black image, this will not be particularly successful for moving images. The paint strokes on each frame will be different, which means the pixels on the edge of the matte will be slightly different from frame to frame, and will cause the edge of the matte to "boil" or "crawl".

A more useful approach is to use a spline drawing tool. Almost all compositing systems will have this facility. With this you draw a shape with spline curves. The control points for these curves can be moved and key framed, so that the curve changes shape as the subject that you are rotoscoping moves. The shape is only rendered as a white on black matte as required. This gives a much better result.

As well as contributing to the compositing process, hand-painted or spline mattes can also contribute to single-image processes. Notably they can be used to mask processes. For example, a mask could be used to localize the effect of a blur filter to one region of an image.

Often the matte resulting from a manual process will be added to those created by automatic processes. For instance, a matte may be pulled from a blue screen. The edge, the important bit, may be extremely good, but the interior may be full of holes, a "thin matte". A rotomatte could be created for the interior of the subject and then added to the original matte to fill in the holes.

4.7.5 Processing Mattes

There are a number of unary processes that are very specific to modifying mattes. Notable processes are *erode*, *dilate*, *median* and *edge detect*. These are often incorporated under the banner of "filters".

The erode process, sometimes known as a minimum filter, decreases the size of a matte by adding an offset, in pixels, from the original edge. You could create two versions of a matte: a solid matte with a coarse edge, and a thin matte with a fine edge. On their own, neither of these mattes would be sufficient. However, the coarse matte could be eroded, and then added to the fine matte.

The dilate process, sometime known as a maximum filter, performs the opposite function to erode, and increases the size of a matte by offsetting the edge pixels.

A median process is only found on some compositing systems. It is most useful for matte repair. If you have a defective matte, i.e. a good edge but black speckles within the outline, a medium process can eradicate the flaws. It does this by detecting black pixels largely surrounded by white, and then filling them in with the value of the surrounding pixels. Of course, if misused, this process can destroy detail. For instance, in a human silhouette the gaps between the fingers could be narrow and sporadic. This could be destroyed by a medium process.

The edge detect process does what it says. It looks for sharp changes of contrast in a scene, and then generates a matte with white pixels at those points. In the case of a blue screen, or an existing matte, this results in an outline of the foreground subject. This outline can be used as a matte for the selective processing of the edge. For instance, a contrast alteration could be made to the edge of a semi transparent matte, without affecting the rest of the image, or a blur could be applied to the edge of a foreground subject, after it has been composited, to better blend it into the scene.

4.7.6 Filtering

Filters can vary from system to system, the tool set often incorporates matte processes and individual color-correction tools. Specific filter operators are *blur*, *sharpen*, *degrain*, *regrain*, *convolve* and *noise*.

Blur

Obviously this blurs the image. Normally it's possible to select the form of the blur filter, for instance "box" or "gaussian". On some compositing systems, you can isolate which channels are blurred. A normal use of a blur filter would be to slightly soften a matte, or to slightly blur the edge of a foreground subject, after it's been composited, so making it sit better in the scene.

Sharpen

This should be used sparingly as it has a tendency to introduce obviously synthetic artifacts into the image. For interlaced video, this can throw the fields into sharp relief.

Color Plates

Plate 4.4 Blue screen stage photography.

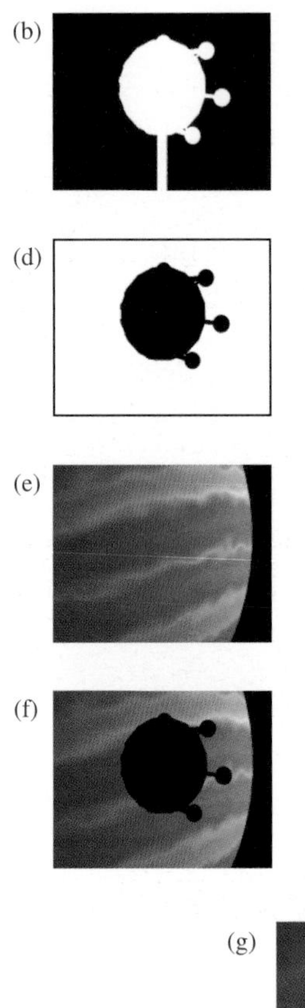

Plate 4.5 The front light/back light technique. (a) Stage photography: matte pass and beauty pass. (b) Matte element (negative). (c) Beauty element. (d) Positive matte with stand removed by garbage matte. (e) Background element. (f) Positive matte making a "hole" in the background element. (g) Composite with the beauty element printed into the "hole" in the background element.

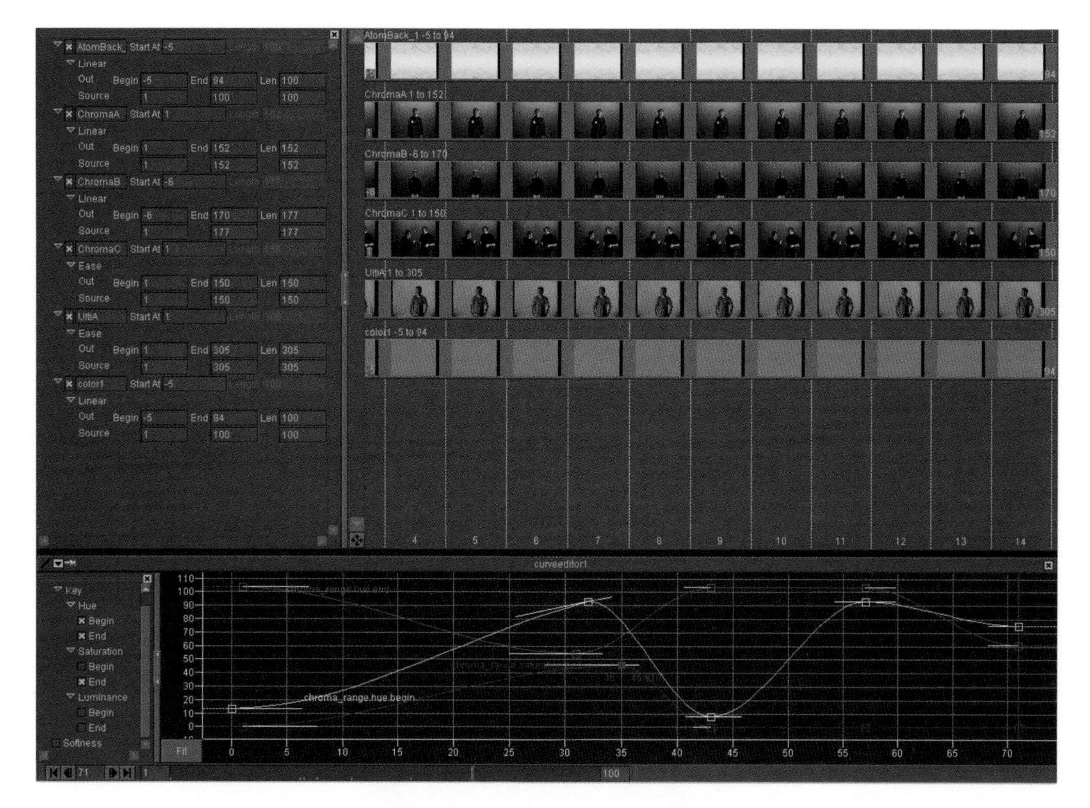

Plate 4.6 Timeline style layout in Rayz.

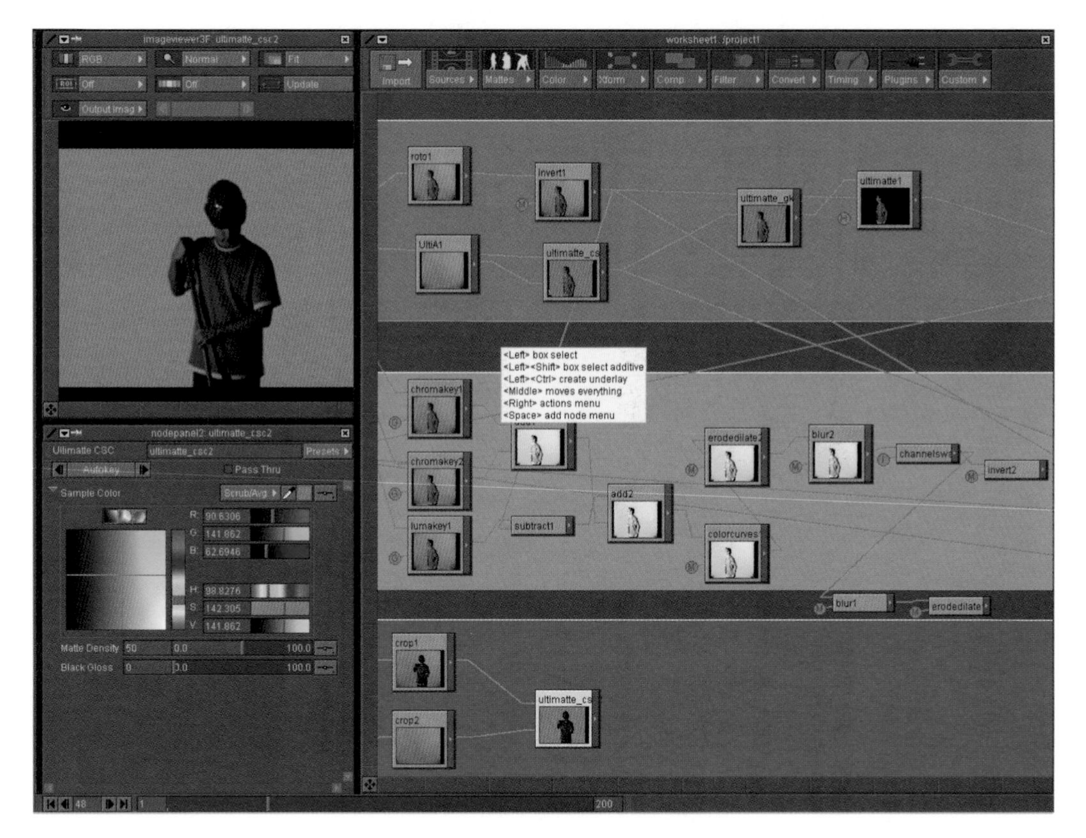

Plate 4.7 Tiled worksheet style layout in Rayz.

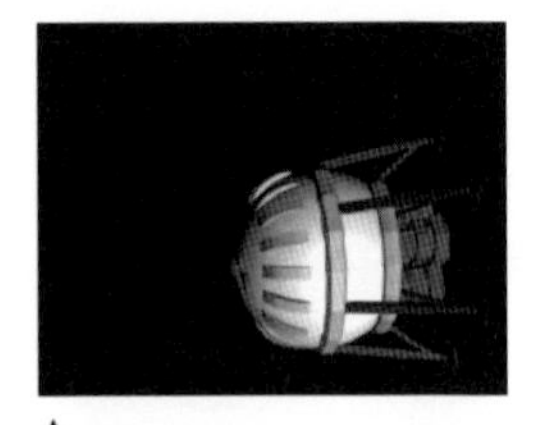

A

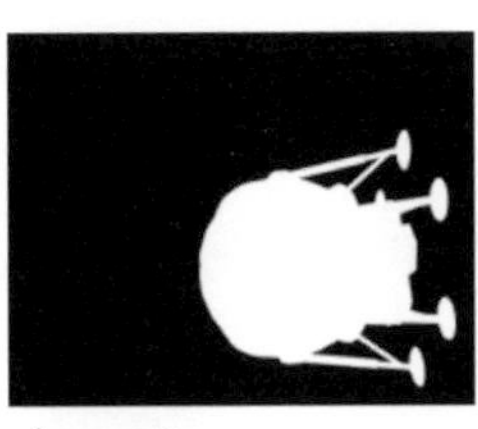

A matte

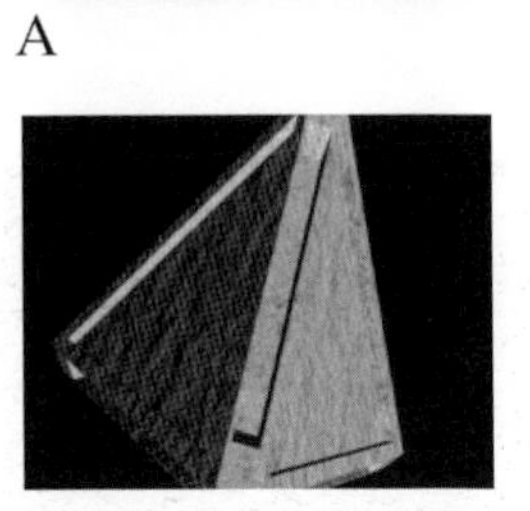

B

B matte

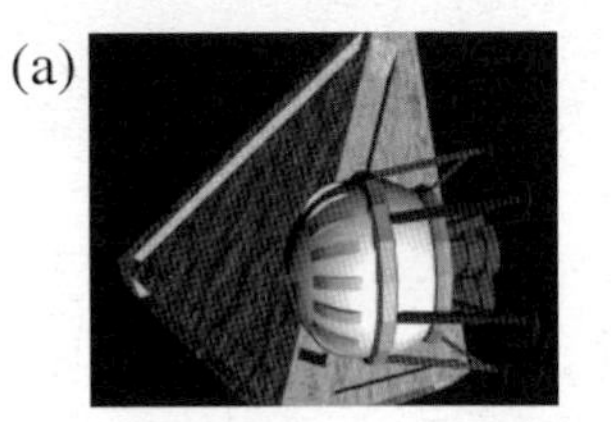

(a) A over B with A's matte

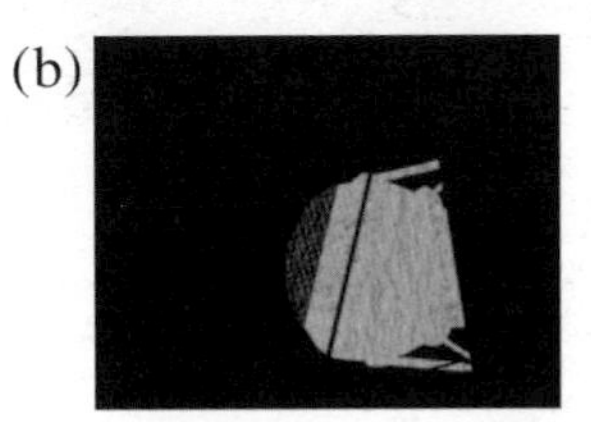

(b) B inside A's matte

(c) B outside A's matte

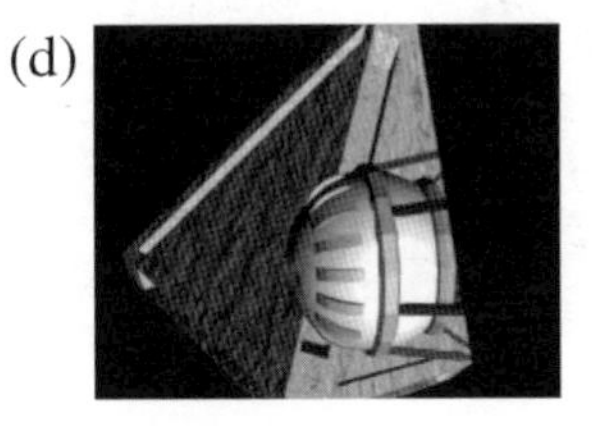

(d) A atop B within B's matte

Plate 4.8 Methods of image combination.

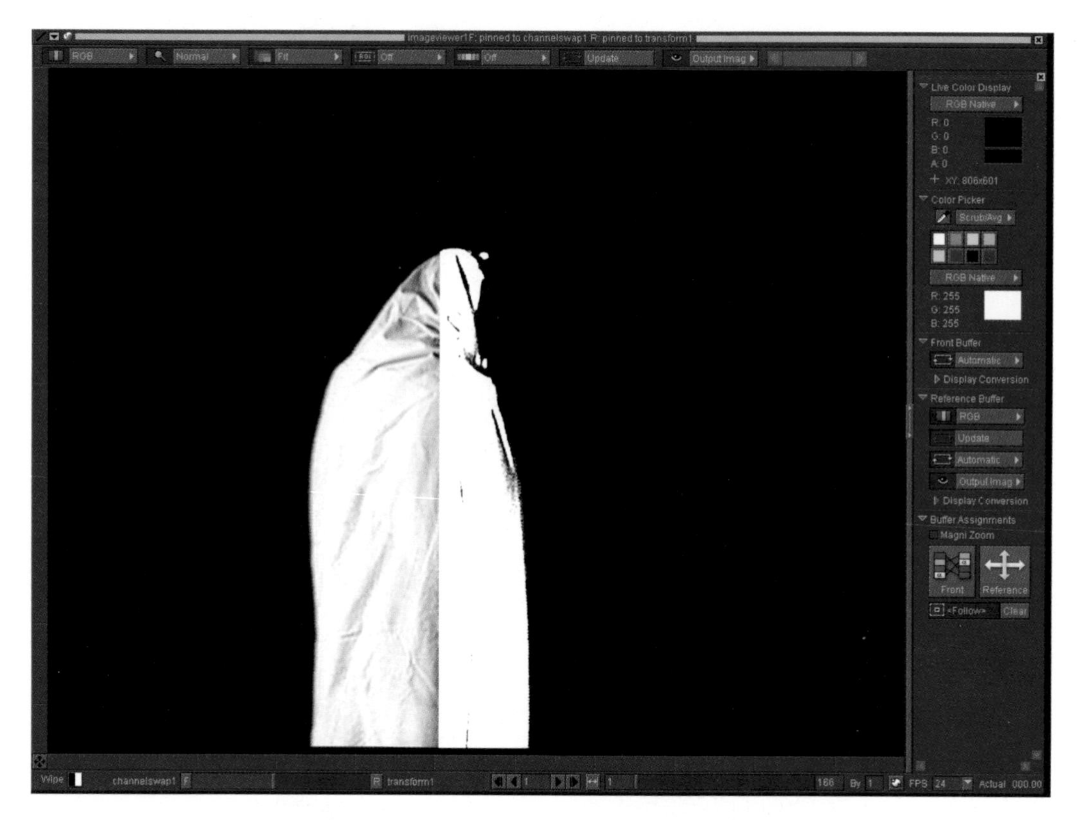

Plate 4.9 A Luma Key in Rayz with original element.

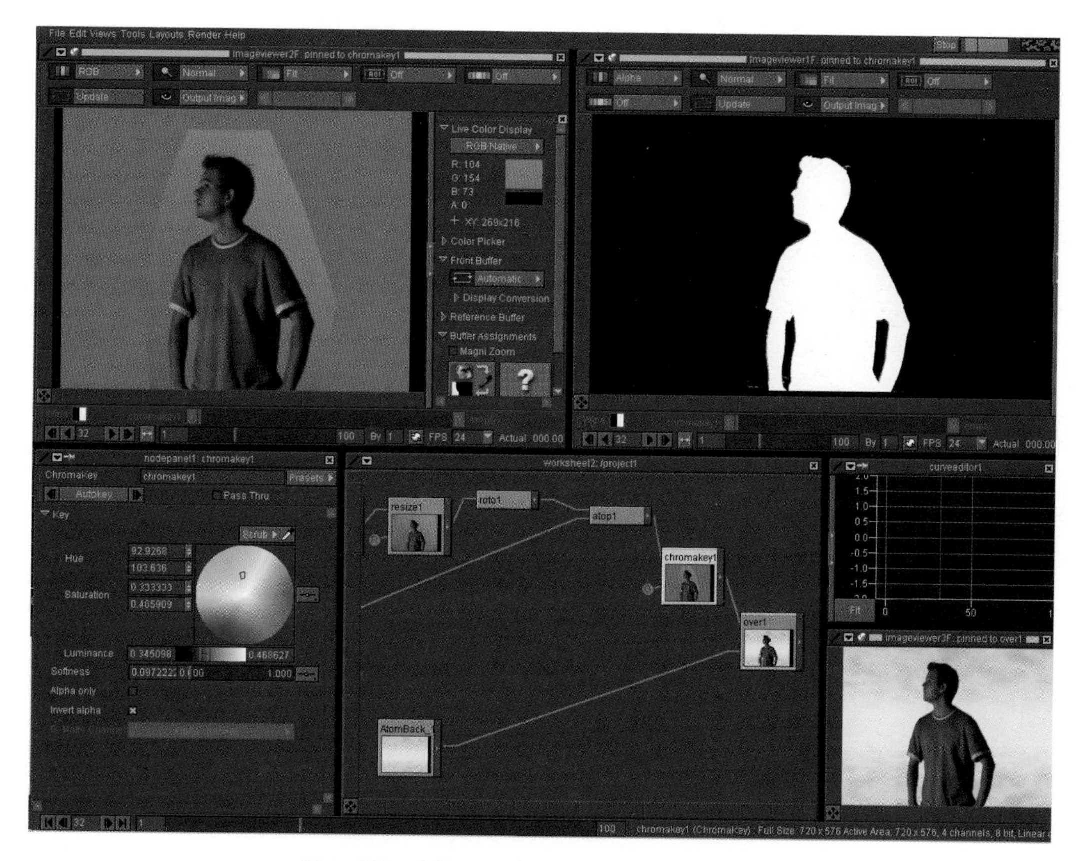

Plate 4.10 A Chroma Key in Rayz with original element.

Plate 4.11 An Ultimatte composite setup in Rayz.

Plate 4.12 A color correction setup in Rayz.

Plate 6.1 Joe's Playtime. Image courtesy of Kevan Shorey.

Plate 6.2 Beyond the see. Image courtesy of David Fish.

Plate 6.3 A flatpack project. Image courtesy of John Haddon.

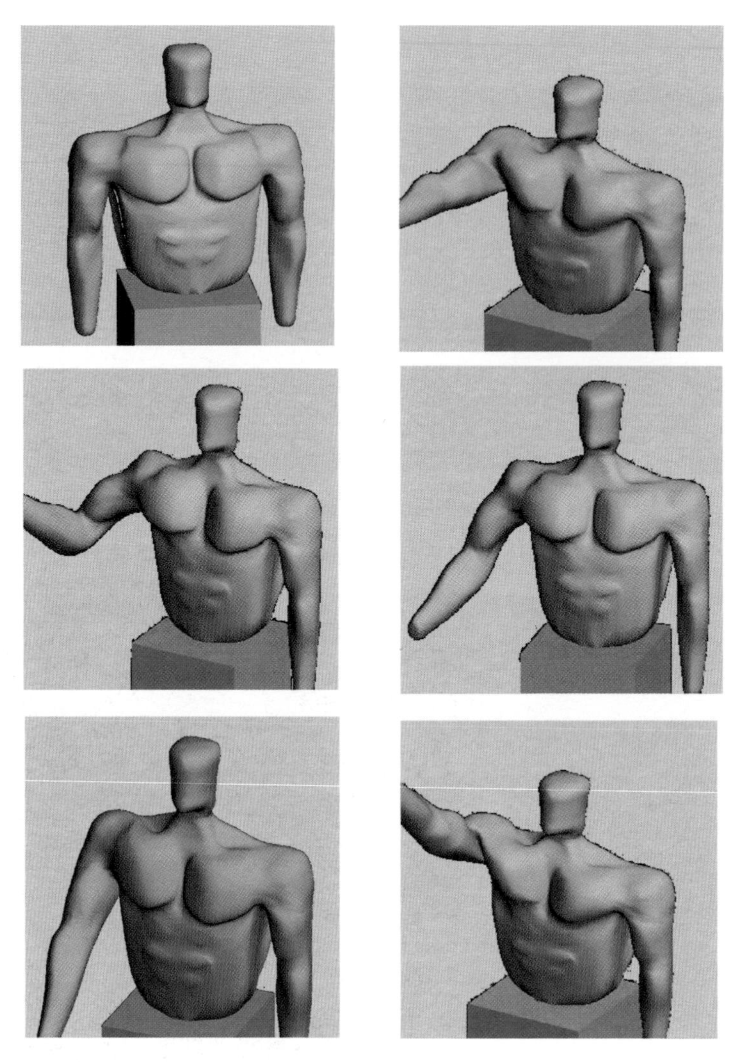

Plate 6.17 Animation character.

Degrain

A degrain filter will remove film grain from an image. This uses a mathematically accurate model of the film grain, so it does not simply blur the image to obscure the grain. However, there is a slight tendency for it to soften the image. This is more pronounced with sensitive high-speed films with larger grain structures.

Regrain

A regrain filter does the opposite. It modifies an image using a mathematically accurate model of film grain, so making the image appear as if it was originally shot on film. Normally this filter will have settings for different films. For instance, Kodak 5245 at 50 ASA or Kodak 5297 at 250 ASA. It is also normal for the filter to allow custom modification if the grain doesn't quite match the presets. For instance, if 5297 were processed as 750 ASA.

Noise

This adds noise to an image. It can normally be applied to the luminance component, and sometimes to the saturation and hue components. This can improve synthetic elements, such as CGI or digital paintings, if they look artificial and don't sit happily in the frame with real elements. With compositing systems that don't have regrain tools, this is often used for "distressing" elements without film grain, to remove their synthetic appearance. One problem with this is that if overapplied or too course it will cause the image to boil.

Convolve

This is basically an editable box filter. You can choose a grid size of 3 × 3 or above, and then type numbers into the squares. These numbers are the weighting of the filter, which defines how it will modify an image. This can be anything from simple blurs or edge enhancements to more complex image manipulation. There are often many presets for filters which can be edited. This often creates fairly dramatic effects that are not appropriate to realistic composites but can be useful for surreal or graphical sequences.

4.7.7 Color Correction

Color correction, or color modification, is sometimes subtle, sometimes dramatic, depending on the purpose. As well as "macro" color correction tools, outlined as follows, a compositing system will often have individual tools, such as contrast, brightness and hue shift grouped under "filters". In the *Rayz* compositing system there is a single "macro" color correction tool, which allows different types of color correction to be layered (see Figure 4.12). Other systems will have these color correctors as different tiles.

RGB

The simplest color corrector allows you to vary the red, green or blue content of the image, effectively allowing you to tint or wash the whole image. This can have aesthetic benefits, in allowing you to subtly alter the mood of the image. For instance, adding 5% more blue could make the image appear colder. Dropping the red and green could make the image appear as if shot at night.

Brightness/Contrast

Another simple color corrector. This either increases or decreases the brightness of an image, or does the same for the contrast. In some cases channels can be isolated. So the green contrast could be increased leaving the blue and red channels untouched or vice versa.

Hue Shift/Saturation/Value

These three tools are normally batched together and will often be referred to as HSV. The most interesting function, and arguably the most rarely used, is the hue shift. This "rotates" the colors of an image. For example, in the hue range, purple may be clockwise from red and green may be clockwise from blue. So a clockwise hue shift would change the reds in an image to purple, the blues to green, and so forth.

The most useful of this palette of tools is saturation. This allows the color intensity to be dropped or raised. Slight alterations will subtly alter the mood of the piece. Significant raises in saturation can create quite surreal effects. A significant decrease in saturation will make the image monochrome. There is often a color corrector or filter specifically for monochrome. The effect of the value parameter is similar to that of the brightness tool.

Color Curve

This is the power tool of the color correction toolset. It's a graph mapping input pixel value against output pixel value. There are normally lines for luminance, red, green and blue. In normal configuration these are straight lines going from bottom left to top right of the graph. The end points can be moved and control points can be added to the line. These can be moved, turning the line into a curve. These curves can modulate all the color aspects, changing brightness, contrast, saturation and inverting the colors. For any sophisticated or subtle color manipulation this is probably the best solution.

Telecine

A rare color corrector, but also a very useful one. This is based on a telecine, color-grading console. This will normally have three trackerballs, corresponding to highlights, mid-tones and shade areas. Different rotation directions on the trackerball correspond to different hues. So the rotation of the trackerball controls the tinting and intensity in the three tonal areas of the image. In a compositing system such as *Cineon*, this is simulated, allowing a pointer to be moved within a hue disk. This is probably best used for fine control of stylized images.

4.7.8 Image Transformations

The ability to move, or transform, an image is very important. On the one hand it can be simply used to reframe a finished image, on the other, it can provide a number of "over-riding" moves to the foreground subject in a blue screen shoot, providing any cropped subject edges are not translated into frame. The main transformations are those found in CGI: scale, rotate and translate. There are also some "macro" tools such as 3D DVE, tracking and stabilization.

Scale

Scale, or scaling, is the ability to change the size of an element. This can be performed on the vertical, horizontal or both axes. When done equally on both axes, it is equivalent to "zooming in" or "zooming out" of an element. A good example of the use of this would be to resize imported images, for instance, CGI rendered at 1K to be composited with film scanned at 2K. A more advanced usage of this would be to change the size of a blue screen element so that the subject appears to move towards, or away from, the viewer.

Rotate

This function can perform corrective rotations to a misaligned element, or it can add over-riding animation where there was none before. One classic example of the latter, from the realm of traditional effects, is *2001: A Space Odyssey*. In this, high-resolution transparencies were taken of miniatures, and these were put under a rostrum camera. The transparency was then reshot as the rostrum was moved, to give the impression of the miniature moving.

There are two obvious issues to consider in rotating a digital element. If the unrotated subject crosses the edge of the frame, that part of the element has to be hidden during the rotation, by zooming in, or scaling up.

The next issue is, which point do you rotate about? The obvious point, if you are making a correction, is the center of the image. However, if you are adding an over-riding camera move, this could make the subject swing around the center point of the image, which is not normally desirable. Thus, the pivot point about which the image rotates, needs to be moved to the center of the subject. Often, a simple rotation will swing the edges of the original image into the viewable area, this will have to be suppressed with a garbage matte.

Translation

Put simply, this means moving the image. Again this can take the form of a corrective manipulation of an image, or an over-riding move. A corrective manipulation would be moving an element around until it has the correct composition in the image. An over-riding move would be putting motion onto the image for creative purposes. This could take on the appearance of a subject translation, or an over-riding camera move.

An example of a subject translation would be shooting a blue screen element of someone holding an umbrella. You could then add translation so that the subject appears to descend through the frame, hanging from the umbrella.

The most common kind of over-riding camera move, would be enhancing the composition of a scene. This is sometimes referred to as "re-racking" (a term from telecine). For instance, in a background plate there could be a tree at the edge of the frame. The image could be translated to the side to move the tree out of frame. The side effect of this is that there will be a black stripe down the opposite side of the frame. To remove this from the frame, you need to zoom in, or "rack in".

A more creative use of the over-riding camera move would be to add tracks, dollies and elevations where there are none. To do this you need to shoot the image wide, then zoom in to create the correct framing. This means you have a lot of surrounding image to track or elevate into. One major implication of this is that the resolution of the reshot image will be considerably reduced. This is a major problem for video, as it is effectively a fixed resolution. It is possible to overcome this in film, by scanning the original film at a higher resolution. For instance, the negative could be scanned at 4K, then you could zoom into 2K, leaving a full-frame width and height for over-riding camera moves.

3D-DVE

This is a special case of image transformation. Everything we've considered so far has been a two-dimensional transformation. However, it is possible to do three-dimensional transformations in a compositing system. This is commonly referred to as a 3D-DVE (3D Digital Video Effect) or 3D transformation. Simply put it treats the element you're manipulating as a flat plane within a 3D space. Therefore it's possible to move the plane within this 3D space. Of course, translating the plane away from the camera will look identical to zooming out or scaling down. It's only when we start to rotate elements that things start to get interesting. Up until now we've considered rotation on a single axis, that perpendicular to the frame, for the sake of clarity, the z axis. The 3D-DVE allows rotation about the x and y axes, those in the plane of the frame. In effect this will move one edge towards the camera, and the opposite away, causing an apparent deformation of the image.

Tracking

Translation, rotation and scaling, the standard 2D transformations lead us onto some macro concepts, namely tracking and stabilization. Tracking involves reading the translation and rotation of an object within the frame. This tracking data can then be applied to another element, using translates and rotates. This gives the appearance of one element being attached to another.

Stabilization

Stabilization works slightly differently. It's very difficult to do a convincing composite of elements if one or more of them have camera shake and are not motion controlled. If, for instance, one element is hand held, the camera movement will have to be stabilized, or removed. This is done by tracking the motion of a distinctive feature of the image, and then inverting this motion and applying it to the same image. This effectively removes camera shake. One side effect of this is that the edges of the image will wobble in and out of the frame. To suppress this there has to be a zoom in, or scale up of the element.

4.7.9 Image Deformations

Image transformations largely retain the shape of the original image, with the exception of non-uniform scales and 3D-DVEs, which give the appearance of image deformation, or shape change. There are tools specifically for image deformation: *warp* and *morph*. Some compositing systems will incorporate these tools. However, standalone systems, notably *Elastic Reality*, tend to be the tools of choice.

A warp is simply displacing some of the pixels in an image to change its shape, or warp the subject. There are two principal techniques for achieving this. One is to add a grid of control points to the image. When a control point is moved, the pixels close by also move. The further away from the control point the pixel is, the less it moves. The alternative approach is to draw spline curves around the dominant features of the images. So, for instance, on an image of a human face, splines could be drawn around the eyes, and the control points moved to make the eyes larger. This allows for much more subtle and accurate deformations compared with the grid method. The spline method is the one used in Elastic Reality. Of course, many images change over time, so the grid or spline control points have to be adjustable. This brings us to morphing.

Morphing is seamlessly changing one image into another, for instance one person's face into another. A dissolve is performed between the images, during which animated warps are taking place on both image sequences. The features on the face in the first image will be moving to the position of those in the second. Meanwhile, the features in the second image will have been pre-warped to those in the first, and be animating back to their original position. The success of a morph sequence largely depends on how compatible the two image sequences are with one another. To be really successful, the image sequences have to be shot with the objective of morphing them.

4.7.10 Painting and Retouching

While rotomattes, or spline elements can be used for most tasks in a compositing system, sometimes it's necessary to digitally paint onto images. As already observed, this can cause problems with painted areas of the image boiling, or not matching the original areas of the image. For the most part, digital paint will be used for repairing flawed images, for instance scratches on damaged negatives, or removing unwanted components, for instance wire rigs on stunt people. This is called "rig removal". Painting with purely digital paint would stand out, so ruining the effect. The way around this is to clone an area of the image, and paint over the component you want to remove, using the cloned area as a brush. Sometimes this brush can come from an entirely different frame in the sequence.

4.7.11 Time Modifications

So far all the transformations that we've made to images are in the spatial domain. Sequences of images also have a temporal domain. Temporal modification is speeding up or slowing down sequences. The most basic form of speeding up a sequence of images would be to remove every other frame, so doubling the speed of a sequence. To speed a sequence up three times, you could remove every second and third frame. Unfortunately this means that the subject will move further between frames and can cause the image to strobe.

Slowing down a sequence is a rather more involved process, as you need more frames. To halve the speed of a sequence, you need to add a frame between each of the existing ones. So you have to manufacture images where there are none. The simplest method of achieving this is to do a dissolve between the two adjacent frames. So that each in between frame is an equal mixture of the preceding and subsequent frame. Again, it's possible to slow it down to a third or quarter speed, but the genesis of the effect will become more and more obvious as you do.

A more advanced approach to this process is to morph between frames, rather than simply dissolving between frames. This has the advantage that each new frame really is unique, this means that you can stretch time to much greater extremes than with the simple dissolve method.

This image morphing could be carried out with a conventional morphing package such as *Elastic Reality*, which means you need to manually define the feature areas. Alternatively an automatic, image recognition approach, could be employed.

This "time warping" effect is also an essential part of bullet-time rigs. A bullet-time rig is an array of cameras, which can be positioned around a subject. These cameras can then shoot a frame at the same time, or frames in sequence. Images from the array of cameras are "stitched" together to create a time-freeze sequence, in which a camera appears to move around a frozen subject. Unfortunately 35 cameras will only create a second and a half of sequence. To get longer sequences you need to interpolate between the frames, i.e. use a morphing tool to recreate the missing frames.

4.7.12 Output

Although it is possible to output digital effects composites as AVIs or Quicktime Movies, for instance for games, for the most part digital effects will be output to videotape or film. Special considerations apply to both of these output routes.

Output to video has the most limitations. This is largely due to the deficiencies of the standard video signal formats, PAL and NTSC. These are now quite aged formats, and their performance is considerably lower than that of the average computer display. For professional video tape formats such as Digi Beta or D1 the vertical resolution will be 576 lines for PAL or 476 for NTSC. For domestic tape formats such as VHS, the vertical resolution will be almost a third of this.

On a computer display an image to be output might have good color saturation, a large tonal range, high whites and deep blacks, and look very compelling. If you transfer such an image to PAL video, even in a high-end tape format such as D1, and you will have burnt out whites, dark gray tones crushed to black, and bleeding from saturated colors, especially red.

So output to standard video is a compromise. The performance of the video signal is significantly inferior to a computer display signal, so the objective is to get the closest approximation to the original image. Most compositing systems will have a tool to "legalize" video. In other words, it will correct the color levels of the image so that they are "video safe". However, care needs to be taken, as the tonal range will be crushed. Dark gray shades will be resolved to black and light gray shades to white.

One additional consideration for video is the framing. On a television or a monitor, a lot of the edge of the image is masked by the plastic surround. All compositing systems will have a TV framing overlay that can be switched on. The outer box will be "safe framing", in other words, everything within it will be visible on screen. The inner box is "safe titling".

Film is a lot more forgiving than video, in terms of image quality. Arguably, depending on the film stock used, it has a greater tonal range than a computer monitor, and it handles extremely saturated colors in a more forgiving way. The best route for outputting digital images to film is via the *Cineon* 10-bit logarithmic file format, as this is designed to match the recording characteristics of the film. An alternative is the Tiff 16 file format, which while in a linear color space has the "definition" to cover the full tonal range of the film.

While, in terms of color and tonal range, there is no beating viewing the final image projected on film, compositors can move some way towards the final look by adjusting their computer monitors, with its high resolution and color depth, to more closely approximate the film color space.

4.8 Conclusion

Digital effects is here to stay. While traditional optical effects were in a lot of cases very successful, they were very limited and very difficult to achieve. They have been decisively superseded by digital techniques.

Currently miniatures, in combination with digital techniques, are holding their own. But it's only a matter of time before CGI supersedes them. Aesthetically, it's open to debate whether CGI looks more pleasing than physical miniatures, but the efficiency and flexibility of CGI will inevitably win the day.

Digital techniques are still in their early days – the digital effects industry is only a decade old. The next 10–20 years of development will bring higher levels of realism and allow higher levels of creativity. Soon the only limit will be imagination.

Bibliography

Brinkmann, R. *The Art and Technique of Digital Compositing*. Morgan Kaufmann, 1999.
Fielding, R. *The Technique of Special Effects Cinematography*. Focal Press, 1985.
Masson, T. *CG101: A Visual Effects Industry Primer*. New Riders, 1999.
Perisic, Z. *Visual Effects Cinematography*. Focal Press, 2000.

5 Cubic Polynomial Curves and Surfaces

Peter Comninos

Introduction

The purpose of this chapter is to introduce the reader to some of the most important concepts, mathematical techniques and algorithms relating to three-dimensional (3D) cubic polynomial curves and surfaces. These techniques are used in most current animation systems to represent the motion of computer-generated objects and to model their surfaces.

This chapter has evolved out of lecture notes used to introduce final-year students of our Bachelor of Arts program in Computer Animation to various aspects of polynomial curves, surfaces and volumes. The material included in this chapter is only a small part of the material to which our students are exposed. They are presented with a much more general and detailed discussion of all aspects of polynomial functions. Owing to the space limitations imposed by the format of this book, we shall limit our discussion to cover only cubic polynomial curves and surfaces.

This chapter is a staged introduction to these topics and assumes no prior knowledge of the subject matter. The material presented here was not written for mathematicians, but for computer animators and technical directors who wish to become expert users of existing software tools or to become toolmakers and write their own software tools. Despite this, however, a certain degree of mathematical rigor is essential in order to explain and illustrate the underlying concepts and techniques at a level that leads to the efficient implementation of computer programs that utilize this technology. Students undertaking this unit have a good level of understanding of the fundamental computer graphics mathematical and programming techniques, which are considered a prerequisite for the material presented here.

It is hoped that the material presented in this chapter will introduce readers to this fascinating topic and hopefully encourage them to further their study of this field of mathematics. Accompanying the presentation of the theory is a set of simple but efficient routines written in the C programming language, which we hope will help illuminate the underlying mathematics and will encourage the reader to experiment by writing their own software tools.

5.1 The Parametric Representation

One of the simplest ways we can represent a 3D curve would be to approximate it by a collection of connected straight line segments, where each line segment is the chord of the corresponding curve segment. Such a collection of line segments is frequently referred to as a polyline. Provided that the curve segments are sufficiently short we would get a reasonable approximation to the curve. Such a polyline can then be seen as a first-degree, piecewise linear approximation of the curve. As we shall see below, each line segment can be represented in parametric form and a point on a line segment can be computed by linearly interpolating its two endpoints.

Similarly, a surface can be represented by a collection of connected quadrilaterals arranged as a polygonal mesh, where each quadrilateral approximates a relatively flat patch of the curved surface. Such a polygonal mesh can then be seen as a first-degree, piecewise bilinear approximation of the surface. As we shall see below, each quadrilateral can be represented in parametric form and a point on a quadrilateral can be computed by bilinearly interpolating its four corners.

5.1.1 Linear Interpolation

Consider the 3D line defined by the two endpoints of the line segment $\overrightarrow{P_0P_1}$, where $P_0 = [x_0\ y_0\ z_0]$ and $P_1 = [x_1\ y_1\ z_1]$, as shown in Figure 5.1.

In parametric form this line is defined by

$$Q(t) = (1-t)\cdot P_0 + t\cdot P_1 \quad \text{where } 0 \le t \le 1 \tag{5.1}$$

or

$$\begin{aligned} x(t) &= (1-t)\cdot x_0 + t\cdot x_1 \\ y(t) &= (1-t)\cdot y_0 + t\cdot y_1 \quad \text{where } 0 \le t \le 1 \\ z(t) &= (1-t)\cdot z_0 + t\cdot z_1 \end{aligned} \tag{5.2}$$

or

$$Q(t) = [x(t)\ y(t)\ z(t)] \quad \text{where } 0 \le t \le 1. \tag{5.3}$$

Thus a parametrically defined 3D line is given by three *univariate linear functions*.

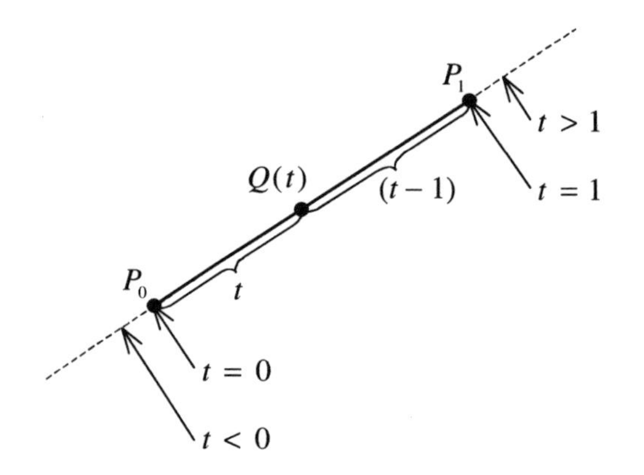

Figure 5.1 The parametric form of a line.

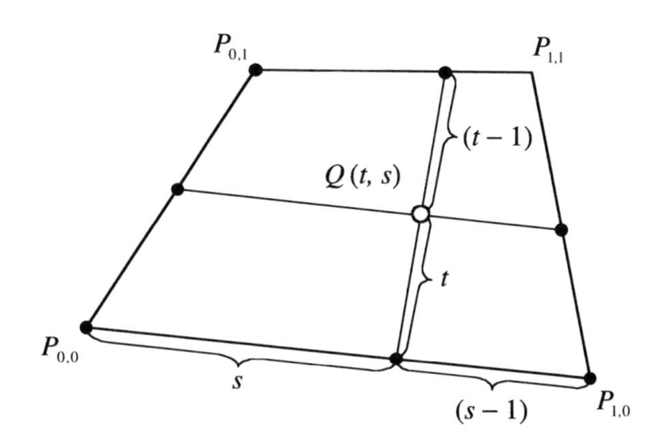

Figure 5.2 The parametric form of a quadrilateral.

5.1.2 Bilinear Interpolation

Consider the 3D quadrilateral defined by the points $P_{0,0}$, $P_{1,0}$, $P_{1,1}$ and $P_{0,1}$ as shown in Figure 5.2. In parametric form this quadrilateral is defined by

$$\begin{aligned} Q(s,t) &= (1-t)\cdot[(1-s)\cdot P_{0,0} + s\cdot P_{1,0}] + t\cdot[(1-s)\cdot P_{0,1} + s\cdot P_{1,1}] \\ &\quad \text{where } 0 \le s,\ t \le 1 \end{aligned} \tag{5.4}$$

or

$$\begin{aligned} x(s,t) &= (1-t)\cdot[(1-s)\cdot x_{0,0} + s\cdot x_{1,0}] + t\cdot[(1-s)\cdot x_{0,1} + s\cdot x_{1,1}] \\ y(s,t) &= (1-t)\cdot[(1-s)\cdot y_{0,0} + s\cdot y_{1,0}] + t\cdot[(1-s)\cdot y_{0,1} + s\cdot y_{1,1}] \\ z(s,t) &= (1-t)\cdot[(1-s)\cdot z_{0,0} + s\cdot z_{1,0}] + t\cdot[(1-s)\cdot z_{0,1} + s\cdot z_{1,1}] \\ &\quad \text{where } 0 \le s,\ t \le 1 \end{aligned} \tag{5.5}$$

or

$$Q(s,t) = [x(s,t)\ \ y(s,t)\ \ z(s,t)] \quad \text{where } 0 \le s,\ t \le 1. \tag{5.6}$$

Thus, a parametrically defined 3D quadrilateral is given by three *bivariate bilinear functions*.

If we use a polyline to represent a 3D curve or a quadrilateral mesh to represent a 3D surface, we will need a large number of line segments or quadrilaterals to obtain a reasonable approximation to the curve or the surface. If we subsequently wished to alter the shape of the curve or the surface interactively, this would be tedious, as we would need to manually reposition a large number of points precisely. So a more compact and easier way to manipulate the representation of piecewise smooth curves and surfaces is required.

The linear approximation functions we have used so far are of degree one. In general we will use *parametric functions* of degree greater than one to represent a curve or a surface.

5.2 Polynomial Parametric Curves

In our quest for higher-degree parametric function representations for curves and surfaces we will utilize polynomial functions. Polynomials are very useful and versatile mathematical tools, which can be evaluated efficiently using computers, can easily be differentiated and integrated, and can easily be pieced together to form spline curves that can approximate any function to any desired accuracy. We write a polynomial function as

$$Q(t) = c_0 \cdot t^0 + c_1 \cdot t^1 + \cdots + c_{d-1} \cdot t^{d-1} + c_d \cdot t^d \tag{5.7}$$

or

$$Q(t) = \sum_{i=0}^{d} c_i \cdot t^i. \tag{5.8}$$

This function represents a polynomial as a *linear combination* of the elementary polynomials $\{t^0, t^1, \ldots, t^{d-1}, t^d\}$. Here $\{c_0, c_1, \ldots, c_{d-1}, c_d\}$ are the coefficients of the elementary polynomials. Sometimes we call this a *monomial* representation of the polynomial.

The *degree* of this polynomial can be d or less depending on the values of the coefficients and the *order* of the polynomial (i.e. the number of its terms) is $k = (d + 1)$.

One or more or even all of the coefficients of a polynomial are allowed to be zero. Consequently, kth-order polynomials are polynomials that have at most degree d. Sometimes we denote a polynomial of order k as P^k. For example, the following are all valid cubic polynomials ($d = 3$, $k = 4$):

$$\begin{aligned}
P_1^4(t) &= 6 + 2t + 3t^2 + 5t^3 \\
P_2^4(t) &= 6 + 2t + 3t^2 + 0t^3 \\
P_3^4(t) &= 6 + 2t + 0t^2 + 0t^3 \\
P_4^4(t) &= 6 + 0t + 0t^2 + 0t^3 \quad \text{(a constant cubic polynomial)} \\
P_5^4(t) &= 0 + 0t + 0t^2 + 0t^3 \quad \text{(the zero cubic polynomial)} \\
P_6^4(t) &= 0 + 0t + 0t^2 + 5t^3.
\end{aligned}$$

A set of polynomials of degree less than or equal to d form a *vector space*, as they are closed under the operations of addition and multiplication by a scalar.

The set of elementary polynomials $\{t^0, t^1, \ldots, t^{d-1}, t^d\}$ forms a *basis* for the *vector space* of polynomials, which is known as the *power basis* and constitutes only one of an infinite number of bases for the vector space of polynomials.

Normally in computer graphics we use polynomials of degree three, that is *cubic polynomials*. A *quadratic polynomial* (of degree two) does not offer sufficient shape flexibility, as it does not allow us to smoothly blend successive curve segments, owing to the fact that quadratic polynomials only have three geometric constraints whereas a minimum of four are usually required.

For polynomials of degree greater than three (*quadric, quintic*, etc.), there is a trade-off between descriptions that are progressively more cumbersome and computationally inefficient, and shape flexibility.

A polynomial of degree d may have up to $(d - 1)$ turning points if all its roots happen to be real. Thus the greater the degree of a polynomial, the more likely it is to oscillate, which will produce unwanted undulations in the shape of the corresponding parametric curve.

Polynomials of degree less than three cannot be made to pass through specific endpoints with specified derivative properties (tangents). This is important since composite curves and surfaces, made up of distinct curve segments, need to be joined smoothly over their segment boundaries.

5.2.1 The Cubic Polynomial Curve Segment

The cubic polynomials that define a curve segment:

$$Q(t) = [x(t) \ y(t) \ z(t)] \tag{5.9}$$

are of the form:

$$\begin{aligned} x(t) &= c_{x3} \cdot t^3 + c_{x2} \cdot t^2 + c_{x1} \cdot t^1 + c_{x0} \cdot t^0 \\ y(t) &= c_{y3} \cdot t^3 + c_{y2} \cdot t^2 + c_{y1} \cdot t^1 + c_{y0} \cdot t^0 \quad \text{where } 0 \le t \le 1. \\ z(t) &= c_{z3} \cdot t^3 + c_{z2} \cdot t^2 + c_{z1} \cdot t^1 + c_{z0} \cdot t^0 \end{aligned} \tag{5.10}$$

These equations can be rewritten in matrix form as

$$Q(t) = [x(t) \ y(t) \ z(t)] = T \cdot C, \tag{5.11}$$

where

$$T = \begin{bmatrix} t^3 & t^2 & t & 1 \end{bmatrix} \quad \text{is called the } \textit{parameter vector} \tag{5.12}$$

and

$$C = \begin{bmatrix} c_{x3} & c_{y3} & c_{z3} \\ c_{x2} & c_{y2} & c_{z2} \\ c_{x1} & c_{y1} & c_{z1} \\ c_{x0} & c_{y0} & c_{z0} \end{bmatrix} \quad \text{is called the } \textit{coefficients matrix.} \tag{5.13}$$

The coefficients matrix can be seen as a row vector of three column vectors:

$$C = \begin{bmatrix} C_x & C_y & C_z \end{bmatrix}, \tag{5.14}$$

where

$$C_x = \begin{bmatrix} c_{x3} \\ c_{x2} \\ c_{x1} \\ c_{x0} \end{bmatrix}, \quad C_y = \begin{bmatrix} c_{y3} \\ c_{y2} \\ c_{y1} \\ c_{y0} \end{bmatrix}, \quad C_z = \begin{bmatrix} c_{z3} \\ c_{z2} \\ c_{z1} \\ c_{z0} \end{bmatrix}. \tag{5.15}$$

The first derivative, $Q^{(1)}(t)$, of $Q(t)$ is the parametric *tangent vector* of the curve:

$$\begin{aligned} Q^{(1)}(t) &= \frac{\partial}{\partial t} Q(t) \\ \therefore Q^{(1)}(t) &= \left[\frac{\partial}{\partial t} x(t) \quad \frac{\partial}{\partial t} y(t) \quad \frac{\partial}{\partial t} z(t) \right] \\ \therefore Q^{(1)}(t) &= \frac{\partial}{\partial t} T \cdot C \\ \therefore Q^{(1)}(t) &= \begin{bmatrix} 3 \cdot t^2 & 2 \cdot t & 1 & 0 \end{bmatrix} \cdot C \\ \therefore Q^{(1)}(t) &= \begin{bmatrix} 3c_{x3} \cdot t^2 + 2c_{x2} \cdot t + c_{x1} & 3c_{y3} \cdot t^2 + 2c_{y2} \cdot t + c_{y1} & 3c_{z3} \cdot t^2 + 2c_{z2} \cdot t + c_{z1} \end{bmatrix}. \end{aligned} \tag{5.16}$$

If two curve segments join together, the curve is said to have G^0 *geometric continuity* and C^0 *parametric continuity*. If two joining curve segments have (at their joint) tangents of equal direction, the curve is said to have G^1 *geometric continuity*. If two joining curve segments have (at their joint) equal tangents (equal in direction and in magnitude), the curve is said to have C^1 *parametric continuity*. If two joining curve segments have (at their joint) $Q^{(2)}(t) = \frac{\partial^2}{\partial t^2} Q(t) \neq 0$, then the curve is said to have C^2 *parametric continuity*.

If we interpret the parameter t as time and a point moves along a parametric cubic in equal time steps, then:

- $Q(t)$ gives the position of the point,
- $Q^{(1)}(t)$ gives the velocity of the point,
- $Q^{(2)}(t)$ gives the acceleration of the point.

A curve segment $Q(t)$ is defined and its shape is determined by

(1) the constraints on its endpoints;

(2) the tangent vectors at its endpoints;

(3) the continuity between neighboring curve segments.

Each cubic polynomial of Equation (5.10) has four coefficients. Thus we need four constraints so that we can formulate four equations in four unknowns and solve for these unknowns.

5.3 Types of Parametric Curve

As we shall see in later sections, it is convenient to define parametric curves using a set of geometric constraints, the most common of which are *control points* that are arranged into a *control polygon*.

There are two distinct types of parametric curves:

I *interpolating curves* that pass through their control points;

II *approximating curves* that pass close to their control points and approximate the shape of their control polygon.

In this chapter we will examine the following cubic polynomial curves.

(i) Hermite Curves

A Hermite curve segment is defined by two endpoints and two endpoint tangent vectors. The curve segment passes through its endpoints and it is thus an interpolating curve.

(ii) Bézier Curves

A Bézier curve segment is defined by two endpoints and two other points that control the tangent vectors at its endpoints. The curve segment passes through its first and last control point and passes close to its middle control points thus approximating its control polygon.

(iii) B-spline/β-spline (Beta-spline) Curves

A B-spline/β-spline curve segment is defined by four control points. Additionally, β-splines require a number of shape parameters. Both splines have C^1 and C^2 continuity at the joints and pass close to their control points, thus approximating their control polygon.

(iv) Cardinal/Catmull–Rom/Kochanek–Bartels Spline Curves

This is a family of interpolating splines. Similarly to the Hermite curves, from which they are descended, each curve segment is defined by two endpoints and two endpoint tangent vectors. The definition of tangent vectors of a segment requires a number of shape parameters. The curve segment passes through its endpoints.

5.4 Polynomial Coefficients and Geometric Constraints

By examining Equations (5.10), (5.11) and (5.13) we can see that in order to change the shape of the curve segment we must modify the 12 coefficients of the coefficients matrix. Though these coefficients might be meaningful to mathematicians they are of very little use to ordinary users, as they have no geometric significance (except at $t = 0$). To see how the polynomial coefficients of Equation (5.10) relate to geometric constraints, that are meaningful to users, we must rewrite the coefficient matrix of Equation (5.11) as

$$C = M \cdot G, \tag{5.17}$$

where M is a 4×4 basis matrix and G is a four-element geometry vector containing the geometric constraints.

Now,

$$Q(t) = T \cdot M \cdot G \tag{5.18}$$

or

$$Q(t) = [t^3 \; t^2 \; t \; 1] \cdot \begin{bmatrix} m_{00} & m_{01} & m_{02} & m_{03} \\ m_{10} & m_{11} & m_{12} & m_{13} \\ m_{20} & m_{21} & m_{22} & m_{23} \\ m_{30} & m_{31} & m_{32} & m_{33} \end{bmatrix} \cdot \begin{bmatrix} G_0 \\ G_1 \\ G_2 \\ G_3 \end{bmatrix}, \tag{5.19}$$

where each G_i is a row vector representing a point or a tangent:

$$G_i = [G_{ix} \; G_{iy} \; G_{iz}]. \tag{5.20}$$

Thus G can be seen as consisting of three column vectors:

$$G = [G_x \; G_y \; G_z] = \begin{bmatrix} G_{0x} & G_{0y} & G_{0z} \\ G_{1x} & G_{1y} & G_{1z} \\ G_{2x} & G_{2y} & G_{2z} \\ G_{3x} & G_{3y} & G_{3z} \end{bmatrix}, \tag{5.21}$$

where

$$G_x = \begin{bmatrix} G_{0x} \\ G_{1x} \\ G_{2x} \\ G_{3x} \end{bmatrix}, \quad G_y = \begin{bmatrix} G_{0y} \\ G_{1y} \\ G_{2y} \\ G_{3y} \end{bmatrix}, \quad G_z = \begin{bmatrix} G_{0z} \\ G_{1z} \\ G_{2z} \\ G_{3z} \end{bmatrix}. \tag{5.22}$$

Given that

$$Q(t) = [x(t)\ \ y(t)\ \ z(t)]$$

it follows that

$$\begin{aligned} x(t) &= T \cdot M \cdot G_x \\ y(t) &= T \cdot M \cdot G_y \\ z(t) &= T \cdot M \cdot G_z. \end{aligned} \tag{5.23}$$

Thus

$$\begin{aligned} x(t) = {} & (t^3 \cdot m_{00} + t^2 \cdot m_{10} + t \cdot m_{20} + m_{30}) \cdot G_{0x} \\ & + (t^3 \cdot m_{01} + t^2 \cdot m_{11} + t \cdot m_{21} + m_{31}) \cdot G_{1x} \\ & + (t^3 \cdot m_{02} + t^2 \cdot m_{12} + t \cdot m_{22} + m_{32}) \cdot G_{2x} \\ & + (t^3 \cdot m_{03} + t^2 \cdot m_{13} + t \cdot m_{23} + m_{33}) \cdot G_{3x} \\ y(t) = {} & (t^3 \cdot m_{00} + t^2 \cdot m_{10} + t \cdot m_{20} + m_{30}) \cdot G_{0y} \\ & + (t^3 \cdot m_{01} + t^2 \cdot m_{11} + t \cdot m_{21} + m_{31}) \cdot G_{1y} \\ & + (t^3 \cdot m_{02} + t^2 \cdot m_{12} + t \cdot m_{22} + m_{32}) \cdot G_{2y} \\ & + (t^3 \cdot m_{03} + t^2 \cdot m_{13} + t \cdot m_{23} + m_{33}) \cdot G_{3y} \\ z(t) = {} & (t^3 \cdot m_{00} + t^2 \cdot m_{10} + t \cdot m_{20} + m_{30}) \cdot G_{0z} \\ & + (t^3 \cdot m_{01} + t^2 \cdot m_{11} + t \cdot m_{21} + m_{31}) \cdot G_{1z} \\ & + (t^3 \cdot m_{02} + t^2 \cdot m_{12} + t \cdot m_{22} + m_{32}) \cdot G_{2z} \\ & + (t^3 \cdot m_{03} + t^2 \cdot m_{13} + t \cdot m_{23} + m_{33}) \cdot G_{3z}. \end{aligned} \tag{5.24}$$

At first sight, Equations (5.19) and (5.24) appear to be more complex than Equations (5.11) and (5.10), respectively. On closer examination, however, it becomes apparent that Equations (5.19) and (5.24) are more revealing from the geometric point of view. The basis matrix is derived by the curve designer (mathematician) and determines the type of curve we are dealing with, while the geometry vector provides the curve user with a meaningful and intuitive tool for altering the shape of the curve segment.

Equation (5.24) emphasizes that the curve segment is the weighted-sum of the elements of the geometry matrix. Each weight is a cubic polynomial of t, and is called the *basis function*. As we shall see later, these functions are sometimes referred to as the *blending functions*.

The basis functions are given by

$$\mathbf{B} = \mathbf{T} \cdot \mathbf{M}. \tag{5.25}$$

Now Equation (5.18) can be rewritten as

$$Q(t) = B \cdot G, \tag{5.26}$$

where B is a row matrix of four elements:

$$B = [B_0(t)\ \ B_1(t)\ \ B_2(t)\ \ B_3(t)] \tag{5.27}$$

and

$$\begin{aligned} B_0(t) &= (t^3 \cdot m_{00} + t^2 \cdot m_{10} + t \cdot m_{20} + m_{30}) \\ B_1(t) &= (t^3 \cdot m_{01} + t^2 \cdot m_{11} + t \cdot m_{21} + m_{31}) \\ B_2(t) &= (t^3 \cdot m_{02} + t^2 \cdot m_{12} + t \cdot m_{22} + m_{32}) \\ B_3(t) &= (t^3 \cdot m_{03} + t^2 \cdot m_{13} + t \cdot m_{23} + m_{33}), \end{aligned} \tag{5.28}$$

which are functions that blend the geometric constraints (hence the name blending functions).

An alternative representation of a cubic polynomial can now be written as

$$Q(t) = \sum_{i=0}^{3} B_i(t) \cdot G_i. \tag{5.29}$$

In this representation the fact that B_i is a function of parameter t is made explicit.

Observation 1

From the above it should be apparent that the basis matrix M of Equation (5.18) converts the geometric constrains expressed in the given basis into the equivalent geometric constraints expressed in the power basis.

From Equation (5.18) we have

$$\begin{aligned} Q(t) &= T \cdot M \cdot G_B \\ &= T \cdot (M \cdot G_B) \\ &= T \cdot G_P, \end{aligned}$$

where G_B is the geometry vector expressed in the given basis and G_P is the geometry matrix expressed in the power basis (i.e. its coefficients matrix). Thus, the matrix M can now be seen as a change of basis transformation.

Observation 2

Because cubic curves are linear combinations (weighted-sums) of their geometry vector, as seen from Equation (5.24) we say that these curves are invariant under rotation, scaling and translation, a property which is known as *affine invariance*. This means that we can transform a curve by transforming its geometry vector alone and then using it to generate the transformed curve.

Observation 3

Let us consider the polynomial curve of degree $d = 1$.

Here

$$Q(t) = (1-t) \cdot P_0 + t \cdot P_1 = \begin{bmatrix} (1-t) & t \end{bmatrix} \cdot \begin{bmatrix} P_0 \\ P_1 \end{bmatrix} = \begin{bmatrix} t & 1 \end{bmatrix} \cdot \begin{bmatrix} -1 & 1 \\ 1 & 0 \end{bmatrix} \cdot \begin{bmatrix} P_0 \\ P_1 \end{bmatrix}.$$

Alternatively,

$$Q(t) = B_0(t) \cdot P_0 + B_1(t) \cdot P_1 = \begin{bmatrix} B_0(t) & B_1(t) \end{bmatrix} \cdot \begin{bmatrix} P_0 \\ P_1 \end{bmatrix},$$

where

$$\begin{aligned} B_0(t) &= (1-t) \\ B_1(t) &= t \end{aligned}$$

are the basis functions for the linear polynomial function. These results are identical to Equation (5.1).

5.5 Cubic Hermite Curves

The Hermite form of a cubic polynomial curve segment is determined by constraints on the endpoints P_0 and P_3 and the tangent vectors at these endpoints R_0 and R_3 (see Figure 5.3). (The indices 0 and 3, rather 0 and 1 are used for notational consistency with the next section.)

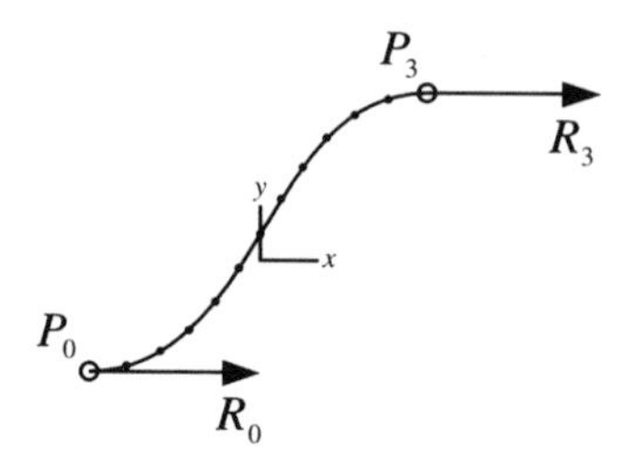

Figure 5.3 A Hermite curve segment.

To determine the *Hermite basis matrix* M_H, which relates the *Hermite geometry vector* G_H to the polynomial coefficients, we write four equations (one for each of the constraints) in the four unknown polynomial coefficients, and solve for these unknowns.

Let G_{Hx} be the x component of the Hermite geometry matrix:

$$G_{Hx} = \begin{bmatrix} P_0 \\ P_3 \\ R_0 \\ R_3 \end{bmatrix}_x . \tag{5.30}$$

From Equations (5.10) and (5.18) we have:

$$\begin{aligned} x(t) &= c_{x3} \cdot t^3 + c_{x2} \cdot t^2 + c_{x1} \cdot t + c_{x0} \\ &= T \cdot C_x \\ &= T \cdot M_H \cdot G_{Hx} \\ &= [t^3 \ t^2 \ t \ 1] \cdot M_H \cdot G_{Hx}. \end{aligned} \tag{5.31}$$

The constraints at $t = 0$ and $t = 1$ substituted into Equation (5.31) give

$$x(0) = P_{0x} = [0 \ 0 \ 0 \ 1] \cdot M_H \cdot G_{Hx} \tag{5.32}$$

$$x(1) = P_{3x} = [1 \ 1 \ 1 \ 1] \cdot M_H \cdot G_{Hx}. \tag{5.33}$$

The tangents at the endpoints are found by differentiating Equation (5.31) to obtain

$$x^{(1)}(t) = [3t^2 \ 2t \ 1 \ 0] \cdot M_H \cdot G_{Hx}. \tag{5.34}$$

So at $t = 0$ and $t = 1$ we have:

$$x^{(1)}(0) = R_{0x} = [0 \ 0 \ 1 \ 0] \cdot M_H \cdot G_{Hx} \tag{5.35}$$

$$x^{(1)}(1) = R_{3x} = [3 \ 2 \ 1 \ 0] \cdot M_H \cdot G_{Hx}. \tag{5.36}$$

Equations (5.32), (5.33), (5.35) and (5.36) can be rewritten in matrix form as

$$\begin{bmatrix} x(0) \\ x(1) \\ x^{(1)}(0) \\ x^{(1)}(1) \end{bmatrix} = \begin{bmatrix} P_0 \\ P_3 \\ R_0 \\ R_3 \end{bmatrix} = G_{Hx} = \begin{bmatrix} 0 & 0 & 0 & 1 \\ 1 & 1 & 1 & 1 \\ 0 & 0 & 1 & 0 \\ 3 & 2 & 1 & 0 \end{bmatrix} \cdot M_H \cdot G_{Hx}. \tag{5.37}$$

From (5.37) we have

$$M_H = \begin{bmatrix} 0 & 0 & 0 & 1 \\ 1 & 1 & 1 & 1 \\ 0 & 0 & 1 & 0 \\ 3 & 2 & 1 & 0 \end{bmatrix}^{-1} = \begin{bmatrix} 2 & -2 & 1 & 1 \\ -3 & 3 & -2 & -1 \\ 0 & 0 & 1 & 0 \\ 1 & 0 & 0 & 0 \end{bmatrix}, \tag{5.38}$$

which is the Hermite basis matrix. Now we can use M_H to find $Q(t)$ based on the geometry vector G_H:

$$Q(t) = [x(t)\ y(t)\ z(t)] = T \cdot M_H \cdot G_H \tag{5.39}$$

and

$$\begin{aligned} x(t) &= T \cdot M_H \cdot G_{Hx} \\ y(t) &= T \cdot M_H \cdot G_{Hy} \\ z(t) &= T \cdot M_H \cdot G_{Hz}. \end{aligned} \tag{5.40}$$

Expanding Equation (5.39) we obtain

$$Q(t) = (2t^3 - 3t^2 + 1) \cdot P_0 + (-2t^3 + 3t^2) \cdot P_3 + (t^3 - 2t^2 + t) \cdot R_0 + (t^3 - t^2) \cdot R_3. \tag{5.41}$$

Alternatively,

$$Q(t) = H_0^3(t) \cdot G_0 + H_1^3(t) \cdot G_1 + H_2^3(t) \cdot G_2 + H_3^3(t) \cdot G_3, \tag{5.42}$$

where

$$\begin{aligned} H_0^3(t) &= (2t^3 - 3t^2 + 1) \\ H_1^3(t) &= (-2t^3 - 3t^2) \\ H_2^3(t) &= (t^3 - 2t^2 + t) \\ H_3^3(t) &= (t^3 - t^2) \end{aligned} \tag{5.43}$$

and

$$\begin{aligned} G_0 &= P_0 \\ G_1 &= P_3 \\ G_2 &= R_0 \\ G_3 &= R_3. \end{aligned} \tag{5.44}$$

The functions $H_i^3(t)$ are known as the cubic *Hermite basis functions*. A plot of these functions can be seen in Figure 5.4. The vectors G_i are known as the geometric constraints of the Hermite curve.

The cubic Hermite curve segment can now be expressed as follows:

$$Q(t) = \sum_{i=0}^{3} H_i^3(t) \cdot G_i. \tag{5.45}$$

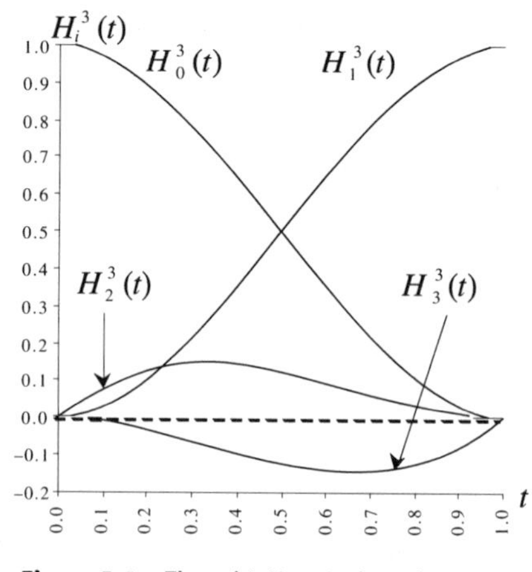

Figure 5.4 The cubic Hermite basis functions.

5.6 Cubic Bézier Curves

Bézier curves are among the most popular polynomial curve representations. These curves were developed independently by Paul de Casteljau around 1959 and shortly afterwards in 1962 by Pierre Bézier. De Casteljau and Bézier worked for the French car manufacturers Citroën and Renault, respectively, and were attempting to develop mathematical techniques to allow them to represent curves and surfaces using the computer. The mathematical technique that underlies Bézier curves and surfaces is that of Bernstein polynomials. De Casteljau was the first to exploit Bernstein polynomials in developing a recursive algorithm for the evaluation of points on the parametric curve. His algorithm is geometric in nature and gives a geometric insight into the process of the creation of the curve. His algorithm appeared in an internal Citroën technical document and was never published. De Casteljau's work went unnoticed until 1975, when W. Boehm obtained copies of the Citroën technical reports. Bézier on the other hand, who developed a CAD system called UNISURF had his work widely published and thus these curves and surfaces bear his name. In 1972, A.R. Forest discovered the connection between Bézier's work and Bernstein polynomials. As we shall discover later, Bézier curves are a special case of uniform B-spline curves.

5.6.1 The Definition of Cubic Bézier Curves

A cubic Bézier curve segment is defined by four control points. The curve segment passes through its first and last control points. The two middle control points are not on the curve but are used to determine the direction and size of its endpoint tangents and therefore can be used to control the shape of the curve.

The start and end tangent vectors are determined by the vectors $3 \cdot (\overline{P_0P_1})$ and $3 \cdot (\overline{P_2P_3})$ and are related to R_0 and R_3 (of the last section) by the equations

$$\begin{aligned} R_0 &= Q^{(1)}(0) = 3 \cdot (P_1 - P_0) \\ R_3 &= Q^{(1)}(1) = 3 \cdot (P_3 - P_2). \end{aligned} \tag{5.46}$$

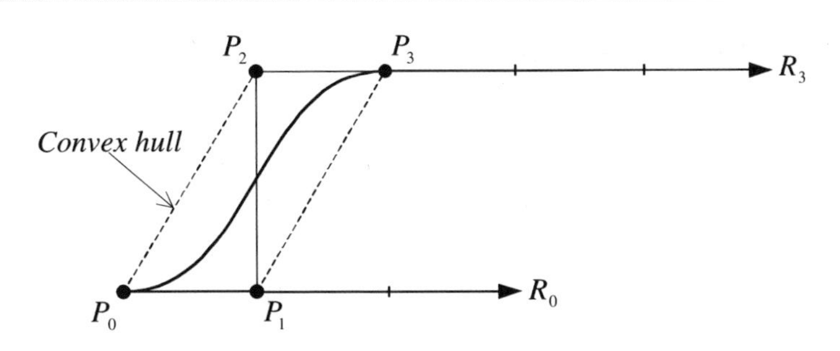

Figure 5.5 A cubic Bézier curve segment.

The cubic Bézier curve interpolates the two end control points and approximates the middle control points.

The Bézier geometry vector is

$$G_B = \begin{bmatrix} P_0 \\ P_1 \\ P_2 \\ P_3 \end{bmatrix}. \tag{5.47}$$

The matrix M_{HB} that defines the relation between the Hermite geometry vector G_H and the Bézier geometry vector G_B is given by

$$G_H = M_{HB} \cdot G_B. \tag{5.48}$$

From Equation (5.46) we have

$$G_H = \begin{bmatrix} P_0 \\ P_3 \\ R_0 \\ R_3 \end{bmatrix} = \begin{bmatrix} 1 & 0 & 0 & 0 \\ 0 & 0 & 0 & 1 \\ -3 & 3 & 0 & 0 \\ 0 & 0 & -3 & 3 \end{bmatrix} \cdot \begin{bmatrix} P_0 \\ P_1 \\ P_2 \\ P_3 \end{bmatrix} = M_{HB} \cdot G_B. \tag{5.49}$$

Recall from Equation (5.39) that $Q(t) = T \cdot M_H \cdot G_H$ and substitute $G_H = M_H \cdot G_B$ to define

$$M_B = M_H \cdot M_{HB}, \tag{5.50}$$

i.e.

$$\begin{aligned} Q(t) &= T \cdot M_H \cdot G_H \\ &= T \cdot M_H \cdot (M_{HB} \cdot G_B) \\ &= T \cdot (M_H \cdot M_{HB}) \cdot G_B \\ &= T \cdot M_B \cdot G_B. \end{aligned} \tag{5.51}$$

Now

$$M_B = M_H \cdot M_{HB} = \begin{bmatrix} -1 & 3 & -3 & 1 \\ 3 & -6 & 3 & 0 \\ -3 & 3 & 0 & 0 \\ 1 & 0 & 0 & 0 \end{bmatrix}, \tag{5.52}$$

which is the *Bézier basis matrix*. Thus,

$$\begin{aligned}
Q(t) &= T \cdot M_B \cdot G_B \\
&= (-t^3 + 3t^2 - 3t + 1) \cdot P_0 + (3t^3 - 6t^2 + 3t + 0) \cdot P_1 \\
&\quad + (-3t^3 + 3t^2 + 0t + 0) \cdot P_2 + (t^3 + 0t^2 + 0t + 0) \cdot P_3 \\
&= t^0 \cdot (1-t)^3 \cdot P_0 + 3t^1 \cdot (1-t)^2 \cdot P_1 + 3t^2 \cdot (1-t)^1 \cdot P_2 + t^3 \cdot (1-t)^0 \cdot P_3 \\
&= (1-t)^3 \cdot P_0 + 3t \cdot (1-t)^2 \cdot P_1 + 3t^2 \cdot (1-t) \cdot P_2 + t^3 \cdot P_3. \qquad (5.53)
\end{aligned}$$

Alternatively,

$$\begin{aligned}
Q(t) &= T \cdot M_B \cdot G_B \\
&= B_B \cdot G_B \\
&= \begin{bmatrix} B_0^3(t) & B_1^3(t) & B_2^3(t) & B_3^3(t) \end{bmatrix} \cdot \begin{bmatrix} P_0 \\ P_1 \\ P_2 \\ P_3 \end{bmatrix} \\
&= B_0^3(t) \cdot P_0 + B_1^3(t) \cdot P_1 + B_2^3(t) \cdot P_2 + B_3^3(t) \cdot P_3 \qquad (5.54)
\end{aligned}$$

where

$$\begin{aligned}
B_0^3(t) &= (1-t)^3 \\
B_1^3(t) &= 3t \cdot (1-t)^2 \\
B_2^3(t) &= 3t^2 \cdot (1-t) \\
B_3^3(t) &= t^3.
\end{aligned} \qquad (5.55)$$

The four polynomials $B_i^3(t)$, which are the weights in Equation (5.54), are called the *Bernstein polynomials*. A plot of these functions is shown in Figure 5.6.

An alternative notation of the cubic Bézier curve is given by

$$Q(t) = \sum_{i=0}^{3} B_i^3(t) \cdot P_i \quad \text{where } 0 \leq t \leq 1. \qquad (5.56)$$

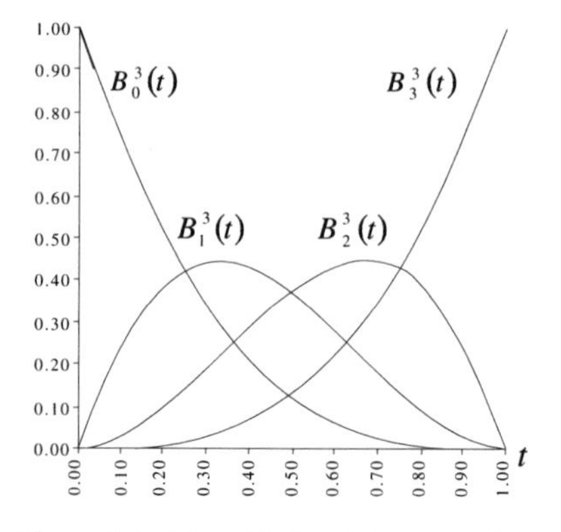

Figure 5.6 The cubic Bernstein polynomials.

The sum of the four Bernstein polynomials is unity everywhere and each polynomial is non-negative everywhere. Thus, $Q(t)$ is the *weighted-average* of the four control points. This condition means that each curve segment is contained in the *convex hull* of the four control points. The convex hull of a 2D curve is a *convex polygon* formed by the four control points. The convex hull of a 3D curve is a *convex polyhedron* (a tetrahedron) formed by the four control points. The convex hull property holds for all cubics defined as the weighted-sum of their control points, provided that their *blending functions* are *non-negative* and that they have a *unit-sum* in the range $0 \leq t \leq 1$. In general, the weighted-sum of n points falls within the convex hull of these points.

5.6.2 The Derivatives of Cubic Bézier Curves

By rewriting Equation (5.51) we obtain

$$
\begin{aligned}
Q(t) &= \begin{bmatrix} t^3 & t^2 & t & 1 \end{bmatrix} \cdot \begin{bmatrix} -1 & 3 & -3 & 1 \\ 3 & -6 & 3 & 0 \\ -3 & 3 & 0 & 0 \\ 1 & 0 & 0 & 0 \end{bmatrix} \cdot \begin{bmatrix} P_0 \\ P_1 \\ P_2 \\ P_3 \end{bmatrix} \\
&= \begin{bmatrix} -t^3+3t^2-3t+1 & 3t^3-6t^2+3t & -3t^3+3t^2 & t^3 \end{bmatrix} \cdot \begin{bmatrix} P_0 \\ P_1 \\ P_2 \\ P_3 \end{bmatrix}.
\end{aligned} \tag{5.57}
$$

From Equation (5.16) we know that the first derivative, $Q^{(1)}(t)$, of $Q(t)$ is given by

$$
\begin{aligned}
Q^{(1)}(t) &= \begin{bmatrix} 3t^2 & 2t & 1 & 0 \end{bmatrix} \cdot \begin{bmatrix} -1 & 3 & -3 & 1 \\ 3 & -6 & 3 & 0 \\ -3 & 3 & 0 & 0 \\ 1 & 0 & 0 & 0 \end{bmatrix} \cdot \begin{bmatrix} P_0 \\ P_1 \\ P_2 \\ P_3 \end{bmatrix} \\
&= \begin{bmatrix} -3t^2+6t-3 & 9t^2-12t+3 & -9t^2+6t & 3t^2 \end{bmatrix} \cdot \begin{bmatrix} P_0 \\ P_1 \\ P_2 \\ P_3 \end{bmatrix}.
\end{aligned} \tag{5.58}
$$

Similarly, the second derivative, $Q^{(2)}(t)$, of $Q(t)$ is given by

$$
\begin{aligned}
Q^{(2)}(t) &= \begin{bmatrix} 6t & 2 & 0 & 0 \end{bmatrix} \cdot \begin{bmatrix} -1 & 3 & -3 & 1 \\ 3 & -6 & 3 & 0 \\ -3 & 3 & 0 & 0 \\ 1 & 0 & 0 & 0 \end{bmatrix} \cdot \begin{bmatrix} P_0 \\ P_1 \\ P_2 \\ P_3 \end{bmatrix} \\
&= \begin{bmatrix} -6t+6 & 18t-12 & -18t+6 & 6t \end{bmatrix} \cdot \begin{bmatrix} P_0 \\ P_1 \\ P_2 \\ P_3 \end{bmatrix}.
\end{aligned} \tag{5.59}
$$

From Equation (5.57), by setting the parameter value to $t = 0$ and $t = 1$ we obtain, respectively,

$$Q(0) = \begin{bmatrix} 1 & 0 & 0 & 0 \end{bmatrix} \cdot \begin{bmatrix} P_0 \\ P_1 \\ P_2 \\ P_3 \end{bmatrix} = P_0 \tag{5.60}$$

and

$$Q(1) = \begin{bmatrix} 0 & 0 & 0 & 1 \end{bmatrix} \cdot \begin{bmatrix} P_0 \\ P_1 \\ P_2 \\ P_3 \end{bmatrix} = P_3. \tag{5.61}$$

The above two equations show that the curve interpolates its first and last control points, as was pointed out at the beginning of the last subsection.

From Equation (5.58), by setting the parameter value to $t = 0$ and $t = 1$ we obtain, respectively,

$$Q^{(1)}(0) = \begin{bmatrix} -3 & 3 & 0 & 0 \end{bmatrix} \cdot \begin{bmatrix} P_0 \\ P_1 \\ P_2 \\ P_3 \end{bmatrix} = 3(P_1 - P_0) \tag{5.62}$$

and

$$Q^{(1)}(1) = \begin{bmatrix} 0 & 0 & -3 & 3 \end{bmatrix} \cdot \begin{bmatrix} P_0 \\ P_1 \\ P_2 \\ P_3 \end{bmatrix} = 3(P_3 - P_2). \tag{5.63}$$

The above two equations show that the tangent vectors at the start and at the end of the curve depend on its first and last control polygon edge, respectively. Compare the above results with Equation (5.46).

From Equation (5.59), by setting the parameter value to $t = 0$ and $t = 1$ we obtain, respectively,

$$Q^{(2)}(0) = \begin{bmatrix} 6 & -12 & 6 & 0 \end{bmatrix} \cdot \begin{bmatrix} P_0 \\ P_1 \\ P_2 \\ P_3 \end{bmatrix} = 6(P_0 - 2P_1 + P_2) = 6(P_2 - P_1) - 6(P_1 - P_0) \tag{5.64}$$

and

$$Q^{(2)}(1) = \begin{bmatrix} 0 & 6 & -12 & 6 \end{bmatrix} \cdot \begin{bmatrix} P_0 \\ P_1 \\ P_2 \\ P_3 \end{bmatrix} = 6(P_1 - 2P_2 + P_3) = 6(P_3 - P_2) - 6(P_2 - P_1). \tag{5.65}$$

The above two equations show that the second derivative at the start and at the end of the curve depend on the first two edges and the last two edges of its control polygon, respectively.

5.6.3 Parametric Continuity Between Two Neighboring Cubic Bézier Curves

Given two cubic Bézier curves $Q_A(t)$ and $Q_B(t)$ defined by the control points $[A_0, A_1, A_2, A_3]$ and $[B_0, B_1, B_2, B_3]$, respectively, let us examine what conditions are required to achieve C^0, C^1 and C^2 parametric continuity between these curves at the joint.

To achieve C^0 continuity we must have

$$Q_B(0) = Q_A(1). \tag{5.66}$$

Using Equations (5.60) and (5.61) we obtain

$$B_0 = A_3, \tag{5.67}$$

which means that the first control point of curve Q_B must be the same as the last control point of curve Q_A (see Figure 5.7).

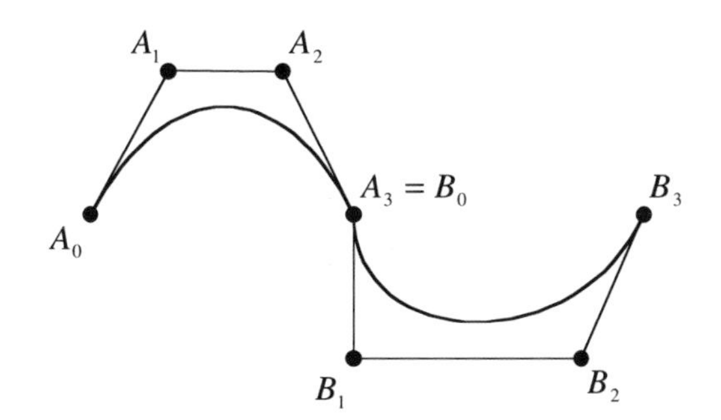

Figure 5.7 Two cubic Bézier curves with C^0 continuity at the joint.

To achieve C^1 continuity we must have

$$Q_B^{(1)}(0) = Q_A^{(1)}(1). \tag{5.68}$$

Using Equations (5.62) and (5.63) we obtain

$$\begin{aligned} 3(B_1 - B_0) &= 3(A_3 - A_2) \\ \therefore\ B_1 - B_0 &= A_3 - A_2. \end{aligned} \tag{5.69}$$

Using Equations (5.67) and (5.69) we obtain

$$\begin{aligned} B_1 - A_3 &= A_3 - A_2 \\ \therefore\ B_1 &= 2A_3 - A_2 \\ \therefore\ B_1 &= A_3 + (A_3 - A_2). \end{aligned} \tag{5.70}$$

From Equation (5.69) we can see that the second control point of curve Q_B must lie on the same line as the last control edge $\overrightarrow{A_2A_3}$ of curve Q_A. Additionally, from Equation (5.70) we can see that point A_3 must lie mid way between points A_2 and B_1 (see Figure 5.8).

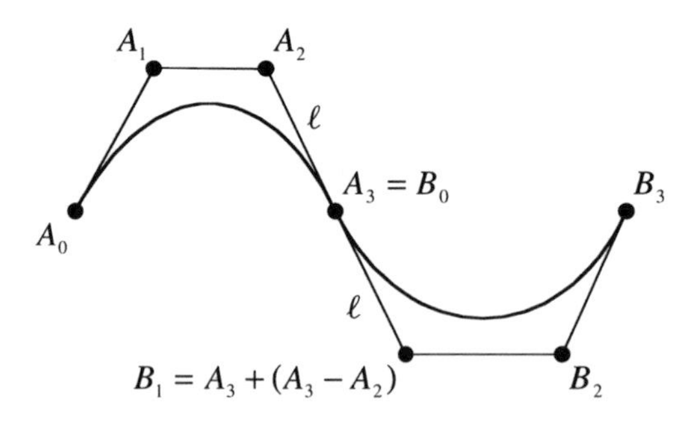

Figure 5.8 Two cubic Bézier curves with C^1 continuity at the joint.

Finally, to achieve C^2 continuity we must have

$$Q_B^{(2)}(0) = Q_A^{(2)}(1). \tag{5.71}$$

Using Equations (5.64) and (5.65) we obtain

$$\begin{aligned} 6(B_2 - B_1) - 6(B_1 - B_0) &= 6(A_3 - A_2) - 6(A_2 - A_1) \\ \therefore\ (B_2 - B_1) - (B_1 - B_0) &= (A_3 - A_2) - (A_2 - A_1). \end{aligned} \tag{5.72}$$

Using Equations (5.67), (5.70) and (5.72) we obtain

$$\begin{aligned} &(B_2 - 2A_3 + A_2) - (2A_3 - A_2 - A_3) = (A_3 - A_2) - (A_2 - A_1) \\ &\therefore\ B_2 - 2A_3 + A_2 - A_3 + A_2 = A_3 - 2A_2 + A_1 \\ &\therefore\ B_2 - 3A_3 + 2A_2 = +A_3 - 2A_2 + A_1 \\ &\therefore\ B_2 = 4A_3 - 4A_2 + A_1 \\ &\therefore\ B_2 = A_1 + 4(A_3 - A_2). \end{aligned} \tag{5.73}$$

From Equation (5.72) we can see that the vector difference between the second and first control polygon edges of curve Q_B must be equal to the vector difference between the third and second control polygon edges of curve Q_A. Alternatively, from Equation (5.73) we can see that control points A_1, A_2, $A_3 = B_0$, B_1 and B_2 must form a convex hull or must be collinear to maintain C^2 parametric continuity across the joining curves (see Figure 5.9).

From the above discussion it should be apparent that the restrictions imposed on the vertices B_0, B_1 and B_2 are quite severe and make it impractical to join cubic Bézier curve segments in order to describe more complex curves. Consequently, to use Bézier curves for practical design purposes it became necessary to introduce higher-degree Bézier curves.

5.7 B-spline Curves

B-splines have a long history. They were originally used as one-dimensional functions in probability and statistical computations. Nicolai Ivanovich Lobachevsky (1792–1856) was the first to introduce the idea of B-spline functions, which were constructed as a convolution of probability functions that were defined in relation to a special knot sequence.

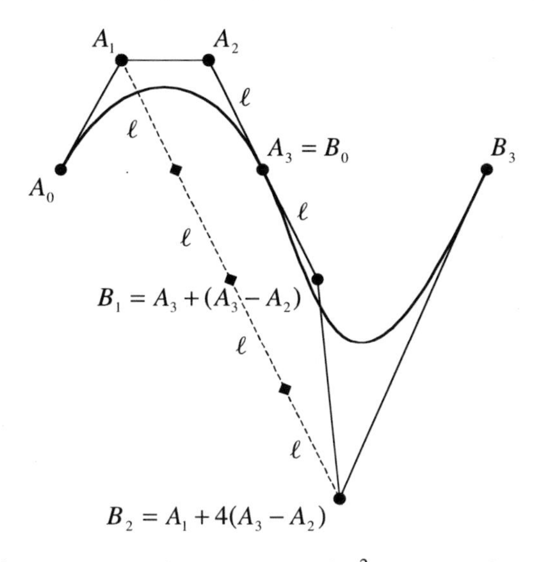

Figure 5.9 Two cubic Bézier curves with C^2 continuity at the joint.

In 1946, I.J. Shoenberg used B-splines for statistical data smoothing and it was his paper, on the approximation of equidistant data by analytic functions, that gave rise to modern spline approximation theory.

M.G. Cox (1971) and C. de Boor (1972) independently developed the recurrence relation (recursive definition) of B-splines, which is useful for numerical computations and is an important theoretical tool in the development of B-splines.

In 1974, W. Gordon and R. Riesenfeld for the first time used B-spline theory to design curves and surfaces. They also combined the theory of Bézier and B-spline curves and showed that Bézier curves are a special case of the more powerful B-splines.

A B-splines curve consists of a number of curve segments, the polynomial coefficients of which depend on just a few control points. This allows for local control. Thus, moving a control point only affects a small segment of the curve. The time needed to compute the polynomial coefficients of a B-spline is thus greatly reduced. Cubic B-splines have C^2 parametric continuity and do not interpolate their control points.

There are four distinct types of B-splines:

- uniform non-rational B-splines (UNRBS);
- uniform rational B-splines (URBS);
- non-uniform non-rational B-splines (NUNRBS);
- non-uniform rational B-splines (NURBS).

Uniform curves have *knots* that are spaced at equal intervals of the parameter t. Conversely, *non-uniform* curves have knots that are unequally spaced.

Curves defined in 3D space by 3D control points using a non-rational polynomial formulation are known as non-rational curves. Such curves are unable to represent conic sections precisely and can only approximate them. As conic sections are frequently used in computer graphics it became necessary to introduce rational formulations of curves, which were capable of representing them exactly. Such curves are defined in homogeneous

4D space by a set of 4D control points and are then represented in 3D by a projective transformation.

We will limit our discussion to cubic B-splines and we will adopt a staged (progressive refinement) approach. First, we will introduce cubic B-splines, then we will examine how these can be pieced together to form more complex curves.

5.7.1 Uniform Non-rational Cubic B-splines

As mentioned earlier we can construct a B-spline curve by piecing together a number of consecutive curve segments each one of which is represented by a polynomial of degree $d = 3$ and order $k = 4$. Each curve segment does not necessarily pass through its control points but may do so under certain conditions, as we shall see later. Each pair of consecutive curve segments exhibits C^2 parametric continuity at their joint. These continuity constraints are achieved by the sharing of control points between neighboring curve segments.

5.7.1.1 Definition of a Cubic B-spline Curve Segment

Cubic B-splines approximate the shape of their defining control polygon. In general, the control polygon consist of $(m + 1)$ control points $P_0, P_1, \ldots, P_m$ (where $m \geq 3$) and it is approximated by a set of $(m - k + 2) = (m - 2)$ cubic polynomial curve segments $Q_0, Q_1, \ldots Q_{m-3}$.

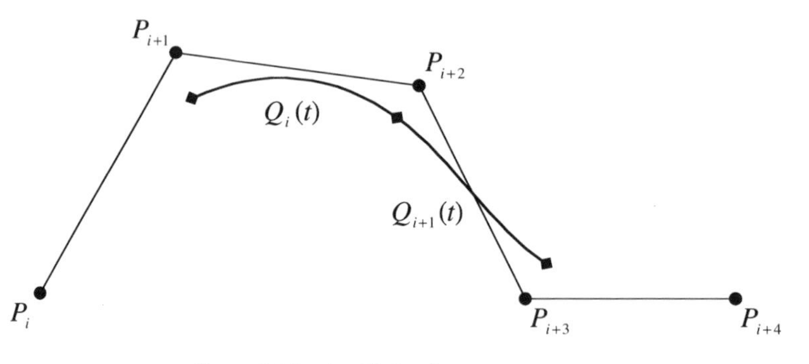

Figure 5.10 A cubic B-spline curve segment.

Given two consecutive curve segments Q_i and Q_{i+1}, as depicted in Figure 5.10, we observe that:

- curve segment Q_i is controlled by the points: $P_i, P_{i+1}, P_{i+2}, P_{i+3}$;
- curve segment Q_{i+1} is controlled by the points: $P_{i+1}, P_{i+2}, P_{i+3}, P_{i+4}$;
- curve segments Q_i and Q_{i+1} share three control points: $P_{i+1}, P_{i+2}, P_{i+3}$.

A single segment of the B-spline curve is defined as

$$Q_i(t) = \sum_{r=0}^{3} B_r(t) \cdot P_{i+r} \quad \text{for } 0 \leq t \leq 1 \quad \text{where } i = \{0, \ldots, m-3\}, \tag{5.74}$$

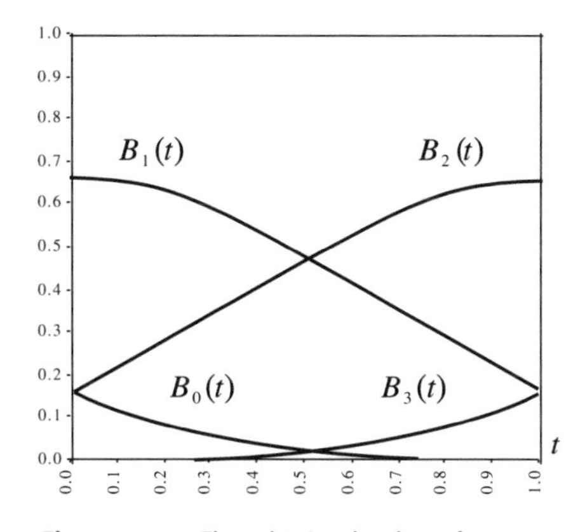

Figure 5.11 The cubic B-spline basis functions.

where

$$
\begin{aligned}
B_0(t) &= \frac{-t^3+3t^2-3t+1}{6} = \frac{(1-t)^3}{6} \\
B_1(t) &= \frac{3t^3-6t^2+4}{6} \\
B_2(t) &= \frac{-3t^3+3t^2+3t+1}{6} \\
B_3(t) &= \frac{t^3}{6}.
\end{aligned}
\tag{5.75}
$$

The functions $B_i(t)$ are the *cubic B-spline basis functions.* A plot of these functions can be seen in Figure 5.11.

A cubic B-spline curve segment can also be expressed in matrix form:

$$Q_i(t) = T \cdot M_{BS} \cdot G_{BS} \tag{5.76}$$

or

$$Q_i(t) = [t^3\ t^2\ t\ 1] \cdot \frac{1}{6} \cdot \begin{bmatrix} -1 & 3 & -3 & 1 \\ 3 & -6 & 3 & 0 \\ -3 & 0 & 3 & 0 \\ 1 & 4 & 1 & 0 \end{bmatrix} \cdot \begin{bmatrix} P_i \\ P_{i+1} \\ P_{i+2} \\ P_{i+3} \end{bmatrix}$$
$$\text{for } 0 \le t \le 1 \quad \text{where } i = \{0, \ldots, m-3\}, \tag{5.77}$$

where M_{BS} is the *B-spline basis matrix* and G_{BS} is the *B-spline geometry vector.*

The derivation of the basis functions (5.75) and the basis matrix in Equation (5.76) are beyond the scope of this exposition.

5.7.1.2 Parametric Continuity Between Two Neighboring Cubic B-spline Curve Segments

At the beginning of this section we stated that the cubic B-spline exhibits C^0, C^1 and C^2 parametric continuity at the curve segment joints. For this to be true the following constraints must be satisfied:

$$\begin{aligned} Q_i(1) &= Q_{i+1}(0) \\ Q_i^{(1)}(1) &= Q_{i+1}^{(1)}(0) \\ Q_i^{(2)}(1) &= Q_{i+1}^{(2)}(0). \end{aligned} \tag{5.78}$$

Let us verify that this is so.

Equation (5.74) can be rewritten as

$$Q_i(t) = B_0(t) \cdot P_i + B_1(t) \cdot P_{i+1} + B_2(t) \cdot P_{i+2} + B_3(t) \cdot P_{i+3} \tag{5.79}$$

or

$$\begin{aligned} Q_i(t) = &\tfrac{1}{6}(-t^3 + 3t^2 - 3t + 1) \cdot P_i + \tfrac{1}{6}(3t^3 - 6t^2 + 4) \cdot P_{i+1} \\ &+ \tfrac{1}{6}(-3t^3 + 3t^2 + 3t + 1) \cdot P_{i+2} + \tfrac{1}{6}(t^3) \cdot P_{i+3}. \end{aligned} \tag{5.80}$$

The first and second derivatives of $Q_i(t)$ are given by

$$\begin{aligned} Q_i(t) = &\tfrac{1}{6}(-3t^2 + 6t - 3) \cdot P_i + \tfrac{1}{6}(9t^2 - 12t) \cdot P_{i+1} + \tfrac{1}{6}(-9t^2 + 6t + 3) \cdot P_{i+2} \\ &+ \tfrac{1}{6}(3t^2) \cdot P_{i+3} \end{aligned} \tag{5.81}$$

and

$$\begin{aligned} &Q_i^{(2)}(t) = \tfrac{1}{6}(-6t + 6) \cdot P_i + \tfrac{1}{6}(18t - 12) \cdot P_{i+1} + \tfrac{1}{6}(-18t + 6) \cdot P_{i+2} + \tfrac{1}{6}(6t) \cdot P_{i+3} \\ &\therefore\ Q_i^{(2)}(t) = (-t + 1) \cdot P_i + (3t - 2) \cdot P_{i+1} + (-3t + 1) \cdot P_{i+2} + (t) \cdot P_{i+3}. \end{aligned} \tag{5.82}$$

Using Equation (5.80) we evaluate $Q_i(1)$ and $Q_{i+1}(0)$:

$$\begin{aligned} Q_i(1) &= \tfrac{1}{6}P_{i+1} + \tfrac{4}{6}P_{i+2} + \tfrac{1}{6}P_{i+3} \\ Q_{i+1}(0) &= \tfrac{1}{6}P_{i+1} + \tfrac{4}{6}P_{i+2} + \tfrac{1}{6}P_{i+3} \\ \therefore\ Q_i(1) &= Q_{i+1}(0). \end{aligned}$$

Using Equation (5.81) we evaluate $Q_i^{(1)}(1)$ and $Q_{i+1}^{(1)}(0)$:

$$\begin{aligned} Q_i^{(1)}(1) &= -\tfrac{1}{2}P_{i+1} + \tfrac{1}{2}P_{i+3} \\ Q_{i+1}^{(1)}(0) &= -\tfrac{1}{2}P_{i+1} + \tfrac{1}{2}P_{i+3} \\ \therefore\ Q_i^{(1)}(1) &= Q_{i+1}^{(1)}(0). \end{aligned}$$

Finally, using Equation (5.82) we evaluate $Q_i^{(2)}(1)$ and $Q_{i+1}^{(2)}(0)$:

$$\begin{aligned} Q_i^{(2)}(1) &= P_{i+1} - 2P_{i+2} + P_{i+3} \\ Q_{i+1}^{(2)}(0) &= P_{i+1} - 2P_{i+2} + P_{i+3} \\ \therefore\ Q_i^{(2)}(1) &= Q_{i+1}^{(2)}(0). \end{aligned}$$

Since all three constraints are satisfied we have verified that the cubic B-spline exhibits C^0, C^1 and C^2 parametric continuity.

5.7.1.3 Local and Global Parameter Values

When dealing with a number of connected cubic curve segments, although each curve segment is defined in its own parameter domain $0 \leq t \leq 1$, we can adjust the parameter so that the parameter domains of all the curve segments are sequential.

We shall adopt the notation t for the *global parameter* across all the segments of the curve and $\hat{t}$ for the *local parameter* across an individual curve segment. The global parameter t is defined in the interval $[0, m-2]$ and the local parameter $\hat{t}$ is defined in the interval $[0, 1]$. Figure 5.12 depicts this arrangement.

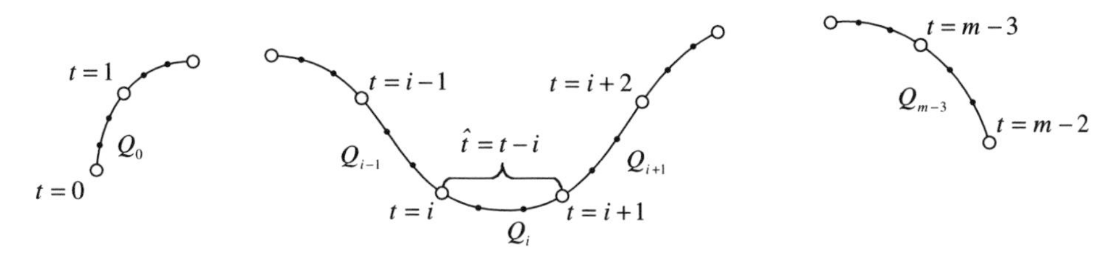

Figure 5.12 The local and global parameters of a multi-segment cubic B-spline curve.

For curve segment Q_i the local parameter is given by $\hat{t} = t - i$.

5.7.1.4 The Relationship Between the Basis Functions and the Blending Functions

Equation (5.74) defines each segment Q_i of the curve in terms of four control points. The basis functions $B_0(\hat{t}), B_1(\hat{t}), B_2(\hat{t}), B_3(\hat{t})$ tell us how the control points $P_i, P_{i+1}, P_{i+2}, P_{i+3}$ are weighted in the calculation of curve segment Q_i. Alternatively, we might express the entire curve as the weighted-sum of all its control points, as follows:

$$Q(t) = \sum_{i=0}^{m} N_i(t) \cdot P_i \quad \text{for } t = [0, m-2], \tag{5.83}$$

where $N_i(t)$ is the blending function for control point P_i. This is a function of the global parameter t and it tells us how the corresponding control point is weighted in the calculation of the entire curve.

Let us examine the relationship that exists between the basis functions of a curve segment and the blending function of a given control point. A control point P_i is used in the evaluation of curve segments Q_{i-3}, Q_{i-2}, Q_{i-1} and Q_i, where it is weighted by the basis functions B_3, B_2, B_1 and B_0 respectively (see Figure 5.13 and Table 5.1). The basis functions associated with a control point P_i in the four consecutive curve segments can be pieced together to form the blending function N_i (see Figure 5.14).

Thus we can define the blending function as

$$N_i(t) = \begin{cases} B_3(\hat{t}_3), & i-3 \leq t < i-2 \\ B_2(\hat{t}_2), & i-2 \leq t < i-1 \\ B_1(\hat{t}_1), & i-1 \leq t < i \\ B_0(\hat{t}_0), & i \leq t < i+1. \end{cases} \tag{5.84}$$

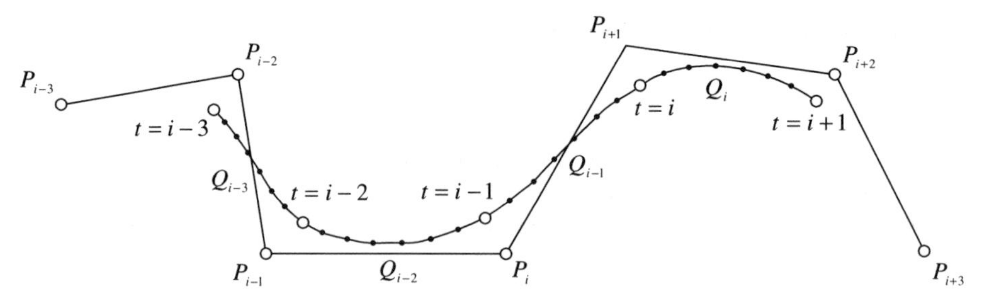

Figure 5.13 The curve segments Q_{i-3}, Q_{i-2}, Q_{i-1} and Q_i.

Table 5.1 The basis function of P_i for curve segments Q_{i-3}, Q_{i-2}, Q_{i-1} and Q_i.

Curve segment	Control points	Basis function of P_i
Q_{i-3}	$P_{i-3}, P_{i-2}, P_{i-1}, P_i$	B_3
Q_{i-2}	$P_{i-2}, P_{i-1}\ P_i, P_{i+1},$	B_2
Q_{i-1}	$P_{i-1}, P_i, P_{i+1}, P_{i+2}$	B_1
Q_i	$P_i, P_{i+1}, P_{i+2}, P_{i+3}$	B_0

Here $N_i(t)$ takes the global parameter t and converts it into the appropriate local parameter $\hat{t}_j$, associated with that segment, as follows: $\hat{t}_j = t - (i - j)$ for $j = 3, 2, 1, 0$.

Observe that N_i is only non-zero in the interval $[i - 3, i + 1]$ and that the control point P_i exerts maximal influence when $t = i - 1$, corresponding to the maximum of the blending function $N_i(t)$.

5.7.1.5 The Effect of Multiple Control Points on a Uniform B-spline

So far we have seen that we can alter the shape of a B-spline curve by altering the positions of its control points. Added shape flexibility can be achieved by placing two or more control points at the same position. By placing more than one control point at the same position we are said to increase the *multiplicity* of a control point, which causes the curve to pass closer to that control point. When the multiplicity of a control point becomes 3, then the curve actually passes through the control point. This extra flexibility, however, is achieved at the expense of geometric continuity. For example, consider a curve defined by five control

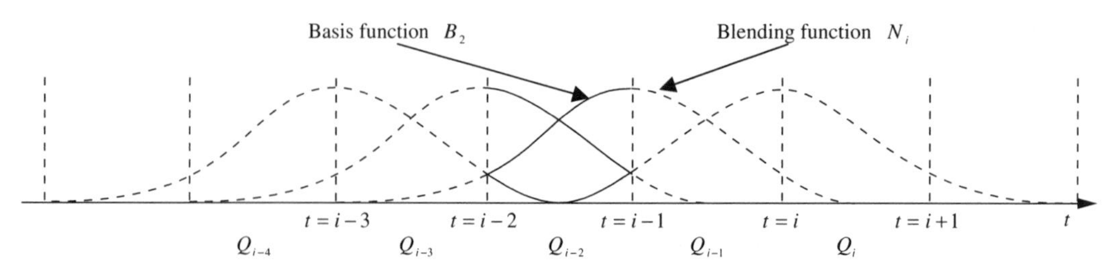

Figure 5.14 The relationship between the basis functions and the blending functions.

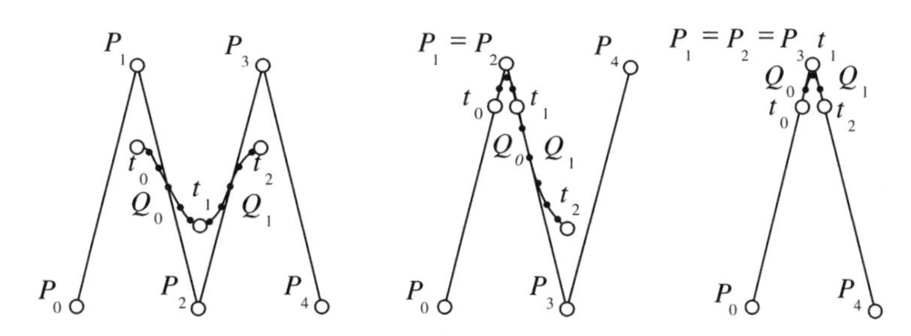

Figure 5.15 The effect of increasing the multiplicity of control point P_1.

points, as depicted in Figure 5.15. Table 5.2 shows what happens to the geometric and the parametric continuities at the joint of the two segments of the curve.

Table 5.2 The effect of increasing the multiplicity of control point P_1 on the geometric continuity between two consecutive curve segments. Observe that when the multiplicity of P_1 becomes 3, both curve segments become linear and lie on the edges of the control polygon.

Multiplicity	Control points	Continuity	Convex hulls
1	$[P_0, P_1, P_2, P_3, P_4]$	G^2, C^2	$(P_0, P_1, P_2, P_3), (P_1, P_2, P_3, P_4)$
2	$[P_0, P_1, P_1, P_3, P_4]$	G^1, C^2	$(P_0, P_1, P_3), (P_1, P_3, P_4)$
3	$[P_0, P_1, P_1, P_1, P_4]$	G^0, C^2	$(P_0, P_1), (P_3, P_4)$

A more detailed description of B-splines and of their properties is beyond the scope of this text.

5.8 Other Intersecting Splines

5.8.1 Beta-splines (β-splines)

In 1981, Brian Barsky completed his PhD studies at Utah University, during the course of which he developed the *beta-spline* (β-spline) curves and surfaces (Barsky 1981). A β-spline has a representation that was specifically designed for computer graphics and computer-aided design. Unlike its predecessors, the derivation of the β-spline was based on fundamental geometric properties rather than abstract algebraic quantities, i.e. it was based on geometric continuity rather than parametric continuity constraints.

In B-splines, from which the β-splines are descended, for a curve to appear smooth we require that neighboring curve segments be C^2 continuous, i.e. we require that the first and second derivatives of the curve be continuous. It turns out that this is overtly constraining and that to obtain a smooth curve it is often sufficient to insist that only its unit tangent vector and its curvature vector be continuous, i.e. to only require that the curve possesses second order geometric continuity G^2. By relaxing the constraints applied to the curve segment joints we gain additional design freedom without having to resort to higher-order polynomials. In particular, for a cubic β-spline we obtain two new degrees of freedom, namely *bias* and *tension*, which act as shape parameters. By selecting appropriate values for these shape parameters a cubic β-spline reduces to a corresponding uniform cubic B-spline. Thus, β-splines are a generalization of uniform B-splines.

β-splines have *local* control with respect to the movement of their control points and may have *global* or *local* control with respect to the change of their shape parameter values. The bias and tension shape parameters associated with a cubic β-spline may be kept constant over the entire curve or may be allowed to vary as we traverse the curve, giving rise to *uniformly, continuously* or *discretely* shaped β-splines. A *uniformly shaped* β-spline has global shape parameters, which are constant for the entire curve. A *continuously shaped* β-spline has local shape parameters, which are functions of the curve parameter and vary over the extent of the curve, i.e. each joint is assigned a distinct value for each shape parameter. A *discretely shaped* β-spline has local shape parameters, which are functions of the curve's *parametric knot vector*. Here each curve segment endpoint has associated with it a *parametric knot* value and each *parametric knot interval* corresponds to a curve segment. These parametric knot intervals may be non-uniformly distributed, as in the case of non-uniform B-splines. Each parametric knot is assigned a distinct value for each shape parameter (Seroussi and Barsky 1992). Our discussion will be limited to uniformly and continuously shaped β-splines.

5.8.1.1 Cubic β-splines

In general, cubic β-splines have the same properties as cubic B-splines but have two additional parameters, namely β_1 and β_2, which provide further control over their shape. β_1 is called the *bias* (or *skew* or *asymmetry*) parameter and β_2 is called the *tension* parameter. To better understand the effect of these parameters on the shape of a cubic β-spline let us consider the single segment cubic β-spline curve illustrated in Figure 5.16.

Let us start by examining the geometric significance and effect of the β_1 shape parameter. In cubic B-splines we insist that two neighboring curve segments meet at a joint with C^2 continuity. C^1 continuity (which is subsumed in C^2 continuity) forces the tangent vectors of the curve segment terminating at the joint and the curve segment starting at the joint to be equal in both direction and size. In β-splines, on the other hand, we only insist on G^2 continuity at the joints. G^1 continuity (which is subsumed in G^2 continuity) forces the tangent vectors of the terminating and the starting curve segments to be equal in direction only but not in size. β_1 is the ratio of the magnitudes of the starting segment tangent vector to the terminating segment tangent vector, as seen in Figure 5.16. Thus, β_1 can be seen as a measure of the influence that the unit tangent vector at the joint has on the two curve segments either side of this joint.

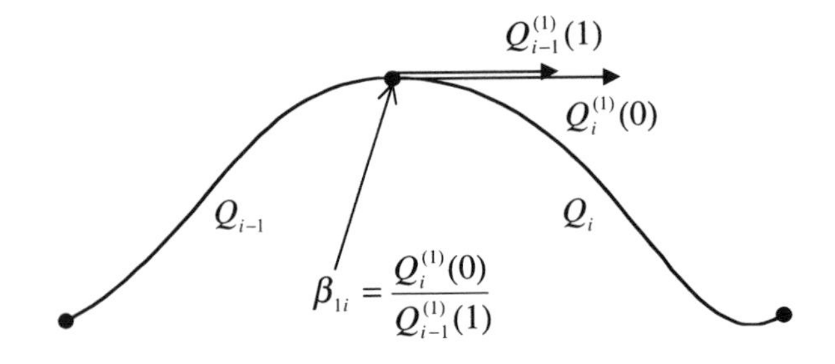

Figure 5.16 The bias parameter at the *i*th joint.

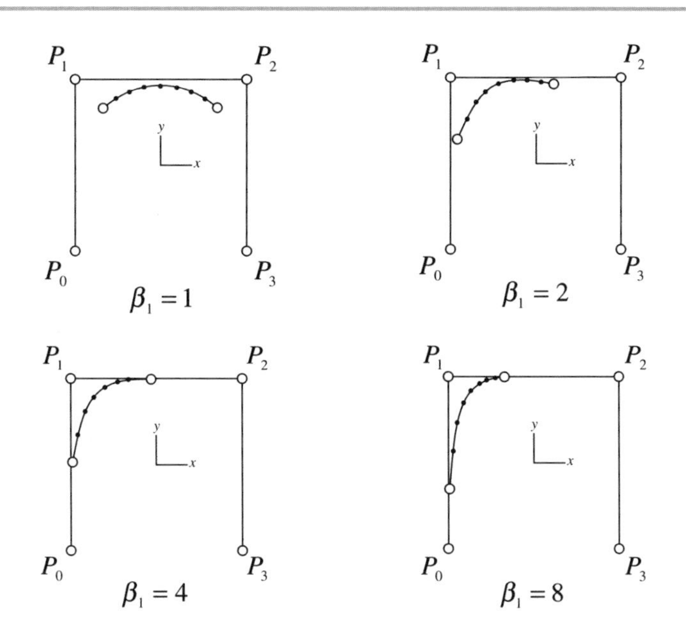

Figure 5.17 Increasing the bias with no tension.

When $\beta_1 = 1$, the two tangents at the joint are of the same size and therefore exert the same influence on the terminating and starting curve segments. As β_1 increases beyond 1, the tangent vector $Q_i^{(1)}(0)$, of the starting segment Q_i, becomes larger and has a greater influence on this segment, forcing the curve to follow the direction of this tangent further and causing the curve segment to be biased with respect to its starting tangent, as seen in Figure 5.17. For values of β_1 less than 1 but greater than 0, the tangent vector $Q_{i-1}^{(1)}(1)$ of the terminating segment Q_{i-1} becomes longer, thus biasing the curve segment with respect to its end tangent, as seen in Figure 5.18.

Figure 5.17 shows the resulting asymmetric skew of a cubic β-spline curve segment as we gradually increase the value of the β_1 shape parameter while keeping $\beta_2 = 0$. Similarly, Figure 5.18 shows the resulting asymmetric skew of the curve segment as we gradually decrease the value of the β_1 shape parameter while keeping $\beta_2 = 0$.

The β_2 shape parameter controls the symmetric tension applied to a curve segment. When $\beta_2 = 0$, the curve is said to be untensed. As the value of $\beta_{2,i}$ increases, it has the effect of attracting the ith joint of the curve towards the $(i + 1)$th control point P_{i+1}. If the same β_2 value is applied to all the joints, the curve uniformly approaches its control polygon and as β_2 approaches ∞, it causes the curve segments to be indistinguishable from their corresponding control edges, as shown in Figures 5.19 and 5.20. Thus, the β_2 shape parameter provides a mechanism for modeling tension in the curve and can be interpreted as a measure of the *tightness* or *looseness* of the curve.

Let us now examine how cubic β-splines are defined. In our discussion we adopt a notation that is slightly different from that used by Barsky and his co-authors in various papers. This is done for consistency with the notation we have adopted in the definition of B-spline curves.

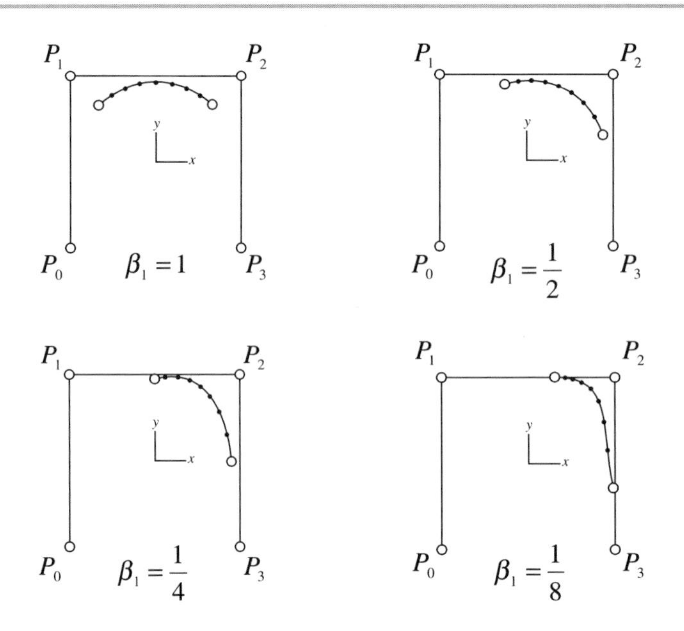

Figure 5.18 Decreasing the bias with no tension.

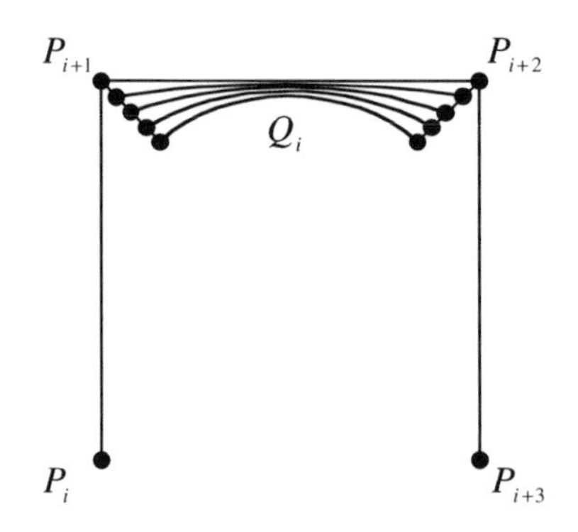

Figure 5.19 Increasing the tension parameter at both joints symmetrically.

5.8.1.2 Uniformly Shaped β-splines

We will start by examining the formulation of uniformly shaped cubic β-splines, where the shape parameters β_1 and β_2 are equal for all joints of the curve.

Given $(m+1)$ control points $\{P_i\}_{i=0}^{m}$, we can define a cubic β-spline curve consisting of $(m-2)$ curve segments $\{Q_i\}_{i=0}^{m-3}$. Each β-spline curve segment is given by

$$Q_i(t) = \sum_{r=0}^{3} B_r(t; \beta_1, \beta_2) \cdot P_{(i+r)} \quad \text{where } 0 \le i \le (m-3) \wedge 0 \le t < 1 \tag{5.85}$$

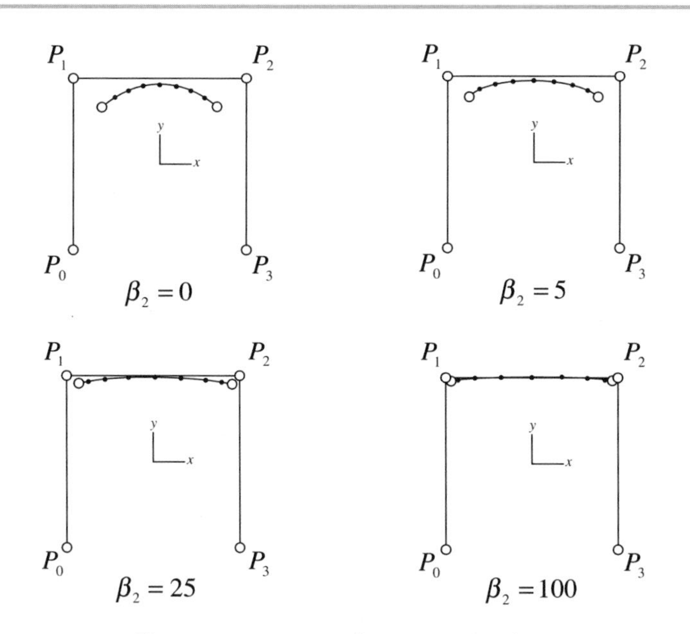

Figure 5.20 Increasing the tension with no bias.

where $B_r(t;\ \beta_1, \beta_2)$ is the rth β-spline basis function, which is a cubic polynomial given by

$$\begin{aligned}
B_0(t;\ \beta_1, \beta_2) &= \frac{1}{\gamma}\left[\left(2\beta_1^3\right) - \left(6\beta_1^3\right)\cdot t + \left(6\beta_1^3\right)\cdot t^2 - \left(2\beta_1^3\right)\cdot t^3\right] \\
B_1(t;\ \beta_1, \beta_2) &= \frac{1}{\gamma}\left[\left(\beta_2 + 4\beta_1^2 + 4\beta_1\right) + \left(6\beta_1^3 - 6\beta_1\right)\cdot t - \left(3\beta_2 + 6\beta_1^3 + 6\beta_1^2\right)\cdot t^2\right. \\
&\quad \left. + \left(2\beta_2 + 2\beta_1^3 + 2\beta_1^2 + 2\beta_1\right)\cdot t^3\right] \\
B_2(t;\ \beta_1, \beta_2) &= \frac{1}{\gamma}\left[2 + (6\beta_1)\cdot t + \left(3\beta_2 + 6\beta_1^2\right)\cdot t^2 - \left(2\beta_2 + 2\beta_1^2 + 2\beta_1 + 2\right)\cdot t^3\right] \\
B_3(t;\ \beta_1, \beta_2) &= \frac{1}{\gamma}\left(2t^3\right)
\end{aligned} \tag{5.86}$$

with

$$\gamma = 2\beta_1^3 + 4\beta_1^2 + \beta_2 + 2 \wedge \gamma \neq 0. \tag{5.87}$$

The matrix representation of a cubic β-spline curve segment is given by

$$Q_i(t) = T \cdot M_\beta \cdot G_\beta \quad \text{where } 0 \le i \le (m-3) \wedge 0 \le t < 1 \tag{5.88}$$

or

$$Q_i(t) = [t^3\ \ t^2\ \ t^1\ \ 1] \cdot M_\beta \cdot \begin{bmatrix} P_i \\ P_{i+1} \\ P_{i+2} \\ P_{i+3} \end{bmatrix}, \tag{5.89}$$

where the β-spline basis matrix M_β is given by

$$M_\beta = \frac{1}{\gamma} \cdot \begin{bmatrix} -2\beta_1^3 & 2(\beta_2 + \beta_1^3 + \beta_1^2 + \beta_1) & -2(\beta_2 + \beta_1^2 + \beta_1 + 1) & 2 \\ 6\beta_1^3 & -3(\beta_2 + 2\beta_1^3 + 2\beta_1^2) & 3(\beta_2 + 2\beta_1^2) & 0 \\ -6\beta_1^3 & 6(\beta_1^3 - \beta_1) & 6\beta_1 & 0 \\ -2\beta_1^3 & \beta_2 + 4(\beta_1^2 + \beta_1) & 2 & 0 \end{bmatrix}. \qquad (5.90)$$

When $\beta_1 = 1$ (no bias) and $\beta_2 = 0$ (no tension) then: $\gamma = 2 \cdot 1 + 4 \cdot 1 + 4 \cdot 1 + 0 + 2 = 12$ and

$$M_\beta = \frac{1}{12} \cdot \begin{bmatrix} -2 & 6 & -6 & 2 \\ 6 & -12 & 6 & 0 \\ -6 & 0 & 6 & 0 \\ 2 & 8 & 2 & 0 \end{bmatrix}$$

From Equation (5.77) we recall that the B-spline basis matrix is:

$$M_{BS} = \frac{1}{6} \cdot \begin{bmatrix} -1 & 3 & -3 & 1 \\ 3 & -6 & 3 & 0 \\ -3 & 0 & 3 & 0 \\ 1 & 4 & 1 & 0 \end{bmatrix}.$$

Thus, $M_\beta = M_{BS}$ and the cubic β-spline reduces to a uniform cubic B-spline.

In uniformly shaped β-splines the same values of the shape parameters β_1 and β_2 are associated with all joints between two consecutive curve segments Q_{i-1} and Q_i. With this arrangement the β shape parameters can only exert global control on the shape of the curve.

5.8.1.3 Continuously Shaped β-splines

In continuously shaped β-splines the shape parameters β_1 and β_2 can assume distinct values for each joint between two consecutive curve segments Q_{i-1} and Q_i, as seen in Figure 5.21. With this arrangement the β parameters can exert local control on the shape of the curve.

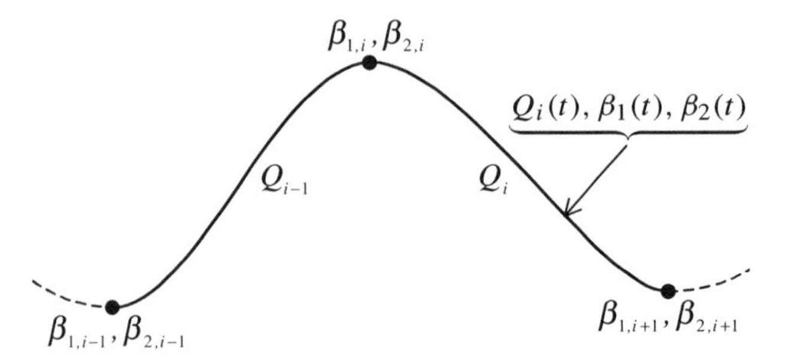

Figure 5.21 The distinct β parameters of a continuously or discretely shaped β-spline.

5.8.1.4 Interpolating Distinct β Parameters

In order to be able to evaluate a point on a continuously shaped β-spline for any given value of the parameter we must be able to calculate a set of values for the β parameters, which are now functions of t, namely $\beta_1(t)$ and $\beta_2(t)$. This can be done by interpolating the β parameters of the joints of the curve segment (Barsky and Beatty 1983).

Let us assume that we are in the process of generating curve segment $Q_i(t)$ which lies between the ith and $(i+1)$th joints. Recall that the value of t varies between 0 and 1 as we span the curve segment. Let, $\beta_{1,i}$ and $\beta_{1,i+1}$ be the bias parameters and $\beta_{2,i}$ and $\beta_{2,i+1}$ be the tension parameters at these joints, respectively. As we span the curve segment $Q_i(t)$ from $t=0$ to $t=1$, the values of the $\beta_1(t)$ and $\beta_2(t)$ parameters must be interpolated from $\beta_{1,i}$ to $\beta_{1,i+1}$ and from $\beta_{2,i}$ to $\beta_{2,i+1}$, respectively (see Figure 5.22). Great care must be taken when performing this interpolation. Barsky and Beatty suggest that in order to preserve geometric continuity between two consecutive curve segments Q_{i-1} and Q_i we must use the following *quintic Hermite interpolation* scheme:

$$\beta_j(t) = H\left(\beta_{j,i-1}, \beta_{j,i}, t\right) = \beta_{j,i-1} + \left(\beta_{j,i} - \beta_{j,i-1}\right) \cdot \left(10t^3 - 15t^4 + 6t^5\right) \quad \text{for } j = 1, 2. \tag{5.91}$$

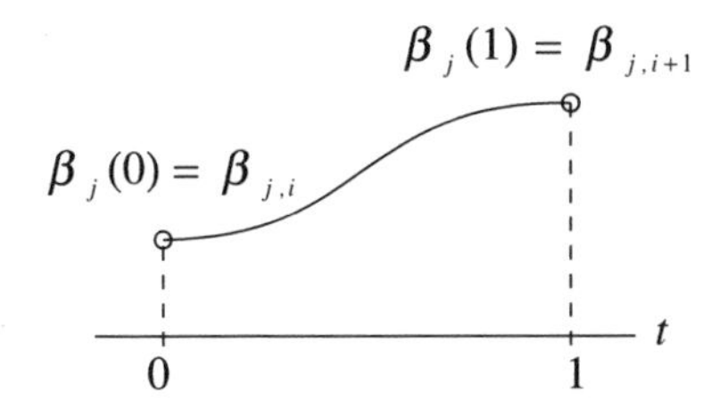

Figure 5.22 The quintic Hermite interpolation scheme.

5.8.2 Beta-2-splines (β_2-splines)

β_2-splines are a special case of β-splines without bias, i.e. $\beta_1 = 1$. The β_2-spline basis matrix is a simplification of the β-spline basis matrix of (5.90) and is given by

$$M_{\beta_2} = \frac{1}{\beta_2 + 12} \cdot \begin{bmatrix} -2 & 2(\beta_2+3) & -2(\beta_2+3) & 2 \\ 6 & -3(\beta_2+4) & 3(\beta_2+2) & 0 \\ -6 & 0 & 6 & 0 \\ 2 & \beta_2+8 & 2 & 0 \end{bmatrix}. \tag{5.92}$$

5.8.3 Kochanek–Bartels Splines

In this subsection we present a brief introduction to the Kochanek–Bartels splines (Kochanek and Bartels 1984; Bartels et al. 1987). Our discussion includes some results and observations not present in either of the references.

In 1984, Doris Kochanek and Richard Bartels developed a method of using cubic interpolating splines in key-frame animation. They used three control parameters that allow the animator to change the *tension*, *continuity* and *bias* of these splines.

Kochanek–Bartels splines are a general class of interpolating cubic splines that includes the *cardinal* and *Catmull–Rom* splines as a proper subset. Unlike other splines Kochanek–Bartels splines are formulated in terms of a set of *Hermite basis functions.* A curve Q interpolating $(m+1)$ control points (CVs) $\{P_i\}_{i=0}^{m}$ is composed of m curve segments $\{Q_i\}_{i=0}^{m-1}$. Each curve segment, Q_i, interpolates the control points P_i and P_{i+1} and requires the tangents R_i and R_{i+1} (at the curve segment endpoints) for its definition. Q_i is defined by

$$Q_i(u) = U \cdot M_H \cdot G_i \quad \text{for } 0 \le u \le 1 \tag{5.93}$$

or

$$Q_i(u) = \begin{bmatrix} u^3 & u^2 & u & 1 \end{bmatrix} \cdot \begin{bmatrix} 2 & -2 & 1 & 1 \\ -3 & 3 & -2 & -1 \\ 0 & 0 & 1 & 0 \\ 1 & 0 & 0 & 0 \end{bmatrix} \cdot \begin{bmatrix} P_i \\ P_{i+1} \\ R_i \\ R_{i+1} \end{bmatrix} \quad \text{for } 0 \le u \le 1, \tag{5.94}$$

where U is the parameter vector, M_H is the *Hermite basis matrix* and G_i is the *geometry vector* for curve segment Q_i.

Alternatively, the curve segment is given by

$$Q_i(u) = \sum_{j=0}^{3} H_j^3(u) \cdot G_{i_j} \quad \text{for } 0 \le u \le 1, \tag{5.95}$$

where G_{i_j} is the jth component of the geometry vector G_i and $H_j^3(u)$ is the jth cubic *Hermite basis function*. These functions are defined as follows:

$$\begin{aligned} H_0^3(u) &= 2u^3 - 3u^2 + 1 \\ H_1^3(u) &= -2u^3 + 3u^2 \\ H_2^3(u) &= u^3 - 2u^2 + u \\ H_3^3(u) &= u^3 - u^2. \end{aligned} \tag{5.96}$$

Let us now look at how we select appropriate values for the tangent vectors R_i and R_{i+1}.

For cardinal splines the tangent vector at P_i is given by

$$R_i = a \cdot (P_{i+1} - P_{i-1}) = a \cdot (P_i - P_{i-1}) + a \cdot (P_{i+1} - P_i). \tag{5.97}$$

Catmull–Rom splines are a special case of cardinal splines with $a = \frac{1}{2}$. Their tangent vector at P_i is the average of the *source chord* $\overline{P_{i-1}P_i}$ and the *destination chord* $\overline{P_iP_{i+1}}$ and is given by

$$R_i = \tfrac{1}{2} \cdot (P_{i+1} - P_{i-1}) = \tfrac{1}{2} \cdot (P_i - P_{i-1}) + \tfrac{1}{2} \cdot (P_{i+1} - P_i). \tag{5.98}$$

The tangent vectors of the first and last control points of a Catmull–Rom spline, $\{R_0, R_m\}$, may be specified by the user or may be set arbitrarily to the zero tangent vector $[0, 0, 0]$. Kochanek–Bartels splines use the Catmull–Rom tangent vector as their default, when the

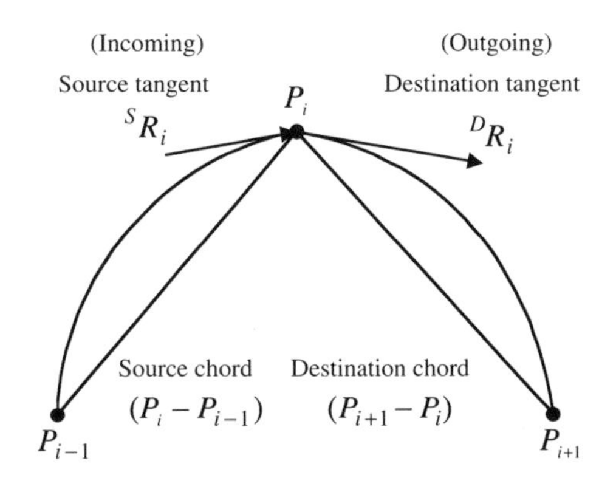

Figure 5.23 The Kochanek–Bartels spline tangent vectors.

tension, continuity and bias parameters are set to their default zero values. Kochanek-Bartels have replaced the tangent vector R_i by the *(incoming) source tangent vector* $^S R_i$ and the *(outgoing) destination tangent vector* $^D R_i$ (see Figure 5.23). The curve segment Q_i can now be defined by the four geometric constraints P_i, P_{i+1}, $^D R_i$ and $^S R_{i+1}$.

The source and destination tangent vectors at control point P_i are given by

$$^S R_i = \frac{(1-t)\cdot(1-c)\cdot(1+b)}{2}\cdot(P_i - P_{i-1}) + \frac{(1-t)\cdot(1+c)\cdot(1-b)}{2}\cdot(P_{i+1} - P_i) \tag{5.99}$$

$$^D R_i = \frac{(1-t)\cdot(1+c)\cdot(1+b)}{2}\cdot(P_i - P_{i-1}) + \frac{(1-t)\cdot(1-c)\cdot(1-b)}{2}\cdot(P_{i+1} - P_i), \tag{5.100}$$

where t is the tension parameter, c is the continuity parameter and b is the bias parameter. Each parameter has a default value of zero. Each control point P_i has one such set of parameters that determines the size of the tangent vectors $^S R_i$ and $^D R_i$.

Let us now examine the significance of these three parameters. We will do so by examining each parameter in isolation.

5.8.3.1 The Tension Parameter

The tension parameter t controls how sharply the curve bends at a point P_i. It is implemented as a scale factor that changes the length of both the source and destination tangent vectors, at a given control point, equally:

$$^S R_i = {}^D R_i = \frac{(1-t)}{2}\cdot(P_i - P_{i-1}) + \frac{(1-t)}{2}\cdot(P_{i+1} - P_i). \tag{5.101}$$

With the default value $t = 0$, the tangent vectors are formed as the average of the two adjacent chords, and we obtain the default curve.

Increasing the value of the tension parameter has the effect of reducing the length of the tangent vectors, thus tightening the curve. When $t = 1$, the lengths of the tangent vectors reduce to zero (see Figures 5.24 and 5.25, and Figure 5.30 in Section 5.8.3.4). Increasing the

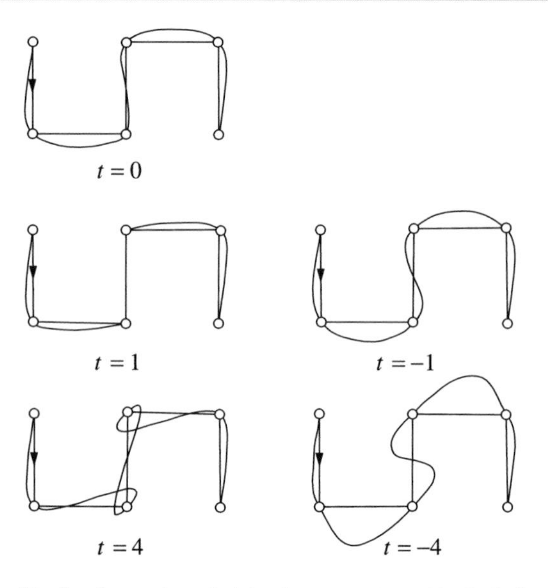

Figure 5.24 The middle two CVs of each curve have their tension parameters set to the indicated values. All other CVs have their parameters set to the default values.

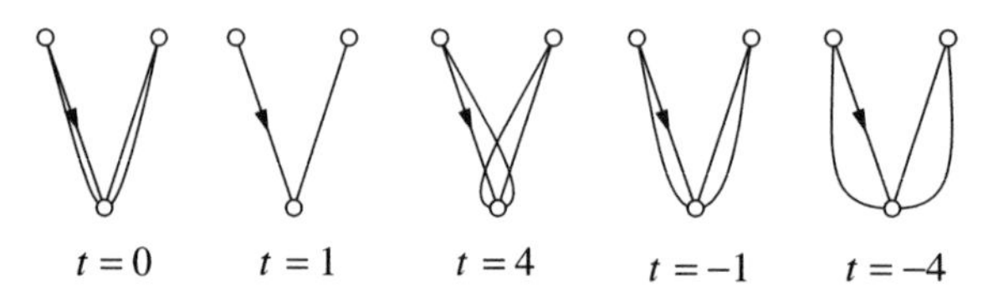

Figure 5.25 The middle CV of each curve has its tension parameter set to the indicated value. All other CVs have their parameters set to the default values.

value of this parameter further introduces a loop in the curve at P_i, as seen in Figures 5.24 and 5.25.

Reducing the value of the tension parameter has the effect of increasing the length of the tangent vectors, thus loosening or slackening the curve. When $t = -1$, the lengths of the tangent vectors double (see Figures 5.24 and 5.25). Decreasing the value of this parameter further introduces a flat section in the curve at P_i, as seen in Figures 5.24 and 5.25.

If two consecutive control points have $t = 1$, then the curve segment joining them reduces to their chord (see Figures 5.24 and 5.30).

5.8.3.2 The Continuity Parameter

The continuity parameter c controls the degree of continuity (smoothness of transition) between two consecutive curve segments at a point P_i. Mathematically speaking the derivative of a spline is either continuous or discontinuous; the user however would like to have more control over the continuity than a simple binary switch. Discontinuities in a path are often required to simulate animation effects such as punching and bouncing. The

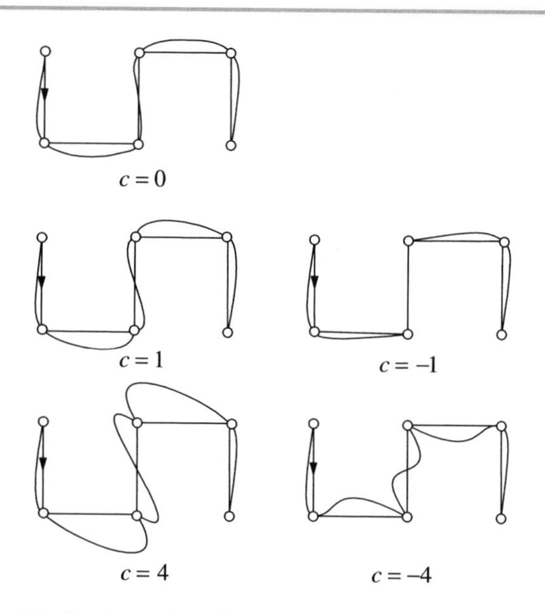

Figure 5.26 The middle two CVs of each curve have their continuity parameters set to the indicated values. All other CVs have their parameters set to the default values.

continuity parameter allows the source and destination tangent vectors to differ from each other:

$$^{S}R_i = \frac{(1-c)}{2} \cdot (P_i - P_{i-1}) + \frac{(1+c)}{2} \cdot (P_{i+1} - P_i) \tag{5.102}$$

$$^{D}R_i = \frac{(1+c)}{2} \cdot (P_i - P_{i-1}) + \frac{(1-c)}{2} \cdot (P_{i+1} - P_i). \tag{5.103}$$

With the default value $c = 0$, we obtain $^{S}R_i = {}^{D}R_i$, which produces a spline with tangent vector continuity at P_i.

As we increase $|c|$, the source and the destination tangent vectors at P_i diverge increasingly and the two neighboring curve segments are increasingly perceived, by the user, as more discontinuous (see Figures 5.26 and 5.27).

When $c = -1$, the source tangent vector $^{S}R_i$ becomes equal to the source chord $\overline{P_{i-1}P_i}$ and the destination tangent vector $^{D}R_i$ becomes equal to the destination chord $\overline{P_iP_{i+1}}$,

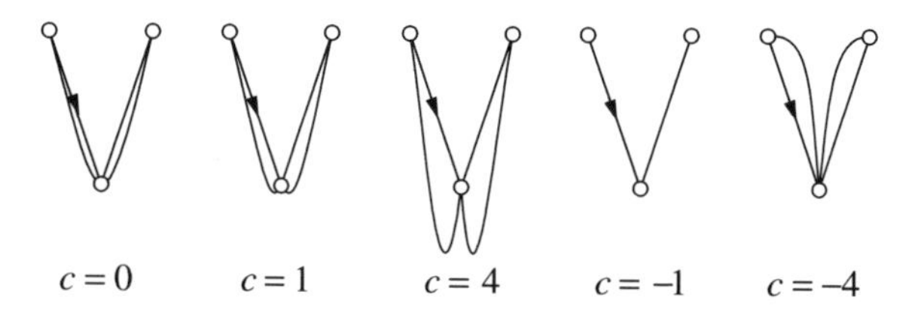

Figure 5.27 The middle CV of each curve has its continuity parameter set to the indicated value. All other CVs have their parameters set to the default values.

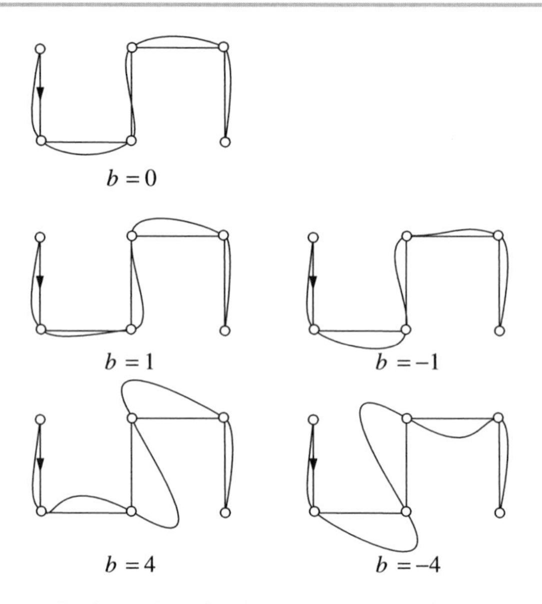

Figure 5.28 The middle two CVs of each curve have their bias parameters set to the indicated values. All other CVs have their parameters set to the default values.

introducing a sharp corner in the curve at P_i, as part of the source curve closely shadows the source chord and part of the destination curve shadows the destination chord.

If two consecutive control points have $c = -1$, then the curve segment joining them reduces to their chord (see Figures 5.26 and 5.30).

When $c = 1$, the source tangent vector ${}^{S}R_i$ becomes equal to the destination chord $\overline{P_iP_{i+1}}$ and the destination tangent vector ${}^{D}R_i$ becomes equal to the source chord $\overline{P_{i-1}P_i}$, introducing a sharp corner in the opposite direction (see Figures 5.26 and 5.27).

Increasing $|c|$ above 1 exaggerates the sharpness of the corner at P_i and introduces undulations to both curve segments meeting at this point (see Figures 5.26 and 5.27).

5.8.3.3 The Bias Parameter

The bias parameter b controls the direction of the path of the curve as it passes through the point P_i. Both source and destination tangent vectors are formed by averaging the source and destination chords, but the bias causes different weights to be used when averaging these chords:

$$ {}^{S}R_i = {}^{D}R_i = \frac{(1+b)}{2} \cdot (P_i - P_{i-1}) + \frac{(1-b)}{2} \cdot (P_{i+1} - P_i). \tag{5.104} $$

When $b = 0$, the two chords are weighted equally, and we obtain the default curve.

As the value of the bias parameter increases, the source chord becomes increasingly dominant in the determination of both tangent vectors. When $b = 1$, both tangent vectors are completely determined by the source chord, which causes the curve path to *overshoot* the control point. If the curve is interpreted as an animation path, then this allows us to *follow through* an action (see Figures 5.28 and 5.29).

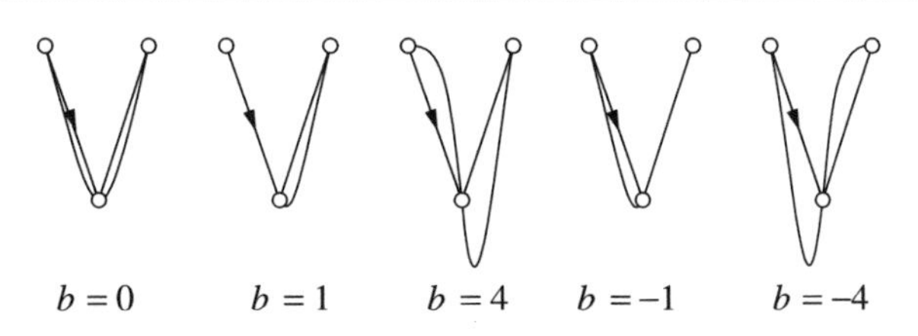

Figure 5.29 The middle CV of each curve has its bias parameter set to the indicated value. All other CVs have their parameters set to the default values.

As the value of the bias parameter decreases, the destination chord becomes increasingly dominant in the determination of both tangent vectors. When $b = -1$, both tangent vectors are completely determined by the destination chord, which causes the curve path to *undershoot* the control point. If the curve is interpreted as an animation path, then this allows us to *exaggerate* an action (see Figures 5.28 and 5.29).

Increasing $|b|$ above 1 exaggerates these effects, as can be seen in Figures 5.28 and 5.29.

5.8.3.4 Continuity Versus Tension

As mentioned earlier, if two consecutive control points have their parameters set to $t = 1$ and $c = 0$ or $t = 0$ and $c = -1$, then the curve segments joining these points reduce, on both occasions, to their chord. If these two distinct curve segments were sampled at equal parametric intervals however, they would yield points in space that are distributed quite differently, as shown in Figure 5.30. Observe that the curve on the left (case $t = 1$ and $c = 0$) shows an uneven distribution of sample points in its middle curve segment, while in the corresponding curve segment on the right (case $t = 0$ and $c = -1$) the sample points are evenly distributed.

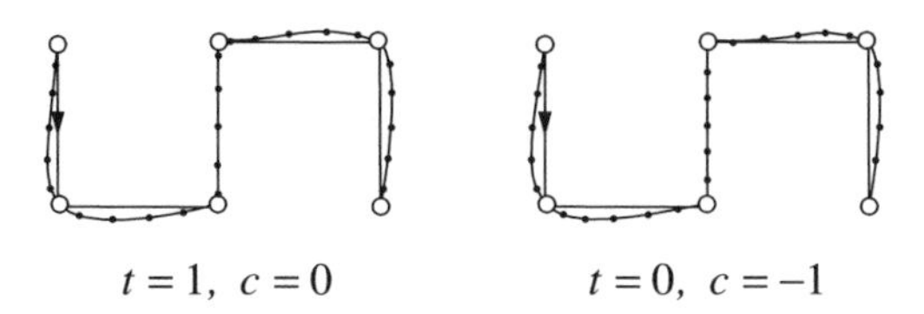

Figure 5.30 The middle two CVs on both curves are set to the indicated values. All other CVs have their parameters set to the default values.

5.8.3.5 Adjustments for the Parameter Step Size

If we assume default continuity ($c = 0$) at the control point P_i, the spline segment between P_{i-1} and P_i should join smoothly with the segment between P_i and P_{i+1}. Equations (5.99) and (5.100) for the incoming and outgoing tangent vectors ${}^{S}R_i$ and ${}^{D}R_i$ can only guarantee a smooth joint between these two curve segments if an equal number of samples is taken on both segments (i.e. an equal number of points is generated on both segments). This might be undesirable if the curve is used to interpolate key-frames, as it implies that there are an equal number of frames between the $(i - 1)$th and the ith key-frames, as well as the ith and the $(i+1)$th key-frames. To allow for different sample rates on the two curve segments meeting at P_i we must adjust the incoming and outgoing tangent vectors as follows:

$$
\begin{aligned}
{}^{D}\hat{R}_i &= {}^{D}R_i \cdot \frac{S_{i-1}}{\frac{1}{2}(S_{i-1}+S_i)} = {}^{D}R_i \cdot \frac{2 \cdot S_{i-1}}{(S_{i-1}+S_i)} \\
{}^{S}\hat{R}_i &= {}^{S}R_i \cdot \frac{S_i}{\frac{1}{2}(S_{i-1}+S_i)} = {}^{S}R_i \cdot \frac{2 \cdot S_i}{(S_{i-1}+S_i)},
\end{aligned}
\tag{5.105}
$$

where S_{i-1} is the number of samples (steps) between P_{i-1} and P_i, S_i is the number of samples (steps) between P_i and P_{i+1}, and $\frac{1}{2}(S_{i-1}+S_i)$ is the average number of samples (steps) between P_{i-1} and P_{i+1}.

5.8.3.6 Experimental Results

Kochanek and Bartels found in experiments that most animators, when asked to modify the control parameters until a spline passing through a set of key positions looked natural to them, left the continuity and bias parameters at their default values but reduced the tension parameter in the range -0.1 to -0.4. They further found that although there were differences between animators, individual animators consistently chose the same values for the three parameters.

5.9 Rational Cubic Polynomial Curves

In geometric modeling the two most important non-linear mathematical representations used to model curves and surfaces are the *implicit form* and the *polynomial form*.

The implicit form can represent concisely and exactly conic sections and primitive quadratic surfaces, such as spheres, cylinders and cones, but cannot represent free-form curves and surfaces, which are of great importance in geometric modeling.

Integral (i.e. non-rational) parametric polynomial forms, such as the Bézier and the B-spline forms, are ideally suited to representing and easily manipulating free-form curves and surfaces, but are incapable of representing exactly conic sections and primitive quadratic surfaces, which are also of great importance to geometric modeling.

The desire to develop a unified mathematical representation that was capable of dealing both with conic sections and quadratic primitives as well as with free-form curves and surfaces led to the development of rational polynomial parametric representations. Rational polynomial curves and surfaces were first introduced into the literature of the computer-aided design field by Steven Coons (Coons 1967).

In this section we examine very briefly rational cubic polynomial curves, with particular emphasis on the *rational Bézier* and *rational B-spline* forms.

Rational curves have two main advantages, over their non-rational counterparts, that have ensured their longevity and success. They are capable of representing conic sections exactly and they are invariant under *central projection* transformations (perspective transformations). Thus, the perspective projection of a rational curve of dimension n is itself a rational curve of dimension $(n-1)$.

The rational counterparts of non-rational cubic polynomial curves are developed using geometric constraints expressed in *homogeneous coordinates*.

5.9.1 Homogeneous Coordinates

In 3D Euclidean space both points and vectors are represented by three components and can only be distinguished by their context. They are, however, quite different. A vector has a magnitude and a direction, but no fixed position, while a point has a position but no magnitude or direction. In homogeneous space we have a way of distinguishing between points and vectors. In this space we represent a 3D point by the 4D homogeneous point $[w \cdot x \quad w \cdot y \quad w \cdot z \quad w]$ and a 3D vector by the homogeneous point $[x \quad y \quad z \quad 0]$:

$$P^h = \begin{cases} \begin{bmatrix} w \cdot x & w \cdot y & w \cdot z & w \end{bmatrix}, & w \neq 0 \\ \begin{bmatrix} x & y & z & 0 \end{bmatrix}, & w = 0. \end{cases} \tag{5.106}$$

The original point/vector P may be retrieved from its homogeneous representation P^h by using the *projective map* H, which is defined as

$$H([X \quad Y \quad Z \quad W]) = \begin{cases} [X/W \quad Y/W \quad Z/W], & W \neq 0 \\ \text{direction}\ ([X \quad Y \quad Z]), & W = 0. \end{cases} \tag{5.107}$$

Thus, the projective map H uses the origin as the center of projection to project the homogeneous point/vector P^h onto the hyperplane $W = 1$ in order to retrieve the 3D point/vector P:

$$P = H(P^h). \tag{5.108}$$

5.9.2 Rational Cubic Polynomial Curve Formulation

As we have seen in earlier sections, a non-rational 3D curve is defined by a set of 3D control points:

$$P_i = \begin{bmatrix} x_i & y_i & z_i \end{bmatrix}$$

and the curve itself is a vector-valued function of the form:

$$Q(t) = \begin{bmatrix} x(t) & y(t) & z(t) \end{bmatrix}.$$

A rational 3D curve is defined by a set of 4D homogeneous control points:

$$P_i^h = \begin{bmatrix} w_i \cdot x_i & w_i \cdot y_i & w_i \cdot z_i & w_i \end{bmatrix} \quad \text{where } w_i \neq 0, \tag{5.109}$$

which give rise to a homogeneous vector-valued function of the form

$$Q^h(t) = \begin{bmatrix} X(t) & Y(t) & Z(t) & W(t) \end{bmatrix}. \tag{5.110}$$

Applying the projective map to this function we obtain the 3D representation of the rational curve:

$$Q(t) = H(Q^h(t))$$

$$\therefore\ Q(t) = \begin{bmatrix} \frac{X(t)}{W(t)} & \frac{Y(t)}{W(t)} & \frac{Z(t)}{W(t)} \end{bmatrix} \tag{5.111}$$

or

$$Q(t) = \left[x(t) \ y(t) \ z(t)\right], \tag{5.112}$$

where

$$\begin{aligned} x(t) &= \frac{X(t)}{W(t)} \\ y(t) &= \frac{Y(t)}{W(t)} \\ z(t) &= \frac{Z(t)}{W(t)}. \end{aligned} \tag{5.113}$$

Any non-rational cubic polynomial curve can be transformed into its corresponding rational version by promoting its 3D vector-valued function into a 4D homogeneous vector-valued function by the addition of $W(t) = 1$. Thus,

$$Q(t) = [x(t) \ y(t) \ z(t)] \quad \Rightarrow \quad Q^h(t) = [x(t) \ y(t) \ z(t) \ 1]. \tag{5.114}$$

This can be achieved by expressing the geometric constraints that define the curve in homogeneous form. Care should be taken when doing this mapping, as 3D points and vectors should be treated differently, as we have seen in Section 5.9.1. A geometric constraint that represents a point should be mapped as

$$P_i = \left[x_i \quad y_i \quad z_i\right] \quad \Rightarrow \quad P_i^h = \left[x_i \quad y_i \quad z_i \quad 1\right]. \tag{5.115}$$

While a geometric constraint that represents a vector (such as a tangent) should be mapped as

$$P_i = \left[x_i \quad y_i \quad z_i\right] \quad \Rightarrow \quad P_i^h = \left[x_i \quad y_i \quad z_i \quad 0\right]. \tag{5.116}$$

Using the above mappings we can derive rational representations for any cubic polynomial curve.

Care should be taken when evaluating rational curve functions, as two special circumstances may cause the evaluation to yield a value of $W(t) = 0$, which produces a point at infinity; an asymptotic result that leads to loss of geometric intuition, not to mention a division by zero in the drawing routine. These circumstances may occur either when some w_i are allowed to be negative and some positive, or when some of the basis functions of the curve are negative for some portion of the parameter range over which the curve is defined. In such cases, we must prevent a division by zero in the drawing routine by setting $W(t)$ to an arbitrary value.

Rational curves are useful for three reasons:

(1) They are invariant under rotation, scaling, translation and perspective transformations of their control points. Non-rational curves are not invariant under the perspective transformation.

(2) They can define any of the conic sections precisely, unlike non-rational curves.

(3) They provide extra control through the weights w_i. When the value of w_i is increased above 1 the curve is pulled towards the control point P_i and when its value is decreased below 1 the curve moves away from the control point. When all the weights w_i are set to 1 the rational curve behaves like its non-rational equivalent. Thus, rational cubic polynomial curves are a generalization of their non-rational equivalents.

5.10 Parametric Surfaces

The motivation that led various researchers to develop mathematical representations for free-form curves also led them to develop related representations for free-form curved surfaces. In general, a mathematical representation of a shape, be it a curve or a surface, has two distinct forms, an analytical form and a synthetic form.

The analytical form of a shape representation is ideally suited to describing shapes that can be measured and quantified. The objective of such a representation may be to achieve an accurate fit to a given set of data points, to minimize the number of measurements needed to describe the shape accurately, to represent the shape in a concise form or to ease the computation of various shape attributes such as the surface area of a shape, its volume or the curvature of its surface.

The synthetic form of a shape representation, on the other hand, is desirable during the shape design process, where a designer interactively creates and modifies a shape, examining and refining the design until the desired form is achieved. The resulting shape model may then be used to render images of the object or to generate instructions for a numerically controlled milling machine that can manufacture the precise shape of the object.

These two forms of the shape representation have led to the development of two distinct philosophies that have influenced the design of surface description and construction methods.

The first philosophy is descended from the work of Steven Coons (1964, 1967) known as the "*little red book*" – and William Gordon (1969), where the surface is generated from a set of known geometric data. The user (designer) thinks in terms of *feature curves* rather than surfaces. These feature curves interpolate known 3D data points and the surface is generated by interpolating a network of feature curves. This approach appeals to users who rely on known geometric data used as constraints in the determination of the surface shape. Such surfaces are collectively known as *interpolating surfaces* and are usually constructed using a *transfinite surface* representation.

The second philosophy is descended from the work of Paul de Casteljau (1963, 1986) and Pierre Bézier (1966, 1967, 1986), where the construction of a surface is guided by a set of geometric data. The Bézier and B-spline methods use a net of control vertices to generate a surface. This approach appeals to users (designers) who rely on visual and aesthetic factors to determine the shape of the surface. Such surfaces are collectively known as *approximating surfaces* and are constructed using a *tensor-product surface* representation. Some approximating surfaces can also be constructed using a tensor-product surface representation.

Roughly speaking, a tensor-product surface can be constructed by a free-form *generator* curve which, as it moves and deforms in space, describes the surface. If we trace a point (at a fixed parameter value) on the generator curve, we see that it describes a curve called the *directrix* curve, since it describes the direction of movement of this point. Such a surface can be seen as being composed of a collection of connected curved patches described by the curve segments of the generator curve and the corresponding directrix curves. A transfinite surface, on the other hand, is defined as a collection of connected surface patches, each of which is defined by four curves describing its boundaries and by the four second-derivative vectors of the surface patch at its corners, which are known as the *twist vectors*. In both the tensor-product and transfinite cases, a surface can be pieced together in a patchwork-quilt-like fashion by joining a collection of surface patches under certain continuity constraints. Thus, such surfaces are often referred to *piecewise parametric surfaces*.

Since the early 1960s a variety of surface representations were introduced for interpolating and approximating surfaces. By their method of construction, free-form parametric surfaces can be categorized as tensor-product or transfinite, as integral or rational and as being constructed of rectangular, triangular or n-sided patches.

The earliest recorded example of a free-form parametric surface was a surface based on triangular patches “surface à pôles” (surfaces with control points) by Paul de Casteljau (1959). De Casteljau is also credited with having constructed the first known example of a tensor-product surface using a rectangular array of control points (de Casteljau 1963). Steven Coons (1964, 1967) is known to have developed the first transfinite surface representation emulating a technique known as *lofting* in pre-computer-aided geometric design days.

In our discussion of free-form parametric surfaces we will concentrate on the construction of bicubic patches and surfaces.

5.10.1 Piecewise Parametric Bicubic Surfaces

In this subsection we examine the construction process for *piecewise parametric bicubic surfaces*, which is a natural and straightforward generalization of parametric cubic curves. As we have seen in previous sections, a cubic curve is formed by piecing together successive curve segments under certain continuity constraints. Here we construct a bicubic surface by piecing together rectangular surface patches to form a composite surface in the way that one constructs a patchwork quilt. The surface is formed as the scaled sum of a topologically rectangular array of control points known as the *control mesh*, *control hull* or *control graph*.

Recall that the general form of the parametric cubic curve is given by

$$Q(s) = S \cdot M \cdot G \quad \text{where } 0 \leq s \leq 1. \tag{5.117}$$

S is the parameter vector (a variable), M is a cubic basis matrix (a constant) and G is the geometry vector (a constant).

Starting with $Q(s)$ as our generator curve, we allow the points in its geometry vector G to vary in 3D space along some path that is parameterized on t, resulting in a function of two parameters:

$$Q(s, t) = S \cdot M \cdot G(t) \quad \text{where } 0 \leq s, t \leq 1 \tag{5.118}$$

or

$$Q(s, t) = S \cdot M \cdot \begin{bmatrix} G_1(t) \\ G_2(t) \\ G_3(t) \\ G_4(t) \end{bmatrix}. \tag{5.119}$$

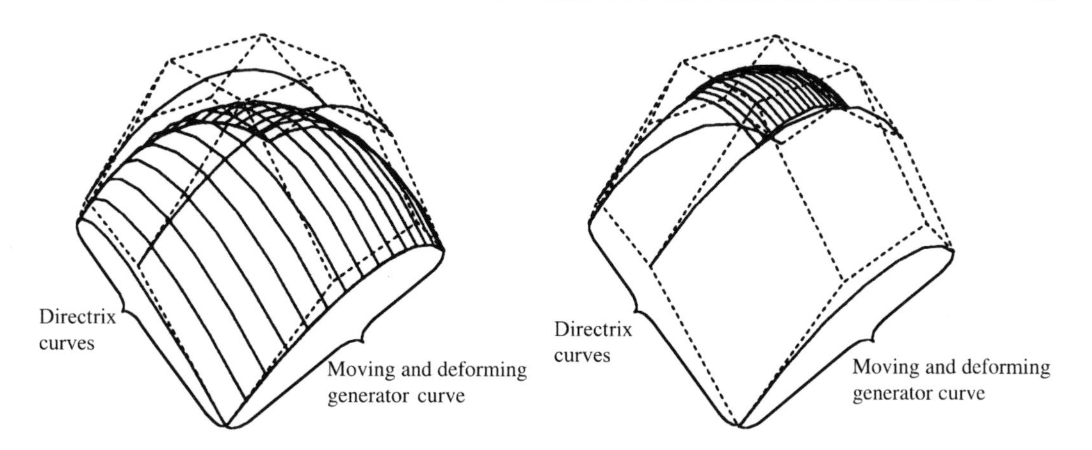

Figure 5.31 The directrix and *generator* curves of a cubic Bézier (left) and B-spline (right) surface patch.

Now, for a fixed value of the parameter t, say $t = t_1$, this new function $Q(s, t_1)$ is a curve, since $G(t_1)$ is constant. Allowing t to assume some new value, say $t = t_2$, such that $\partial t = (t_2 - t_1)$ is very small, $Q(s, t_2)$ is a slightly different curve. Repeating this operation a large number of times for values of $t \in [0, 1]$, we define a family of curves arbitrarily close to each other. The set of all such curves defines a surface patch. If the directrix curves $G_i(t)$ are themselves cubics, then the surface patch is said to be a *parametric bicubic surface patch* (see Figure 5.31).

Each curve $G_i(t)$ can be expressed as

$$G_i(t) = T \cdot M \cdot \mathbf{G}_i \tag{5.120}$$

where

$$\mathbf{G}_i = \begin{bmatrix} G_{i1} \\ G_{i2} \\ G_{i3} \\ G_{i4} \end{bmatrix}. \tag{5.121}$$

Thus $\mathbf{G}_i$ is the geometry vector of the directrix curve $G_i(t)$.

Given the matrix identity $(A \cdot B \cdot C)^T = C^T \cdot B^T \cdot A^T$, we can transpose Equation (5.120) to obtain

$$G_i(t)^T = \mathbf{G}_i^T \cdot M^T \cdot T^T \tag{5.122}$$

$$\therefore\ G_i(t)^T = \begin{bmatrix} G_{i1} & G_{i2} & G_{i3} & G_{i4} \end{bmatrix} \cdot M^T \cdot T^T. \tag{5.123}$$

Since $G_i(t)^T$ and $G_i(t)$ are two different ways of denoting the same point, using Equation (5.122) we can rewrite Equation (5.119) as

$$Q(s, t) = S \cdot M \cdot \begin{bmatrix} G_1(t) \\ G_2(t) \\ G_3(t) \\ G_4(t) \end{bmatrix} = S \cdot M \cdot \begin{bmatrix} \mathbf{G}_1^T \cdot M^T \cdot T^T \\ \mathbf{G}_2^T \cdot M^T \cdot T^T \\ \mathbf{G}_3^T \cdot M^T \cdot T^T \\ \mathbf{G}_4^T \cdot M^T \cdot T^T \end{bmatrix}$$

$$\therefore\ Q(s,t) = S \cdot M \cdot \begin{bmatrix} \mathbf{G}_1^T \\ \mathbf{G}_2^T \\ \mathbf{G}_3^T \\ \mathbf{G}_4^T \end{bmatrix} \cdot M^T \cdot T^T.$$

Using Equation (5.123) we can rewrite the above result as

$$Q(s,t) = S \cdot M \cdot \begin{bmatrix} G_{11} & G_{12} & G_{13} & G_{14} \\ G_{21} & G_{22} & G_{23} & G_{24} \\ G_{31} & G_{32} & G_{33} & G_{34} \\ G_{41} & G_{42} & G_{43} & G_{44} \end{bmatrix} \cdot M^T \cdot T^T \tag{5.124}$$

or

$$Q(s,t) = S \cdot M \cdot \mathbf{G} \cdot M^T \cdot T^T \quad \text{where } 0 \le s, t \le 1. \tag{5.125}$$

Which written separately for the x, y, z components, gives:

$$\begin{aligned} x(s,t) &= S \cdot M \cdot \mathbf{G}_x \cdot M^T \cdot T^T \\ y(s,t) &= S \cdot M \cdot \mathbf{G}_y \cdot M^T \cdot T^T \\ z(s,t) &= S \cdot M \cdot \mathbf{G}_z \cdot M^T \cdot T^T, \end{aligned} \tag{5.126}$$

where $\mathbf{G}$ is the geometry matrix of the patch (i.e. a square control mesh of $(4 \times 4) = 16$ control points), $\mathbf{G}_x$, $\mathbf{G}_y$, $\mathbf{G}_z$ are the x, y, z components of the geometry matrix, respectively, and M is a cubic basis matrix. It should be observed that no restrictions are placed on this basis matrix. We could even select different basis matrices for the s and t parametric directions. Thus, Equation (5.125) could be rewritten as

$$Q(s,t) = S \cdot M_s \cdot \mathbf{G} \cdot M_t^T \cdot T^T \quad \text{where } 0 \le s, t \le 1 \tag{5.127}$$

and where M_s and M_t are the basis matrices in the s and t parametric directions, respectively.

To deal with rational surface patches we assume that each control point is of the form:

$$G_{ij} = \begin{bmatrix} w_{ij} \cdot x_{ij} & w_{ij} \cdot y_{ij} & w_{ij} \cdot z_{ij} & w_{ij} \end{bmatrix}$$

and by extending Equation (5.127) to deal with the fourth coordinate, we obtain:

$$\begin{aligned} X(s,t) &= S \cdot M \cdot \mathbf{G}_x \cdot M^T \cdot T^T \\ Y(s,t) &= S \cdot M \cdot \mathbf{G}_y \cdot M^T \cdot T^T \\ Z(s,t) &= S \cdot M \cdot \mathbf{G}_z \cdot M^T \cdot T^T \\ W(s,t) &= S \cdot M \cdot \mathbf{G}_w \cdot M^T \cdot T^T. \end{aligned} \tag{5.128}$$

Applying the projective map to these functions, as we have done in the case of rational curves, we obtain the 3D representation of the rational surface patch as

$$\begin{aligned} x(s,t) &= \frac{X(s,t)}{W(s,t)} \\ y(s,t) &= \frac{Y(s,t)}{W(s,t)} \\ z(s,t) &= \frac{Z(s,t)}{W(s,t)}. \end{aligned} \tag{5.129}$$

5.10.1.1 Bicubic Bézier Surface Patches

By setting the basis matrix in Equation (5.125) to the cubic Bézier basis matrix M_B, we derive the formulation for a bicubic Bézier surface patch:

$$Q(s, t) = S \cdot M_B \cdot \mathbf{G}_B \cdot M_B^T \cdot T^T \quad \text{where } 0 \leq s, t \leq 1 \tag{5.130}$$

or in component form:

$$\begin{aligned} x(s, t) &= S \cdot M_B \cdot \mathbf{G}_{Bx} \cdot M_B^T \cdot T^T \\ y(s, t) &= S \cdot M_B \cdot \mathbf{G}_{By} \cdot M_B^T \cdot T^T \\ z(s, t) &= S \cdot M_B \cdot \mathbf{G}_{Bz} \cdot M_B^T \cdot T^T, \end{aligned} \tag{5.131}$$

where $\mathbf{G}_B$ is the Bézier geometry matrix, i.e. a square control mesh of $(4 \times 4) = 16$ control points, $\mathbf{G}_{Bx}$, $\mathbf{G}_{By}$, $\mathbf{G}_{Bz}$ are the x, y, z components of the Bézier geometry matrix, respectively, and M_B is the cubic Bézier basis matrix, which is defined as

$$M_B = \begin{bmatrix} -1 & 3 & -3 & 1 \\ 3 & -6 & 3 & 0 \\ -3 & 3 & 0 & 0 \\ 1 & 0 & 0 & 0 \end{bmatrix}.$$

Figure 5.32 depicts a bicubic Bézier surface patch generated in this manner.

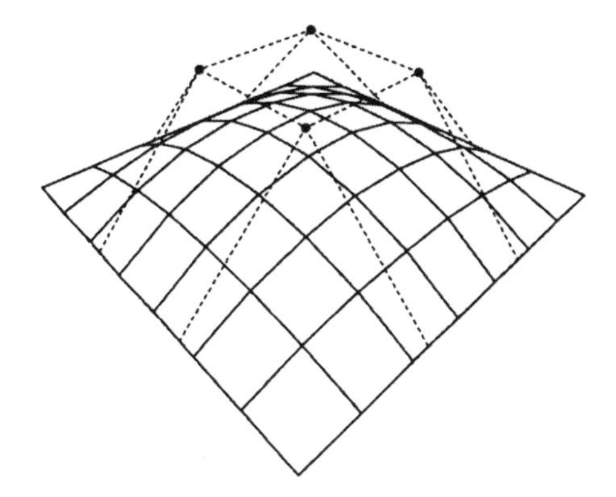

Figure 5.32 A bicubic Bézier surface patch.

5.10.1.2 Bicubic B-spline Surface Patches

Similarly, by setting the basis matrix in Equation (5.125) to the cubic B-spline basis matrix M_{BS}, we derive the formulation for a bicubic B-spline surface patch:

$$Q(s, t) = S \cdot M_{BS} \cdot \mathbf{G}_{BS} \cdot M_{BS}^T \cdot T^T \quad \text{where } 0 \leq s, t \leq 1 \tag{5.132}$$

or in component form:

$$\begin{aligned} x(s,t) &= S \cdot M_{BS} \cdot \mathbf{G}_{BSx} \cdot M_{BS}^T \cdot T^T \\ y(s,t) &= S \cdot M_{BS} \cdot \mathbf{G}_{BSy} \cdot M_{BS}^T \cdot T^T \\ z(s,t) &= S \cdot M_{BS} \cdot \mathbf{G}_{BSz} \cdot M_{BS}^T \cdot T^T, \end{aligned} \tag{5.133}$$

where $\mathbf{G}_{BS}$ is the B-spline geometry matrix, i.e. a square control mesh of $(4 \times 4) = 16$ control points, $\mathbf{G}_{BSx}$, $\mathbf{G}_{BSy}$, $\mathbf{G}_{BSz}$ are the x, y, z components of the B-spline geometry matrix, respectively, and M_{BS} is the cubic B-spline basis matrix, which is defined as

$$M_{BS} = \frac{1}{6} \cdot \begin{bmatrix} -1 & 3 & -3 & 1 \\ 3 & -6 & 3 & 0 \\ -3 & 0 & 3 & 0 \\ 1 & 4 & 1 & 0 \end{bmatrix}.$$

Figure 5.33 depicts a bicubic B-spline surface patch generated in this manner.

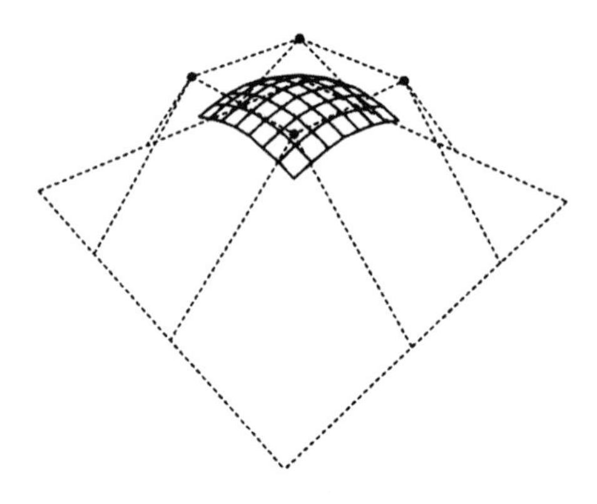

Figure 5.33 A bicubic B-spline surface patch.

5.11 Drawing Cubic Curves and Surfaces

In this section we examine ways of drawing cubic curves, by piecing together a collection of cubic curve segments, and bicubic surfaces, by piecing together a collection of bicubic surface patches. We provide a set of simple functions written in the C programming language. In order to de-couple the curve and surface routines from the graphics package in use we will make the following simplifying assumptions. We will assume that the viewing transformation has been set-up and communicated to the graphics package. Furthermore, we will assume that the graphics package in use has a routine capable of displaying a homogeneous line segment of the form

$$\overrightarrow{P_1^h P_2^h} \quad \text{where } P_i^h = \left[x_i, y_i, z_i, w_i\right].$$

Thus we assume the existence of a function:

```
draw_homogeneous_line(x1, y1, z1, w1, x2, y2, z2, w2);
```

Such a function is equivalent to the following sequence of OpenGL function calls:

```
GlBegin(CL_LINES);
GlVertex4d(x1, y1, z1, w1);
GlVertex4d(x2, y2, z2, w2);
GlEnd();
```

For brevity we will assume the existence of the following C language macro definition which performs matrix multiplication:

```
/*
 * Macro Matrix_Multiply: M3 = M1 * M2
 */

#define Matrix_Multiply(M1, M2, M3)                      \
{                                                        \
 int r, c, i;                                            \
                                                         \
 for (r = 0; r < 4; r++)                                 \
 for (c = 0; c < 4; c++)                                 \
  {                                                      \
   (M3)[r][c] = 0.0;                                     \
                                                         \
   for (i = 0; i < 4; i++)                               \
   (M3)[r][c] += ((M1)[r][i] * (M2)[i][c]);              \
  }                                                      \
}
```

5.11.1 Drawing a Cubic Curve Segment

We can draw a cubic curve segment using a simple *brute-force* technique or using a more sophisticated and much more efficient method known as the *forward difference* technique. Let us start by examining the brute-force technique, as it is easier to understand and to implement.

5.11.1.1 Drawing a Cubic Curve Segment with the Brute-force Technique

The following C function draws a cubic curve segment defined by four geometric constraints:

$$Q_i(t) = \begin{bmatrix} t^3 & t^2 & t & 1 \end{bmatrix} \cdot M \cdot \begin{bmatrix} x_i & y_i & z_i & w_i \\ x_{i+1} & y_{i+1} & z_{i+1} & w_{i+1} \\ x_{i+2} & y_{i+2} & z_{i+2} & w_{i+2} \\ x_{i+3} & y_{i+3} & z_{i+3} & w_{i+3} \end{bmatrix} \quad \text{for } 0 \le t \le 1.$$

On each iteration of the *for* loop this routine draws one line segment of the curve segment. Points on the curve segment are evaluated at the parameter values: $t = \left\{\frac{0}{n}, \frac{1}{n}, \frac{2}{n}, \ldots, \frac{n}{n}\right\}$.

```
typedef double matrix_3d_t[4][4];


void draw_cubic_curve_segment(M, G, n)

matrix_3d_t M; /* Basis Matrix of the curve                         */
matrix_3d_t G; /* Geometry Vector of the curve                      */
int         n; /* Number of line segments on the curve segment */
{
 matrix_3d_t C; /* Coefficient Matrix */
 double      t1, t2, t3, dt,
             x1, y1, z1, w1,
             x2, y2, z2, w2;
 int         ci;


 dt = 1.0 / n;

/* Compute_Coefficient_Matrix: C = M * G */

Matrix_Multiply(M, G, C);

x1 = C[3][0]; y1 = C[3][1]; z1 = C[3][2]; w1 = C[3][3];

if (!(fabs(w1) > 0) w1 = 1;

for (ci = 1; ci <= n; ci++)
 {
  t1 = ci * dt;
  t2 = t1 * t1;
  t3 = t2 * t1;

  x2 = t3 * C[0][0] + t2 * C[1][0] + t1 * C[2][0] + C[3][0];
  y2 = t3 * C[0][1] + t2 * C[1][1] + t1 * C[2][1] + C[3][1];
  z2 = t3 * C[0][2] + t2 * C[1][2] + t1 * C[2][2] + C[3][2];
  w2 = t3 * C[0][3] + t2 * C[1][3] + t1 * C[2][3] + C[3][3];

  if (!(fabs(w2) > 0) w2 = 1;

  draw_homogeneous_line(x1, y1, z1, w1, x2, y2, z2, w2);

  x1 = x2; y1 = y2; z1 = z2; w1 = w2;
 }
} /* draw_cubic_curve_segment */
```

5.11.1.2 Drawing a Cubic Curve Segment with the Forward Difference Technique

(See the Appendix to this chapter for a more detailed discussion on *forward differences*).

First we build the *precision matrix*, P, from the precision, n, of the curve segment:

$$P = \begin{bmatrix} \frac{6}{n^3} & 0 & 0 & 0 \\ \frac{6}{n^3} & \frac{2}{n^2} & 0 & 0 \\ \frac{1}{n^3} & \frac{1}{n^2} & \frac{1}{n} & 0 \\ 0 & 0 & 0 & 1 \end{bmatrix}.$$

Then we build a matrix F from the product of the *precision matrix*, the *basis matrix* and the *geometry vector*:

$$F = P \cdot M \cdot G.$$

The bottom row of the matrix F identifies the first of the $(n+1)$ sample points on the curve segment. To generate the remaining sample points on the curve segment, the following algorithm is used to iterate the matrix F as a *forward difference matrix*. The third row of F is added to its fourth row, its second row is added to its third row, its first row is added to its second row and finally its fourth row is output as the next sample point on the curve. This process is repeated n times.

Note that if the *precision matrix*, P, were iterated as a *forward difference matrix* it would generate the following vectors:

$$\left[\left(\frac{0}{n}\right)^3 \ \left(\frac{0}{n}\right)^2 \ \left(\frac{0}{n}\right)^1 \ 1\right], \left[\left(\frac{1}{n}\right)^3 \ \left(\frac{1}{n}\right)^2 \ \left(\frac{1}{n}\right)^1 \ 1\right],$$
$$\left[\left(\frac{2}{n}\right)^3 \ \left(\frac{2}{n}\right)^2 \ \left(\frac{2}{n}\right)^1 \ 1\right], \ldots$$

which is the same sequence of vectors as generated by varying $t = \left\{\frac{0}{n}, \frac{1}{n}, \frac{2}{n}, \ldots, \frac{n}{n}\right\}$ in the parameter vector $[t^3 \ \ t^2 \ \ t \ \ 1]$.

By iterating matrix F for each of the parameter values $t = \left\{\frac{0}{n}, \frac{1}{n}, \frac{2}{n}, \ldots, \frac{n}{n}\right\}$ we obtain $[t^3 \ \ t^2 \ \ t \ \ 1] \cdot M \cdot G$, which are the sample points on the curve segment.

The following C function draws a curve segment defined by four geometric constraints. On each iteration of the *for* loop it draws one line segment of the curve segment.

```
typedef double matrix_3d_t[4][4];


void draw_curve_segment_forward_difference(M, G, n)

matrix_3d_t M, /* Basis Matrix of the curve                    */
            G; /* Geometry Vector of the curve                 */
int         n; /* Number of line segments in the curve segment */

{
```

```
/*
 * This function draws a Cubic Curve Segment using Forward Differences.
 */

matrix_3d_t P, /* Precision Matrix          */
            B, /* Precision by Basis Matrix */
            F; /* Forward Difference Matrix */

double      x1, y1, z1, w1,
            x2, y2, z2, w2;
int         r, c, i, s;


/*
 * Setup the Precision Matrix.
 */

P[0][0] = 6.0/(n*n*n); P[0][1] = 0.0;       P[0][2] = 0.0;   P[0][3] = 0.0;
P[1][0] = 6.0/(n*n*n); P[1][1] = 2.0/(n*n); P[1][2] = 0.0;   P[1][3] = 0.0;
P[2][0] = 1.0/(n*n*n); P[2][1] = 1.0/(n*n); P[2][2] = 1.0/n; P[2][3] = 0.0;
P[3][0] = 0.0;         P[3][1] = 0.0;       P[3][2] = 0.0;   P[3][3] = 1.0;

/* Concatenate the precision and basis matrices */

Matrix_Multiply(P, M, B);

/* Compute the forward difference matrix */

Matrix_Multiply(B, G, F);

/* Get the start point on the curve segment */

x1 = F[3][0]; y1 = F[3][1]; z1 = F[3][2]; w1 = F[3][3];

if (!(fabs(w1) > 0) w1 = 1;

for (s = 0; s < n; s++)
 {
  /* Iterate the Forward Difference Matrix */

  for (r = 3; r > 0; r--)
  for (c = 0; c < 4; c++)
   F[r][c] += F[r-1][c];

   x2 = F[3][0];
   y2 = F[3][1];
   z2 = F[3][2];
   w2 = F[3][3];

   if (!(fabs(w2) > 0) w2 = 1;

   draw_homogeneous_line(x1, y1, z1, w1, x2, y2, z2, w2);

  x1 = x2; y1 = y2; z1 = z2; w1 = w2;
 }
} /* draw_curve_segment_forward_difference */
```

5.11.2 Drawing a Cubic Curve

The following C function draws a cubic curve by piecing together a number of cubic curve segments. On each iteration of the outer *for* loop it builds the geometry matrix of the corresponding curve segment and calls the curve segment drawing routine.

```
typedef double vector_4d_t[4];


void draw_cubic_curve(cv_array, ncv, M, nc)

vector_4d_t cv_array[]; /* Control Vertex Array of the curve              */
int         ncv;        /* Number of Control Vertices                     */
matrix_3d_t M;          /* Basis Matrix of the curve                      */
int         n;          /* Number of line segments on each curve segment */

{
 matrix_3d_t G;    /* Geometry Matrix of curve segment       */
 int         ncs,  /* Number of Curve Segments in the Curve */
             ci,   /* Curve Segment Index                    */
             r, c;


 ncs = (ncv - 1) - 4 + 2;

 for (ci = 0; ci < ncs; ci ++)
  {
   for (r = 0; r < 4; r++)
   for (c = 0; c < 4; c++)
    G[r][c]=cv_array[r+ci][c];

   draw_cubic_curve_segment_forward_difference (M, G, n);
  }
} /* draw_cubic_curve */
```

5.11.3 Drawing a Bicubic Surface Patch

The following C function draws a bicubic surface patch by drawing a set of isoparametric cubic curve segments in the *s* and *t* parametric directions.

```
typedef double matrix_3d_t[4][4];
typedef double vector_4d_t[4];
typedef vector_4d_t geometry_matrix_t[4][4];

/* Macro Compute_Coefficient_Matrix: Ci = Ms * Gi * transpose(Mt) */

#define Compute_Coefficient_Matrix(Ms, Mt, Gi, Ci, i) \
{                                                     \
 matrix_3d_t T; /* Temporary Matrix */                \
 int         r1,                                      \
             r2,                                      \
             c1,                                      \
             c2;                                      \
```

```
                                                                \
for (r1 = 0; r1 < 4; r1++)                                      \
for (c2 = 0; c2 < 4; c2++)                                      \
  {                                                             \
   T[r1][c2] = 0;                                               \
                                                                \
   for (c1 = 0; c1 < 4; c1++)                                   \
    T[r1][c2] += ((Ms)[r1][c1] * (Gi)[c1][c2][(i)]);            \
 }                                                              \
                                                                \
for (r1 = 0; r1 < 4; r1++)                                      \
for (c2 = 0; c2 < 4; c2++)                                      \
 {                                                              \
  (Ci)[r1][c2] = 0;                                             \
                                                                \
  for (c1 = 0; c1 < 4; c1++)                                    \
   (Ci)[r1][c2] += ((T)[r1][c1] * (Mt)[c2][c1]);                \
 }                                                              \
}


void draw_bicubic_surface_patch(G, Ms, Mt, ns, nt, nl)

geometry_matrix_t G;  /* Geometry Matrix of the Surface Patch            */
matrix_3d_t       Ms; /* Basis Matrix of the Surface in the s direction */
matrix_3d_t       Mt; /* Basis Matrix of the Surface in the t direction */
int               ns; /* Number of curves of constant s to be drawn     */
int               nt; /* Number of curves of constant t to be drawn     */
int               nl; /* Number of Line Segments in each Curve Segment  */

{
 matrix_3d_t Cx; /* Coefficient Matrix for the x component */
 matrix_3d_t Cy; /* Coefficient Matrix for the y component */
 matrix_3d_t Cz; /* Coefficient Matrix for the z component */
 matrix_3d_t Cw; /* Coefficient Matrix for the w component */

 vector_4d_t Rx,
             Ry,
             Rz,
             Rw;


 int         i, si, ti, li;

 double      ds, s1, s2, s3,
             dt, t1, t2, t3,
             dl,
             x1, y1, z1, w1,
             x2, y2, z2, w2;

 /*
  * Compute the Coefficient Matrices:
  *
  * Cx = Ms * Gx * transpose(Mt)
  * Cy = Ms * Gy * transpose(Mt)
  * Cz = Ms * Gz * transpose(Mt)
```

```
 * Cw = Ms * Gw * transpose(Mt)
 */

Compute_Coefficient_Matrix(Ms, Mt, G, Cx, 0);
Compute_Coefficient_Matrix(Ms, Mt, G, Cy, 1);
Compute_Coefficient_Matrix(Ms, Mt, G, Cz, 2);
Compute_Coefficient_Matrix(Ms, Mt, G, Cw, 3);

ds = 1.0 / (ns - 1);
dt = 1.0 / (nt - 1);
dl = 1.0 / (nl - 1);

/* Draw ns curves of constant s, varying t=[0..1] */

for (si = 0; si < ns; si++)
 {
  s1 = si * ds;
  s2 = s1 * s1;
  s3 = s2 * s1;

  /*
   * Compute the products:
   *
   * Rx = s * Cx
   * Ry = s * Cy
   * Rz = s * Cz
   * Rw = s * Cw
   */

  for (i = 0; i < 4; i++)
   {
    Rx[i] = s3 * Cx[0][i] + s2 * Cx[1][i] + s1 * Cx[2][i] + Cx[3][i];
    Ry[i] = s3 * Cy[0][i] + s2 * Cy[1][i] + s1 * Cy[2][i] + Cy[3][i];
    Rz[i] = s3 * Cz[0][i] + s2 * Cz[1][i] + s1 * Cz[2][i] + Cz[3][i];
    Rw[i] = s3 * Cw[0][i] + s2 * Cw[1][i] + s1 * Cw[2][i] + Cw[3][i];
   }

  x1 = Rx[3]; y1 = Ry[3]; z1 = Rz[3]; w1 = Rw[3];

  if (!(fabs(w1) > 0) w1 = 1;

  for (li = 0; li < nl; li++)
   {
    t1 = li * dl;
    t2 = t1 * t1;
    t3 = t2 * t1;

    x2 = Rx[0] * t3 + Rx[1] * t2 + Rx[2] * t1 + Rx[3];
    y2 = Ry[0] * t3 + Ry[1] * t2 + Ry[2] * t1 + Ry[3];
    z2 = Rz[0] * t3 + Rz[1] * t2 + Rz[2] * t1 + Rz[3];
    w2 = Rw[0] * t3 + Rw[1] * t2 + Rw[2] * t1 + Rw[3];

    if (!(fabs(w2) > 0) w2 = 1;

    draw_homogeneous_line(x1, y1, z1, w1, x2, y2, z2, w2);
```

```
      x1 = x2; y1 = y2; z1 = z2; w1 = w2;
     }
   }

 /* Draw nt curves of constant t, varying s=[0..1] */

 for (ti = 0; ti < nt; ti++)
   {
    t1 = ti * dt;
    t2 = t1 * t1;
    t3 = t2 * t1;

    /*
     * Compute the products:
     *
     * Rx = Cx * t
     * Ry = Cy * t
     * Rz = Cz * t
     * Rw = Cw * t
     */

    for (i = 0; i < 4; i++)
     {
      Rx[i] = Cx[i][0] * t3 + Cx[i][1] * t2 + Cx[i][2] * t1 + Cx[i][3];
      Ry[i] = Cy[i][0] * t3 + Cy[i][1] * t2 + Cy[i][2] * t1 + Cy[i][3];
      Rz[i] = Cz[i][0] * t3 + Cz[i][1] * t2 + Cz[i][2] * t1 + Cz[i][3];
      Rw[i] = Cw[i][0] * t3 + Cw[i][1] * t2 + Cw[i][2] * t1 + Cw[i][3];
    }

    x1 = Rx[3]; y1 = Ry[3]; z1 = Rz[3]; w1 = Rw[3];

    if (!(fabs(w1) > 0) w1 = 1;

    for (li = 0; li < nl; li++)
     {
      s1 = li * dl;
      s2 = s1 * s1;
      s3 = s2 * s1;

      x2 = s3 * Rx[0] + s2 * Rx[1] + s1 * Rx[2] + Rx[3];
      y2 = s3 * Ry[0] + s2 * Ry[1] + s1 * Ry[2] + Ry[3];
      z2 = s3 * Rz[0] + s2 * Rz[1] + s1 * Rz[2] + Rz[3];
      w2 = s3 * Rw[0] + s2 * Rw[1] + s1 * Rw[2] + Rw[3];

      if (!(fabs(w2) > 0) w2 = 1;

      draw_homogeneous_line(x1, y1, z1, w1, x2, y2, z2, w2);

      x1 = x2; y1 = y2; z1 = z2; w1 = w2;
     }
   }
} /* draw_bicubic_surface_patch */
```

5.11.4 Drawing a Bicubic Surface

The following C function draws a bicubic surface by piecing together a number of bicubic surface patches. On each iteration of the two outer *for* loops it builds the geometry matrix of the corresponding surface patch and calls the surface patch drawing routine (see Figures 5.34 and 5.35).

```
typedef vector_4d_t control_grid_t[GRID_ROWS][GRID_COLUMNS];


void draw_bicubic_surface(control_grid, nr, nc, Ms, Mt, ns, nt, nl)

control_grid_t control_grid;/* Control Grid of the Surface              */
int            nr;          /* Number of Rows in Control Grid           */
int            nc;          /* Number of Columns in Control Grid        */
matrix_3d_t    Ms;          /* Basis Matrix in the s direction          */
matrix_3d_t    Mt;          /* Basis Matrix in the t direction          */
int            ns;          /* Number of Curves of constant s           */
int            nt;          /* Number of Curves of constant t           */
int            nl;          /* Number of Line Segments in a Curve Segment */

{
 geometry_matrix_t G;     /* Geometry Matrix of the Surface Patch     */
 int               ncs_s; /* Number of Curve Segments in s Direction */
 int               ncs_t; /* Number of Curve Segments in t Direction */
 int               ci_s;  /* Curve Segment Index      in s Direction */
 int               ci_t;  /* Curve Segment Index      in t Direction */
 int               r,
                   c,
                   i;
 static int        k = 4;

 ncs_s = (nr - 1) - k + 2;
 ncs_t = (nc - 1) - k + 2;

 for (ci_s = 0; ci_s < ncs_s; ci_s ++)
 for (ci_t = 0; ci_t < ncs_t; ci_t ++)
  {
   for (r = 0; r < 4; r++)
   for (c = 0; c < 4; c++)
   for (i = 0; i < 4; i++)
    G[r][c][i]=control_grid[ci_s+r][ci_t+c][i];

  draw_bicubic_surface_patch(G, Ms, Mt, ns, nt, nl);
 }
} /* draw_bicubic_surface */
```

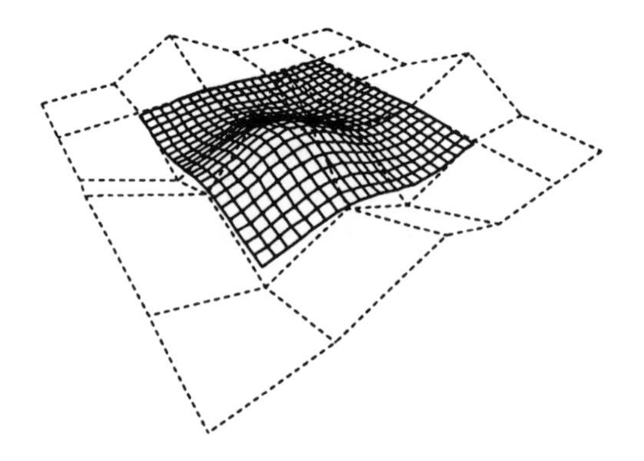

Figure 5.34 A surface drawn by piecing together a set of bicubic B-spline surface patches.

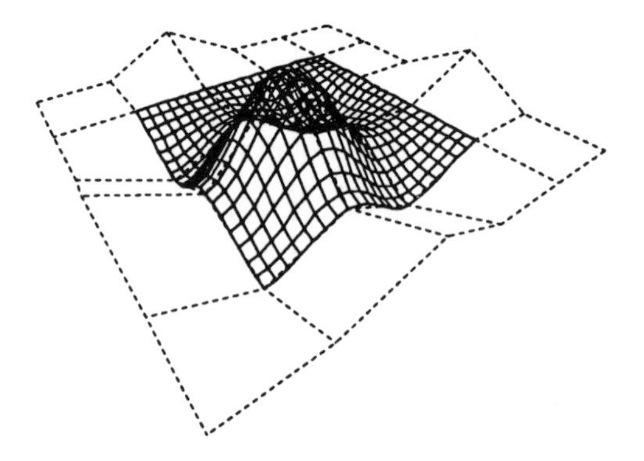

Figure 5.35 A surface drawn by piecing together a set of bicubic cardinal spline surface patches.

Appendix

A.12 Introduction to Forward Differences

Suppose we have a continuous univariate function f. If we evaluate this function at a parameter value t, its value is given by $f(t)$. If we further evaluate this function at parameter value $t+\delta$, where δ is a small positive increment of t, then its value is given by $f(t+\delta)$. The difference between these two function evaluations is called the *forward difference* of function f and is denoted by $\Delta f(t)$, where Δ is called the *forward difference operator*. Thus

$$\Delta f(t) = f(t+\delta) - f(t), \quad \exists\delta > 0, \tag{A.1}$$

where δ is the difference in the parameter value from one evaluation of the function to the next.

Rearranging the terms of Equation (A.1) we obtain

$$f(t+\delta) = f(t) + \Delta f(t). \tag{A.2}$$

This means that, if we know the value of the function $f(t)$ at the parameter value t and we know the value of its forward difference $\Delta f(t)$, then we can obtain the next value of the function $f(t+\delta)$ without having to re-evaluate the function afresh at parameter value $t+\delta$. When the evaluation of the function is computationally expensive, this technique can substantially speed up the evaluation of the function at successive parameter intervals.

If the function f is to be evaluated at $(n+1)$ equal increments δ of the parameter value t, we can simplify the notation by rewriting Equations (A.1) and (A.2) as

$$\Delta f_i = f_{i+1} - f_i \tag{A.3}$$

and

$$f_{i+1} = f_i + \Delta f_i, \tag{A.4}$$

where

$$\left.\begin{array}{lll} f_i = f(t_i),\ \Delta f_i = \Delta f(t_i) & \textit{with} & t_i = i \cdot \delta \\ f_{i+1} = f(t_{i+1}) & \textit{with} & t_{i+1} = (i+1) \cdot \delta \end{array}\right\} \text{ for } \quad i = 0, \ldots, n-1.$$

This formulation is sometimes known as *Newton's form of a polynomial* and is written as

$$\Delta^p f_i = \Delta^{p-1} f_{i+1} - \Delta^{p-1} f_i \tag{A.5}$$

and

$$\Delta^{p-1} f_{i+1} = \Delta^{p-1} f_i + \Delta^p f_i, \tag{A.6}$$

where by definition

$$\Delta^0 f_i = f_i. \tag{A.7}$$

The process of evaluating a function at successive equidistant intervals of its parameter using the above technique is known as *forward differencing* and it is a well-established technique in computer graphics.

A.13 Evaluating a Cubic Polynomial with Forward Differences

Here we wish to evaluate a piecewise cubic polynomial so that we can display the curve it defines as efficiently as possible. The cubic polynomial will be of the form

$$q(t) = at^3 + bt^2 + ct + d \quad \text{with } t \in [0, 1]$$

if the curve represented by $q(t)$ is to be approximated by n line segments, then we need to evaluate $q(t)$ at $(n+1)$ equally spaced values of the parameter t. Thus, t will assume the values

$$t = i \cdot \delta \quad \forall\ i = 0, 1, 2, \ldots, n,$$

where δ is the step size of the parameter t and $n \cdot \delta = 1$, i.e. $\delta = \frac{1}{n}$.

Let us start by looking at a linear polynomial of the form

$$p(t) = ct + d \quad \text{with } t \in [0, 1].$$

Its forward difference is given by

$$\begin{aligned}\Delta^1 p(t) &= p(t+\delta) - p(t)\\ &= c(t+\delta) + d - ct - d\\ &= ct + c\delta - ct\\ \therefore \Delta^1 p(t) &= c\delta. \end{aligned} \tag{A.8}$$

This is the difference between one evaluation of the polynomial (at parameter value t) and its next evaluation (at parameter value $t+\delta$).

Thus

$$\begin{aligned} p(t+\delta) &= p(t) + \Delta^1 p(t)\\ \therefore \quad p(t+\delta) &= p(t) + c\delta. \end{aligned} \tag{A.9}$$

Now we can efficiently generate the values of the polynomial, p_i, at parameter values t_i with the following algorithm, starting with the values $t_0 = 0$, $p_0 = p(0)$ and $\Delta^1 = \Delta^1 p(0)$:

$$\begin{aligned} & t_0 \leftarrow 0\\ & p_0 \leftarrow d\\ & \Delta^1 \leftarrow c\delta \end{aligned}$$

$$\begin{aligned} & \text{for } i = 1 \text{ to } n \text{ do}\\ & \text{begin}\\ & \quad t_i \leftarrow t_{i-1} + \delta\\ & \quad p_i \leftarrow p_{i-1} + \Delta^1\\ & \text{end} \end{aligned} \tag{A.A1}$$

As we would like to generalize this technique to deal with polynomials of higher degree, we proceed with the next step by considering a quadratic polynomial of the form

$$r(t) = bt^2 + ct + d \quad \text{with } t \in [0, 1].$$

Its forward difference will be

$$\begin{aligned}\Delta^1 r(t) &= r(t+\delta) - r(t)\\ &= b(t+\delta)^2 + c(t+\delta) + d - bt^2 - ct - d\\ &= b(t^2 + 2\delta t + \delta^2) + c(t+\delta) - bt^2 - ct\\ &= bt^2 + 2b\delta t + b\delta^2 + ct + c\delta - bt^2 - ct\\ \therefore \Delta^1 r(t) &= (2b\delta)t + (b\delta^2 + c\delta). \end{aligned} \tag{A.10}$$

Thus,

$$\begin{aligned} r(t+\delta) &= r(t) + \Delta^1 r(t)\\ \therefore r(t+\delta) &= r(t) + (2b\delta)t + (b\delta^2 + c\delta). \end{aligned} \tag{A.11}$$

A cursory examination of Equations (A.8) and (A.10) reveals that $\Delta^1 p(t)$ and $\Delta^1 r(t)$ are substantially different. Whereas $\Delta^1 p(t)$ is a constant, $\Delta^1 r(t)$ is a linear polynomial of t.

Unfortunately, this implies that the technique we have used above to evaluate $p(t+\delta)$ cannot be used directly to evaluate $r(t+\delta)$, because $p(t+\delta)$ is evaluated by adding a constant to its previous value (see Equation (A.9)), while $r(t+\delta)$ can only be evaluated by adding a function of t to its previous value (see Equation (A.11)). Thus in the loop of algorithm (A.A1) we can no longer add a constant to p_{i-1} to obtain p_i.

To overcome this problem we could of course pre-compute and store the values of $\Delta^1 r(t)$ at parameter values $t = 0, \delta, 2\delta, \ldots, n\delta$ and then add these values to p_{i-1} inside the loop. We notice, however, that $\Delta^1 r(t)$ is itself a linear polynomial and we already know how to evaluate such a polynomial efficiently for such a sequence of parameter values. We simply compute the starting value $\Delta^1 r(0)$ and then add its forward difference $\Delta^1(\Delta^1 r(0)) = \Delta^2 r(0)$ to $\Delta^1 r(i\delta)$ to obtain $\Delta^1 r(i\delta+\delta)$. Using Equation (A.10) we obtain

$$\Delta^1 r(0) = b\delta^2 + c\delta$$

and we compute the second forward difference of $r(t)$ as

$$\begin{aligned}
\Delta^2 r(t) &= \Delta^1(\Delta^1 r(t)) \\
&= \Delta^1 r(t+\delta) - \Delta^1 r(t) \\
&= (2b\delta)(t+\delta) + (b\delta^2 + c\delta) - (2b\delta)t - (b\delta^2 + c\delta) \\
&= (2b\delta)t + 2b\delta^2 - (2b\delta)t \\
\Delta^2 r(t) &= 2b\delta^2.
\end{aligned} \tag{A.12}$$

Now we can efficiently generate the values of the polynomial, r_i, at parameter values t_i with the following algorithm, starting with the values $t_0 = 0$, $r_0 = r(0)$, $\Delta^1 = \Delta^1 r(0)$ and $\Delta^2 = \Delta^2 r(0)$:

$$\begin{aligned}
& t_0 \leftarrow 0 \\
& r_0 \leftarrow d \\
& \Delta^1 \leftarrow b\delta^2 + c\delta \\
& \Delta^2 \leftarrow 2b\delta^2 \\
& \\
& \text{for } i = 1 \text{ to } n \text{ do} \\
& \text{begin} \\
& \quad t_i \leftarrow t_{i-1} + \delta \\
& \quad r_i \leftarrow r_{i-1} + \Delta^1 \\
& \quad \Delta^1 \leftarrow \Delta^1 + \Delta^2 \\
& \text{end}
\end{aligned} \tag{A.A2}$$

Let us re-examine what we have discovered thus far. When $r(t)$ is a quadratic polynomial, then the value of its first forward difference $\Delta^1 r(t)$ that must be added to $r(t)$ to obtain $r(t+\delta)$ is not constant – it changes value as the parameter value changes from t to $t+\delta$. Fortunately, however, $\Delta^1 r(t)$ is itself easy to recalculate after we have reached $t+\delta$ so as to obtain the increment that will be required to compute the next iteration of the loop (i.e. to move from $t+\delta$ to $t+2\delta$).

Next, we extend this approach to a cubic polynomial. In this case,

$$q(t) = at^3 + bt^2 + ct + d \quad \text{with } t \in [0, 1].$$

Here we need to compute the first, second and third forward differences of $q(t)$:

$$\begin{aligned}
\Delta^1 q(t) &= q(t+\delta) - q(t) \\
&= a(t+\delta)^3 + b(t+\delta)^2 + c(t+\delta) + d - at^3 - bt^2 - ct - d \\
&= a(t^3 + 3t^2\delta + 3t\delta^2 + \delta^3) + b(t^2 + 2t\delta + \delta^2) + c(t+\delta) - at^3 - bt^2 - ct \\
&= at^3 + 3a\delta t^2 + 3a\delta^2 t + a\delta^3 + bt^2 + 2b\delta t + b\delta^2 + ct + c\delta - at^3 - bt^2 - ct \\
\therefore\ \Delta^1 q(t) &= (3a\delta)t^2 + (3a\delta^2 + 2b\delta)t + (a\delta^3 + b\delta^2 + c\delta) \qquad \text{(A.13)}
\end{aligned}$$

$$\begin{aligned}
\Delta^2 q(t) &= \Delta^1 q(t+\delta) - \Delta^1 q(t) \\
&= (3a\delta)(t+\delta)^2 + (3a\delta^2 + 2b\delta)(t+\delta) + (a\delta^3 + b\delta^2 + c\delta) - (3a\delta)t^2 \\
&\quad - (3a\delta^2 + 2b\delta)t - (a\delta^3 + b\delta^2 + c\delta) \\
&= (3a\delta)(t^2 + 2t\delta + \delta^2) + (3a\delta^2 + 2b\delta)(t+\delta) - (3a\delta)t^2 - (3a\delta^2 + 2b\delta)t \\
&= 3a\delta t^2 + 6a\delta^2 t + 3a\delta^3 + (3a\delta^2 + 2b\delta)t + 3a\delta^3 + 2b\delta^2 \\
&\quad - 3a\delta t^2 - (3a\delta^2 + 2b\delta)t \\
\therefore\ \Delta^2 q(t) &= (6a\delta^2)t + (6a\delta^3 + 2b\delta^2) \qquad \text{(A.14)}
\end{aligned}$$

$$\begin{aligned}
\Delta^3 q(t) &= \Delta^2 q(t+\delta) - \Delta^2 q(t) \\
&= (6a\delta^2)(t+\delta) + (6a\delta^3 + 2b\delta^2) - (6a\delta^2)t - (6a\delta^3 + 2b\delta^2) \\
&= (6a\delta^2)t + 6a\delta^3 - (6a\delta^2)t \\
\therefore\ \Delta^3 q(t) &= 6a\delta^3. \qquad \text{(A.15)}
\end{aligned}$$

From Equations (A.13)–(A.15) we see that if we know the values of the functions $q(t)$, $\Delta^1 q(t)$, $\Delta^2 q(t)$ and $\Delta^3 q(t)$ for a given parameter value t, then we can determine their values for $t+\delta$:

$$\begin{aligned}
q(t+\delta) &= q(t) + \Delta^1 q(t) \\
\Delta^1 q(t+\delta) &= \Delta^1 q(t) + \Delta^2 q(t) \\
\Delta^2 q(t+\delta) &= \Delta^2 q(t) + \Delta^3 q(t) \\
\Delta^3 q(t+\delta) &= 6a\delta^3.
\end{aligned}$$

Having obtained the values of these functions for the parameter value $t+\delta$, we use the same equations to determine their values for $t+2\delta$, and so on since the above equations are valid for any t. In particular, to determine the value of these functions for $t' = i\delta$ we write:

$$\begin{aligned}
q(t'+\delta) &= q(t') + \Delta^1 q(t') \\
\Delta^1 q(t'+\delta) &= \Delta^1 q(t') + \Delta^2 q(t') \\
\Delta^2 q(t'+\delta) &= \Delta^2 q(t') + \Delta^3 q(t') \\
\Delta^3 q(t'+\delta) &= 6a\delta^3.
\end{aligned}$$

Thus, the following algorithm will generate the values of the polynomial, q_i, at parameter values t_i, starting with the values $t_0 = 0$, $q_0 = q(0)$, $\Delta^1 = \Delta^1 q(0)$, $\Delta^2 = \Delta^2 q(0)$ and $\Delta^3 = \Delta^3 q(0)$.

The algorithm is:

$$
\begin{aligned}
&t_0 \leftarrow 0\\
&q_0 \leftarrow d\\
&\Delta^1 \leftarrow a\delta^3 + b\delta^2 + c\delta\\
&\Delta^2 \leftarrow 6a\delta^3 + 2b\delta^2\\
&\Delta^3 \leftarrow 6a\delta^3\\
&\\
&\text{for } i = 1 \text{ to } n \text{ do}\\
&\text{begin}\\
&\quad t_i \leftarrow t_{i-1} + \delta\\
&\quad q_i \leftarrow q_{i-1} + \Delta^1\\
&\quad \Delta^1 \leftarrow \Delta^1 + \Delta^2\\
&\quad \Delta^2 \leftarrow \Delta^2 + \Delta^3\\
&\text{end}
\end{aligned}
\tag{A.A3}
$$

where the initial conditions are:

$$
\begin{aligned}
q(0) &= d\\
\Delta^1 q(0) &= a\delta^3 + b\delta^2 + c\delta\\
\Delta^2 q(0) &= 6a\delta^3 + 2b\delta^2\\
\Delta^3 q(t) &= 6a\delta^3.
\end{aligned}
$$

The above initial conditions can be rewritten in matrix form:

$$
F = \begin{bmatrix} \Delta^3 q(0) \\ \Delta^2 q(0) \\ \Delta^1 q(0) \\ q(0) \end{bmatrix} = \begin{bmatrix} 6\delta^3 & 0 & 0 & 0 \\ 6\delta^3 & 2\delta^2 & 0 & 0 \\ \delta^3 & \delta^2 & \delta & 0 \\ 0 & 0 & 0 & 1 \end{bmatrix} \cdot \begin{bmatrix} a \\ b \\ c \\ d \end{bmatrix}. \tag{A.16}
$$

Since $\delta = \frac{1}{n}$, the above equation can be rewritten as

$$
F = \begin{bmatrix} \frac{6}{n^3} & 0 & 0 & 0 \\ \frac{6}{n^3} & \frac{2}{n^2} & 0 & 0 \\ \frac{1}{n^3} & \frac{1}{n^2} & \frac{1}{n} & 0 \\ 0 & 0 & 0 & 1 \end{bmatrix} \cdot \begin{bmatrix} a \\ b \\ c \\ d \end{bmatrix} = P \cdot C, \tag{A.17}
$$

where F is called the *forward difference matrix*, P is called the *precision matrix* and C is the *coefficients matrix*.

From Section 5.4, we know that the coefficients matrix is equal to the product of the *basis matrix* M by the *geometry matrix* G, i.e. $C = M \cdot G$. Substituting C in Equation (A.17) we obtain

$$
F = P \cdot M \cdot G. \tag{A.18}
$$

Thus, the forward difference matrix is defined as the product of the precision, the basis and the geometry matrices. By iterating this matrix as described in Section 5.11 we can compute successive sample points on a curve very efficiently.

Bibliography

Barsky, B.A., The beta-spline: a local representation based on shape parameters and fundamental geometric measures, *PhD Dissertation*, Dept of Computer Science, University of Utah, Salt Lake City, Utah 84112, 1981.

Barsky, B.A., *Computer Graphics and Geometric Modelling Using Beta-splines*, Springer-Verlag, 1988.

Barsky, B.A. and Beatty, J.C., Local control of bias and tension in beta-splines, *ACM Transactions on Graphics*, **2**(2), 109–134, 1983.

Bartels, R.H., Beatty, J.C. and Barsky, B.A., *An Introduction to Splines for use in Computer Graphics Modeling*, Morgan-Kaufman, pp. 422–434, 1987.

Bézier, P., Définition numerique de courbes et surfaces I, *Automátisme*, Vol. XI, pp. 625–632, 1966.

Bézier, P., Définition numerique de courbes et surfaces II, *Automátisme*, Vol. XII, pp. 17–21, 1967.

Bézier, P., *The Mathematical Basis of the Unisurf CAD System*, Butterworths, 1986.

Coons, S.A., Surfaces for computer aided design, *Technical Report*, Project MAC-TR-41, MIT, Cambridge, Massachusetts 02139, 1964.

Coons, S.A., Surfaces for computer aided design of space forms, *Technical Report*, Project MAC-TR-41, MIT, Cambridge, Massachusetts 02139, 1967. (Available as AD-663-504 from the National Information Service, Springfield, Virginia 22161.)

de Casteljau, P., *Outillage Méthodes Calcul*, Citröen, 1959.

de Casteljau, P., *Courbes et Surfaces à Pôles*, Citröen, 1963.

de Casteljau, P., *Shape Mathematics and CAD*, Kogan Page, 1986.

Gordon, W., Spline-blended surface interpolation through curve networks, *Journal of Mathematics and Mechanics*, **18**(10), 931–952, 1969.

Kochanek, D.H.U. and Bartels, R.H., Interpolating splines with local tension, continuity and bias control, *Computer Graphics, SIGGRAPH-84 Conference Proceedings*, **18**(3), 33–41, 1984.

Seroussi, G. and Barsky, B.A., An explicit derivation of discretely shaped beta-spline basis functions of arbitrary order, *Mathematical Methods in Computer-Aided Geometric Design II (Proceedings of the 191-91 Conference on Curves, Surfaces, CAGD and Image Processing)*, Lyche, T. and Schumaker, L.L. (eds.), Academic Press, pp. 667–584, 1992.

6 Smooth Surface Representation over Irregular Meshes

Jian J. Zhang

6.1 Introduction

The previous chapter has described a number of surface representation schemes in surface modeling. These schemes, despite being popular in geometric modeling, do suffer from one important limitation: the control mesh must be structured to have a rectangular topology. This limitation has made it very awkward for many modeling tasks in computer animation, such as the modeling of branches and non-quadrilateral patches.

Some remedies have been proposed to alleviate this limitation. The main methods are:

- trimmed surfaces;
- networks of triangular patches; and
- n-sided patches.

One of the primary methods in overcoming the topological limitation of a traditional surface scheme is "trimming and stitching". This allows two or more surface patches to be connected together with the required topology. Many animation packages, such as Alias|Wavefront Maya and SoftImage|XSI, have implemented it with NURBS surfaces. Although this scheme gets around the problem, it is difficult to maintain the smoothness conditions at the seams, especially in interactive shape modeling and animation.

The second remedy is to blend a non-quadrilateral area with several triangular sub-patches. This approach splits an irregular area into a number of triangular regions and interpolates each region with a triangular surface patch. All the triangular surface patches are then stitched together smoothly. Because of the way it is modeled, when modifying a blending surface, it is often difficult to maintain the tangent continuity between the triangular patches.

The third method is similar to the second except that a complete n-sided patch is constructed and used to blend an irregular area. Since this is an integral patch, internal smoothness is automatically guaranteed. Most such modeling schemes, however, do not permit shape modification after a surface patch has been generated.

Apart from the traditional surfaces, another group of surface generation schemes have recently gained increasing popularity in computer animation. They are known as *subdivision surfaces*. These surfaces differ from the traditional ones in that they do not usually afford a closed-form mathematical expression. The surfaces are actually the result of a repeated subdivision of a polyhedron. Such a polyhedron can have an arbitrary topology.

Unlike the traditional surfaces, the resulting subdivision surfaces are not restricted by the topology of the model. Compared with other methods, which aim to overcome the topological limitation, subdivision surfaces are regarded as being advantageous due to their topological flexibility and are therefore attractive to animators. This advantage is further amplified by many successful applications in computer animation, amongst which is *Geri's Game* by Pixar Animation Studios.

There are a number of schemes that are able to generate a surface by either subdividing an arbitrary polyhedron or by combining subdivision with sub-patch generation operations. In this chapter, we introduce three such schemes.

6.2 Quasi-cubic Subdivision Surfaces

Catmull and Clark (1978) proposed one of the earliest schemes for subdivision surfaces. This scheme takes as input an arbitrary initial *control mesh*, which is a collection of points connected by line segments. The points are usually known as *vertices* and the line segments as *edges*. An area enclosed by a number of edges is called a *face*. This initial control mesh is then refined (subdivided) to create new vertices, edges and faces which at each subdivision step form a new control mesh. By selecting the subdivision rules carefully, this process will converge to a limiting surface. It can be shown that this scheme generates a generalized cubic B-spline surface except at some isolated *extraordinary points*. This is to say that when its topology is quadrilateral, this scheme generates a generalized cubic B-spline surface.

To better understand this modeling scheme, we will derive the subdivision rules by splitting a cubic B-spline surface into smaller patches. This splitting process creates a new control polyhedron at each refinement step.

6.2.1 Cubic B-spline Surface Splitting

A cubic B-spline patch can be expressed as

$$\mathbf{S}(u, v) = \mathbf{U}\,\mathbf{M}\,\mathbf{P}\,\mathbf{M}^T\,\mathbf{V}^T \tag{6.1}$$

where

$$\mathbf{M} = \frac{1}{6}\begin{bmatrix} 1 & 4 & 1 & 0 \\ -3 & 0 & 3 & 0 \\ 3 & -6 & 3 & 0 \\ -1 & 3 & -3 & 1 \end{bmatrix} \tag{6.2}$$

stands for the B-spline basis matrix, and

$$\mathbf{P} = \begin{bmatrix} p_{11} & p_{12} & p_{13} & p_{14} \\ p_{21} & p_{22} & p_{23} & p_{24} \\ p_{31} & p_{32} & p_{33} & p_{34} \\ p_{41} & p_{42} & p_{43} & p_{44} \end{bmatrix} \tag{6.3}$$

represents the control points, and

$$\mathbf{U} = [1\ u\ u^2\ u^3] \qquad \mathbf{V} = [1\ v\ v^2\ v^3]. \tag{6.4}$$

Here $u, v \in [0, 1]$ are the parameters of the surface.

Now let us generate the control points which will correspond to a sub-patch of this patch, and assume that the parameters of this sub-patch are given within the range of $[0, 0.5]$. Taking the same B-spline expression as (6.1), this sub-patch can be given as

$$\mathbf{S}(u_1, v_1) = \mathbf{U}_1\,\mathbf{M}\,\mathbf{P}\,\mathbf{M}^T\,\mathbf{V}_1^T \tag{6.5}$$

where $u_1 = 0.5u$, $v_1 = 0.5v$, $\mathbf{U}_1 = \mathbf{U}\,\mathbf{L}$, $\mathbf{V}_1 = \mathbf{V}\,\mathbf{L}$ and the matrix $\mathbf{L}$ is given by

$$\mathbf{L} = \frac{1}{8}\begin{bmatrix} 8 & 0 & 0 & 0 \\ 0 & 4 & 0 & 0 \\ 0 & 0 & 2 & 0 \\ 0 & 0 & 0 & 1 \end{bmatrix}. \tag{6.6}$$

Thus we have

$$\mathbf{S}(u_1, v_1) = \mathbf{U}\,\mathbf{L}\,\mathbf{M}\,\mathbf{P}\,\mathbf{M}^T\,\mathbf{L}^T\,\mathbf{V}^T. \tag{6.7}$$

Since (6.7) represents part of the same cubic B-spline as (6.1) under a new set of control points $\mathbf{Q}$, we have

$$\mathbf{S}(u, v) = \mathbf{U}\,\mathbf{M}\,\mathbf{Q}\,\mathbf{M}^T\,\mathbf{V}^T. \tag{6.8}$$

Letting (6.7) be equal to (6.8), for any value of u and v we will have,

$$\mathbf{M}\,\mathbf{Q}\,\mathbf{M}^T = \mathbf{L}\,\mathbf{M}\,\mathbf{P}\,\mathbf{M}^T\,\mathbf{L}^T. \tag{6.9}$$

It allows the computation of a new set of control points,

$$\mathbf{Q} = \mathbf{H}_1\,\mathbf{P}\,\mathbf{H}_1^T, \tag{6.10}$$

where

$$\mathbf{H}_1 = \mathbf{M}^{-1}\,\mathbf{L}\,\mathbf{M} = \frac{1}{8}\begin{bmatrix} 4 & 4 & 0 & 0 \\ 1 & 6 & 1 & 0 \\ 0 & 4 & 4 & 0 \\ 0 & 1 & 6 & 1 \end{bmatrix}. \tag{6.11}$$

This gives the relationship between the original control points and the new control points after a patch split. $\mathbf{H}_1$ is hence called the *splitting matrix*.

Figure 6.1 illustrates both the old and the new control meshes of a B-spline patch where $\mathbf{P} = [p_{ij}]$ denotes the old control mesh and $\mathbf{Q} = [q_{ij}]$ denotes the new one. Their relationship is defined by (6.10), i.e. any point q_{ij} of the new mesh can be given as a linear combination of p_{ij} of the old control mesh.

The new control mesh consists of three types of points: face points f, edge points e and vertex points v.

- A new face point f is associated with a face of the old mesh, such as q_{11}, q_{13}, q_{31} and q_{33}.
- A new edge point e is associated with a face and an edge of the old mesh, such as q_{12} and q_{21}.
- A new vertex point v is associated with a vertex and a face of the old mesh, such as q_{22}.

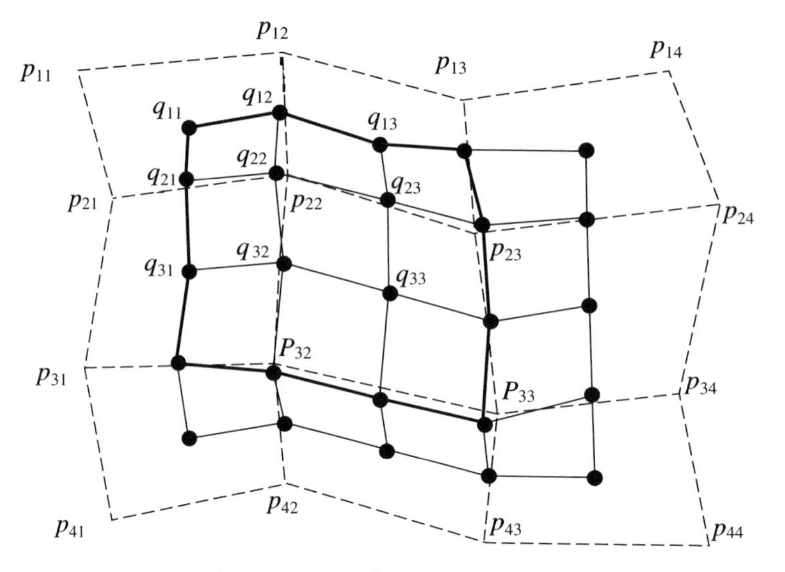

Figure 6.1 Cubic B-spline patch splitting.

Referring to Figure 6.1, we can compute the new control points from (6.10). For example, the new face point

$$q_{11} = \frac{p_{11} + p_{12} + p_{21} + p_{22}}{4}, \tag{6.12}$$

which is the average of the old points that define the face.

Similarly, a new edge point can be derived as

$$q_{12} = \frac{q_{11} + q_{13} + p_{12} + p_{22}}{4}, \tag{6.13}$$

which is the average of the end points of the old edge and new face points of the adjacent faces.

A vertex point is given by

$$q_{22} = \frac{p_{22}}{2} + \frac{(p_{12} + p_{21} + p_{32} + p_{23}) + (q_{11} + q_{13} + q_{31} + q_{33})}{16} \tag{6.14}$$

which is a weighted average of the associated old vertex point, old edge points and the new face points.

Since the new points are derived from the same B-spline expression, they will generate part (a sub-patch) of the same surface. From the properties of B-spline surfaces, if we continue splitting the surface, the control points will get closer and closer to the surface. In other words, this splitting process will converge to the B-spline surface when the number of splits approaches infinity.

In the next section, we will generalize the patch splitting process for an arbitrary control mesh.

6.2.2 Catmull–Clark Subdivision Surfaces

As a natural extension, we can generalize the expressions (6.12)–(6.14) to accommodate control meshes of an arbitrary topology, i.e. faces have any number of sides (Catmull and Clark 1978). The generalization should have (6.12)–(6.14) as a special case when the topology is quadrilateral.

Let us denote a control mesh by M^i, where i denotes the number of applications of the subdivision rules on the mesh. Therefore the initial mesh is M^0, and the subsequent meshes are denoted as $M^1, M^2, \ldots$ generated by the repeated application of the subdivision rules. We still term the points of a mesh face, edge and vertex points. These three types of points and the meshes are illustrated in Figure 6.2 (DeRose et al. 1998).

Assume the current mesh is M^i and we want to create the points of the next subdivision mesh M^{i+1}. From (6.12)–(6.14), the generalized subdivision rules are given as follows (Catmull and Clark 1978).

- A face point f^{i+1} is positioned at the centroid of the corresponding face of M^i, i.e. it is the average of the vertices of the face:

$$f^{i+1} = \text{average of vertices of the corresponding old face.} \tag{6.15}$$

- An edge point e_j^{i+1} is computed by

$$e_j^{i+1} = \frac{v^i + e_j^i + f_{j-1}^{i+1} + f_j^{i+1}}{4}, \tag{6.16}$$

  where the subscripts are taken modulo the valence n (the number of edges incident to the vertex) of the central vertex. So a new edge point is the average of the endpoints of the old edge and two new face points of the faces that share this edge.

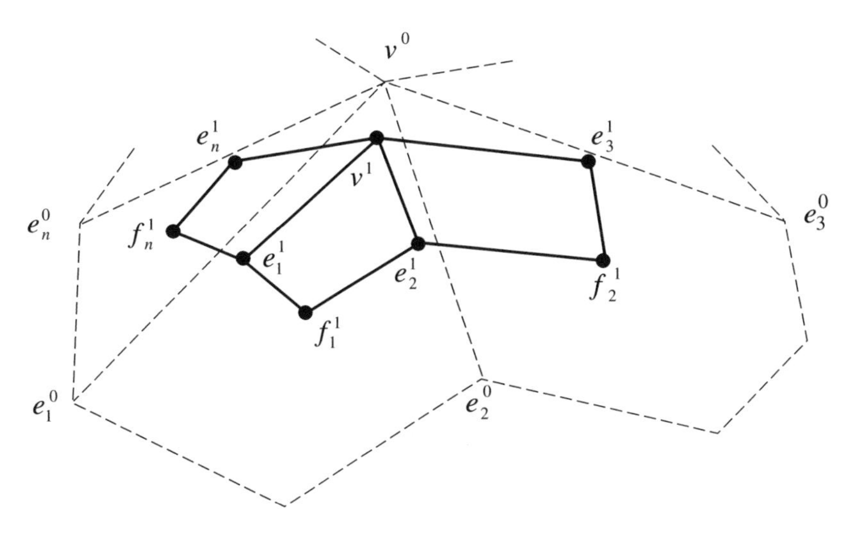

Figure 6.2 Subdivision of an irregular mesh.

- A vertex point v^{i+1} is given by

$$v^{i+1} = \frac{n-2}{n} v^i + \frac{1}{n^2} \sum_j e_j^i + \frac{1}{n^2} \sum_j f_j^{i+1}. \qquad (6.17)$$

Once all the new points are generated, the new mesh M^{i+1} is constructed by:

- connecting a new face point to the new edge points of the edges which bound the old face;
- connecting a new vertex point to the new edge points of all old edges that are incident to the corresponding old vertex point;
- identifying a new face as the enclosure of the new edges.

Vertices of valence 4 are called the *ordinary points*, whereas the others are called the *extraordinary points*. From these rules it can be seen that when $n = 4$, the rules duplicate expressions (6.12)–(6.14) and hence will generate a cubic B-spline surface.

Figure 6.3 shows an initial control mesh, its intermediate subdivisions and the corresponding subdivision surface are generated by the rules (6.15)–(6.17).

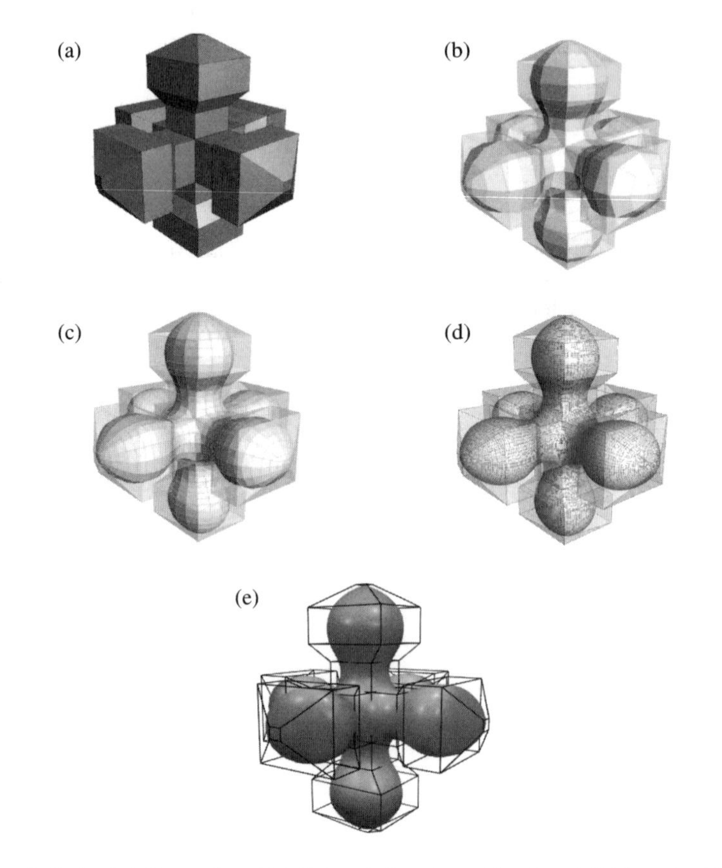

Figure 6.3 Catmull–Clark subdivision surface. (a) Initial control mesh; applying the subdivision rules: (b) once; (c) twice; (d) four times; (e) the final surface.

6.2.3 Properties of Catmull–Clark Surfaces

From the rules and procedure of the Catmull–Clark subdivision, the following properties emerge (Halstead et al. 1993).

- The control mesh, and hence the generated surface, can be of any arbitrary topology consisting of ordinary and extraordinary points.
- The mesh is composed of only quadrilateral faces after the first subdivision.
- A control mesh will generate a cubic B-spline surface except at the extraordinary points. Therefore excluding these isolated points, the surface is tangent and curvature continuous.
- After the first subdivision step, the number of extraordinary points is fixed. It is the number of non-quadrilateral faces of the initial mesh.
- With repeated application of the subdivision rules to a mesh, the influence of the extraordinary points becomes progressively isolated. At these points the surface can be shown to have a well-defined tangent plane, but the curvature is not continuous (Ball and Storry 1988). Therefore a Catmull–Clark subdivision surface is always smooth regardless of its topology.

6.2.4 Surfaces with Sharp and Semi-sharp Edges

Many animation objects and characters exhibit sharp and semi-sharp edges. It is understood that these edges may follow an arbitrary topology or a complicated 3D curve. Therefore they are very difficult to model, not only because with the conventional piecewise smooth surfaces it is often hard to produce such features, but also because of their complexity and variety. Although the Catmull–Clark subdivision surface scheme is capable of constructing a surface over an irregular mesh, the current rules are not flexible enough to create creases on an overall smooth surface.

To get around this problem, DeRose et al. (1998) proposed some complementary rules to the above subdivision procedure. These rules are a generalization of the original Catmull–Clark subdivision scheme, so as to allow the creation of subdivision surfaces having both smooth and sharp features. For example, Figure 6.4 illustrates a Catmull–Clark subdivision surface with semi-sharp edges. It uses the same control mesh as that of Figure 6.3, but with a slight modification of the subdivision rules applied. The original rules (*smooth rules*), and the new rules (*sharp rules*) are applied to a mesh at different steps to achieve a required result. For example, a semi-sharp crease may be obtained by applying the sharp rules at the first few subdivision steps and the smooth rules at the final stage to generate the final surface.

6.2.4.1 Creation of Sharp Edges on Subdivision Surfaces

With this new set of subdivision rules, the user can tag any edge of a mesh as being sharp. Starting from an old mesh M^i, these subdivision rules will generate the face, edge and vertex points of the new mesh M^{i+1}, taking into account the indications of sharp edges.

- *Face points*. Same as the original rule, the face points are always placed at the face centroid regardless of whether an edge is tagged as sharp or not.

Figure 6.4 Catmull–Clark subdivision surface using modified rules.

- *Edge points.* If there is a tagged edge $v^i e^i_j$, the corresponding new edge point of M^{i+1} is positioned at the midpoint of the edge, i.e.

$$e^{i+1}_j = \frac{v^i + e^i_j}{2}. \tag{6.18}$$

- *Vertex points.* A new vertex point is computed according to the number of sharp edges incident at the corresponding vertex point of M^i. There are three situations.

(1) One sharp edge is incident at the corresponding vertex point. This vertex is called a *dart.* The new vertex point is computed using the original rule (6.17).

(2) Two sharp edges, $v^i e^i_j$ and $v^i e^i_k$, are incident at the corresponding vertex point. Such a vertex is called a *crease vertex.* A new rule will be used for its creation, which is

$$v^{i+1} = \frac{e^i_j + 6v^i + e^i_k}{8}. \tag{6.19}$$

(3) Three or more sharp edges are incident at the corresponding vertex point. This is called a *corner.* A vertex can also be made a corner by directly tagging it. The corresponding new vertex is then positioned using the corner rule, which prevents the corner point from moving,

$$v^{i+1} = v^i. \tag{6.20}$$

Figure 6.5 illustrates the applications of these modified rules and their effects on the subdivision surfaces.

6.2.4.2 Creation of Semi-sharp Edges on Subdivision Surfaces

Semi-sharp edges can be created by assigning a not necessarily integer sharpness factor to an edge. Let us assume that a sequence of edges, $e_1, e_2, e_3, \ldots,$ of a mesh have an associated

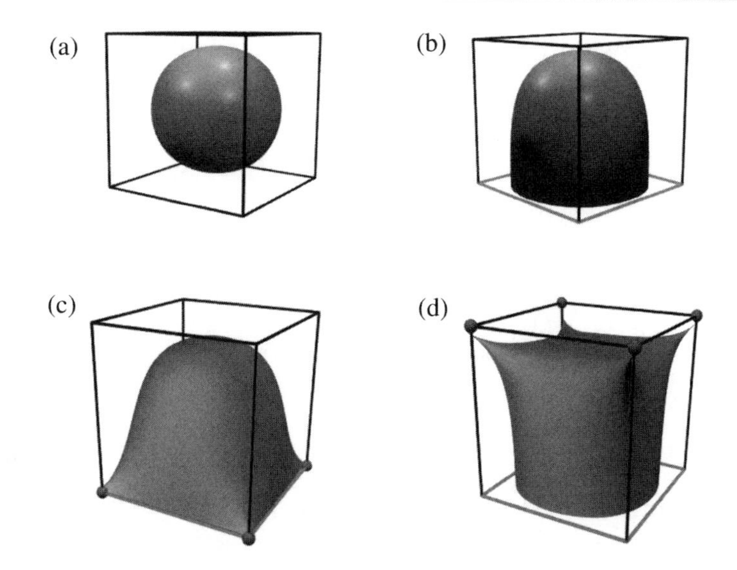

Figure 6.5 Subdivision surfaces resulting from sharp edges. (a) Subdivision using smooth rules; (b) the bottom edges are marked as sharp edges; (c) the bottom edges are marked as sharp edges and the bottom vertices are marked as corners; (d) the bottom edges are marked as sharp edges and the top vertices are marked as corners.

sharpness. Adopting the notation of DeRose et al. (1998), the sharpness factor of such an edge e_i is expressed as $e_i \cdot s$, where $e_i \cdot s = 0$ indicates an smooth edge and $e_i \cdot s \geq 1$ a sharp edge, whereas $0 < e_i \cdot s < 1$ specifies a semi-sharp edge.

To compute the new face, edge and vertex points, the following rules are used:

- *Face points.* As usual, the face points are always placed at the face centroid.
- *Edge points.* A new edge point is created depending on whether the edge is smooth, sharp or semi-sharp.
  - For a smooth edge, the edge point is computed using the smooth rule (6.16).
  - For a sharp edge, the edge point is computed using the sharp rule (6.18).
  - For a semi-sharp edge, a weighted average of both the sharp and smooth rules is used. Let a be the edge point computed using the smooth rule and b using the sharp rule, then the edge point should be given by

$$(1 - e \cdot s)a + e \cdot sb. \tag{6.21}$$

- *Vertex points.* A new vertex point is created depending on the number and the smoothness of the edges associated with the corresponding vertex.
  - There is one or no sharp edges incident at the corresponding vertex point. The new vertex point is computed using the original rule (6.17).
  - There are three or more sharp edges incident at the corresponding vertex point. The new vertex point is computed using the corner rule (6.20).
  - There are two sharp edges incident at the corresponding vertex point. A new vertex point is computed depending on the average sharpness $v \cdot s$ of these two edges. If $v \cdot s > 1$, the new vertex point is computed using crease vertex rule (6.19). However, if

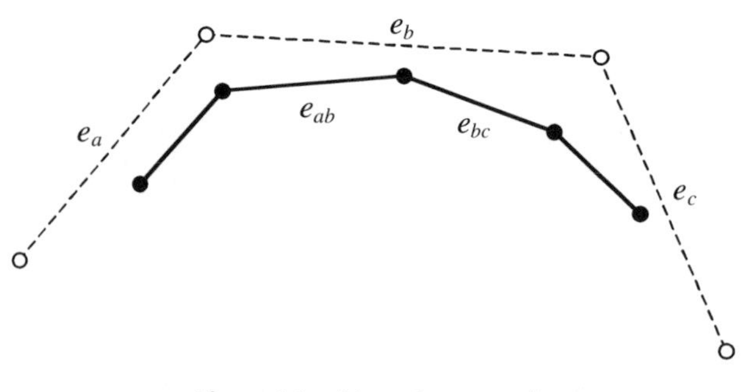

Figure 6.6 Edges of a new mesh.

$0 < v \cdot s < 1$, a weighted average of the crease vertex rule and the smooth rule should be used. Let c be the vertex point computed using the smooth rule (6.17) and d using the crease vertex rule (6.19), then the new vertex point should be given by

$$(1 - v \cdot s)c + v \cdot sd \tag{6.22}$$

Since each subdivision makes the faces and edges smaller/shorter, suggested by DeRose et al. (1998), the sharpness factor of each new edge should also be reduced accordingly. If e_a, e_b, e_c are three adjacent edges of mesh M^i (Figure 6.6), and e_{ab} and e_{bc} are the new edges in the new mesh M^{i+1} corresponding to edge e_b, their sharpness factors are calculated as follows:

$$\begin{aligned} e_{ab \cdot s} &= \max\left(\frac{e_{a \cdot s} + 3e_{b \cdot s}}{4} - 1, 0\right) \\ e_{bc \cdot s} &= \max\left(\frac{3e_{b \cdot s} + e_{c \cdot s}}{4} - 1, 0\right). \end{aligned} \tag{6.23}$$

Figure 6.7 shows two models generated by applying the subdivision rules to the same initial mesh as that of Figure 6.3. But the edges are assigned different sharpness factors, hence the difference of the shapes.

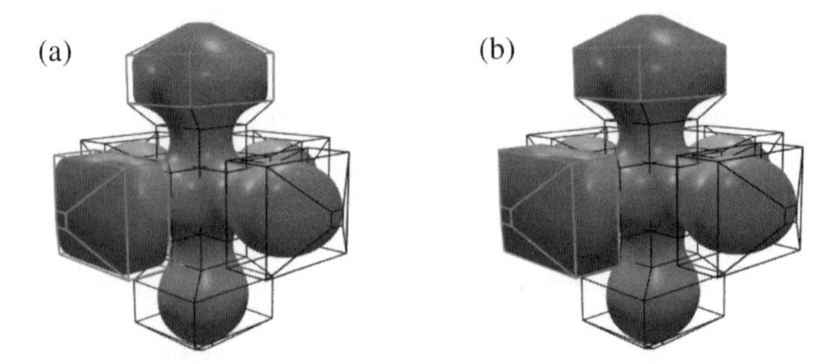

Figure 6.7 Effects of sharpness factors. (a) Edge sharpness = 0.2 for the edges of the top and left-hand blocks; (b) edge sharpness = 1.0 for the edges of the top and left-hand blocks.

6.2.5 Other Comments

To generate a Catmull–Clark subdivision surface, the most direct and simplest method is to apply the subdivision rules repeatedly to a mesh until a predetermined precision is reached. Then the surface is approximated by the polygons obtained after the subdivision steps. Nevertheless, although a surface can be approximated using a collection of subdivision polygons, some of the properties of the scheme are exploitable for the better use of the algorithms and improving overall efficiency and quality.

Apart from the neighborhood of extraordinary points and sharp features, a Catmull–Clark subdivision surface is covered by cubic B-spline surface patches. Therefore, during each subdivision, more B-spline patches are identified, which can be dealt with as ordinary B-spline surfaces.

For those faces that cannot be covered by B-spline patches, a method for computing the position and normal of a given point on the surface was proposed by Halstead et al. (1993).

Color plates 6.1–6.3 show some animation characters produced using Catmull–Clark subdivision surfaces.

6.3 Quadratic Subdivision Surfaces

The Catmull–Clark subdivision surfaces are mostly cubic except in areas near the extraordinary points. Therefore excluding the extraordinary points, they are curvature continuous. This scheme has now gained popularity in computer animation. Among many different subdivision schemes, there is another one that has also been well received by the geometric modeling community, this is the subdivision scheme proposed by Doo and Sabin (1978).

The Doo–Sabin scheme, unlike the Catmull–Clark subdivision scheme, produces a smooth quadratic surface throughout. This feature has been regarded as an advantage due to its continuity consistency over the surface.

6.3.1 Doo–Sabin Subdivision Surfaces

Like the Catmull–Clark scheme, it also takes as input a control mesh and produces a smooth surface in the limit. Let us denote an arbitrary control mesh of the subdivision by M^i, where i denotes the number of applications of the subdivision rules on the mesh. Therefore the initial mesh is M^0, and subsequent meshes are denoted by $M^1, M^2, \ldots$ generated by repeated application of the subdivision rules. This sequence of meshes will finally converge to a smooth quadratic surface.

The Doo–Sabin scheme also creates new points at each subdivision, from which the new faces are formed. To see the subdivision process more clearly, let us first look at the subdivision of a regular quadratic B-spline patch (Joy 1999) and then generalize the derived rules for the subdivision of irregular meshes.

6.3.2 Quadratic B-spline Surface Splitting

A quadratic B-spline patch can be expressed as

$$\mathbf{S}(u, v) = \mathbf{U}\,\mathbf{M}\,\mathbf{P}\,\mathbf{M}^T\mathbf{V}^T, \tag{6.24}$$

where

$$\mathbf{M} = \frac{1}{2}\begin{bmatrix} 1 & 1 & 0 \\ -2 & 2 & 0 \\ 1 & -2 & 1 \end{bmatrix} \tag{6.25}$$

represents the B-spline basis matrix, and

$$\mathbf{P} = \begin{bmatrix} p_{11} & p_{12} & p_{13} \\ p_{21} & p_{22} & p_{23} \\ p_{31} & p_{32} & p_{33} \end{bmatrix} \tag{6.26}$$

represents the control points, and

$$\mathbf{U} = [1 \;\; u \;\; u^2] \qquad \mathbf{V} = [1 \;\; v \;\; v^2], \tag{6.27}$$

where $u, v \in [0, 1]$ are the parametric variables of the surface.

Let us now generate the control points that will correspond to a sub-patch of this patch. Consider the reparameterization of the surface by setting $u_1 = 0.5u$ and $v_1 = 0.5v$, substituting the new parameters into (6.24), the sub-patch is given by

$$\begin{aligned} \mathbf{S}(u_1, v_1) &= \mathbf{U}_1\,\mathbf{M}\,\mathbf{P}\,\mathbf{M}^T\,\mathbf{V}^T \\ &= \mathbf{U}\,\mathbf{L}\,\mathbf{M}\,\mathbf{P}\,\mathbf{M}^T\,\mathbf{L}^T\,\mathbf{V}_1^T \end{aligned} \tag{6.28}$$

where the matrix $\mathbf{L}$ is

$$\mathbf{L} = \frac{1}{4}\begin{bmatrix} 4 & 0 & 0 \\ 0 & 2 & 0 \\ 0 & 0 & 1 \end{bmatrix}. \tag{6.29}$$

Since (6.24) represents part of the same quadratic B-spline surface as (6.28) does under a new set of control points $\mathbf{Q}$, we have

$$\mathbf{S}(u, v) = \mathbf{U}\,\mathbf{M}\,\mathbf{Q}\,\mathbf{M}^T\,\mathbf{V}^T. \tag{6.30}$$

Equating (6.24) and (6.28), for any value of u and v we have,

$$\mathbf{M}\,\mathbf{Q}\,\mathbf{M}^T = \mathbf{L}\,\mathbf{M}\,\mathbf{P}\,\mathbf{M}^T\,\mathbf{L}^T. \tag{6.31}$$

This allows the computation of a new set of control points,

$$\mathbf{Q} = \mathbf{H}_1\,\mathbf{P}\,\mathbf{H}_1^T, \tag{6.32}$$

where

$$\mathbf{H}_1 = \mathbf{M}^{-1}\,\mathbf{L}\,\mathbf{M} = \frac{1}{4}\begin{bmatrix} 3 & 1 & 0 \\ 1 & 3 & 0 \\ 0 & 3 & 1 \end{bmatrix}. \tag{6.33}$$

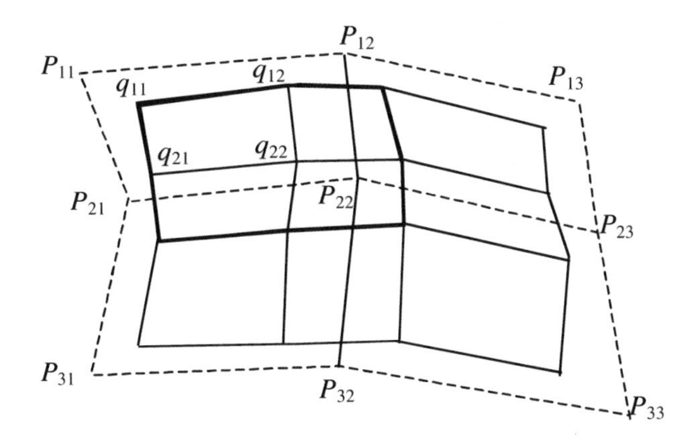

Figure 6.8 Quadratic B-spline patch splitting.

Similar to the cubic case explained in the last section, $\mathbf{H}_1$ is also called the *splitting matrix* for quadratic B-spline patches. This matrix codes the relationship between the original control points and the new ones after a patch split.

The new and old control points of a quadratic B-spline patch are depicted in Figure 6.8, where $\mathbf{P} = [p_{ij}]$ denotes the old control mesh and $\mathbf{Q} = [q_{ij}]$ denotes the new one. Their relationship is defined by (6.32), i.e. any point q_{ij} of the new mesh is a linear combination of p_{ij} of the old control mesh. Expanding the elements of the matrix, we have

$$\mathbf{Q} = \mathbf{H}_1\,\mathbf{P}\,\mathbf{H}_1^T = \frac{1}{16}\begin{bmatrix} q_{11} & q_{12} & q_{13} \\ q_{21} & q_{22} & q_{23} \\ q_{31} & q_{32} & q_{33} \end{bmatrix}, \tag{6.34}$$

where

$$\begin{aligned}
q_{11} &= \tfrac{1}{16}[3(3p_{11}+p_{21})+(3p_{12}+p_{22})] \\
q_{12} &= \tfrac{1}{16}[(3p_{11}+p_{21})+3(3p_{12}+p_{22})] \\
q_{13} &= \tfrac{1}{16}[3(3p_{12}+p_{22})+(3p_{13}+p_{23})] \\
q_{21} &= \tfrac{1}{16}[3(p_{11}+3p_{21})+(p_{12}+3p_{22})] \\
q_{22} &= \tfrac{1}{16}[(p_{11}+3p_{21})+3(p_{12}+3p_{22})] \\
q_{23} &= \tfrac{1}{16}[3(p_{12}+3p_{22})+(p_{13}+3p_{23})] \\
q_{31} &= \tfrac{1}{16}[3(3p_{21}+p_{31})+(3p_{22}+p_{32})] \\
q_{32} &= \tfrac{1}{16}[(3p_{21}+p_{31})+3(3p_{22}+p_{32})] \\
q_{33} &= \tfrac{1}{16}[3(3p_{22}+p_{32})+(3p_{23}+p_{33})].
\end{aligned} \tag{6.35}$$

6.3.3 Doo–Sabin Subdivision Rules

From the splitting process it is clear that the new points are computed only on a face-by-face basis, meaning that the vertices of a new face are only dependent on those of the corresponding old face. Let $V^i = (v_1^i, \ldots, v_n^i)^T$ be the vertices of an n-sided face of mesh M^i and $V^{i+1} = (v_1^{i+1}, \ldots, v_n^{i+1})^T$ be the vertices of the corresponding new face of mesh M^{i+1}. Then the subdivision rule can be stated as follows:

$$V^{i+1} = \sum_{j=1}^{n} \alpha_{ij} V^i, \tag{6.36}$$

where coefficients α_{ij} are defined by

$$\alpha_{ij} = \frac{n+5}{4n} \quad \text{for } i = j \tag{6.37}$$

$$\alpha_{ij} = \frac{3 + 2\cos[2\pi(i-j)/n]}{4n} \quad \text{for } i \neq j. \tag{6.38}$$

Once all new points are created, a new mesh M^{i+1} can be generated by connecting them into faces. A mesh consists of three types of faces: *F-faces*, *E-faces* and *V-faces*, as illustrated by Figure 6.9 (Nasri 1987).

F-face

For each face F in the old mesh, a new face is constructed by connecting all the new points of the corresponding old face F. A new face so produced is called an F-face (Figure 6.9a).

E-face

For each edge shared by two faces F_1 and F_2, a new face is constructed by connecting the new points of face F_1 and F_2 corresponding to the end points of the edge. A new face so produced is called an E-face (Figure 6.9b). Clearly, every E-face will be four-sided.

V-face

Except at the boundaries, for each n-valent vertex V of a mesh where n faces meet, a new face is constructed by connecting all the new points of the surrounding faces, which corresponds to vertex V. This new face is called a V-face (Figure 6.9c).

From these subdivision rules, it can be observed that a face of a mesh will produce a sequence of F-faces that in the end converge to its centroid in the limit. Therefore the centroid of each face is definitely on the limit surface.

Figure 6.10 illustrates the generation of a Doo–Sabin subdivision surface.

6.3.4 Properties of Doo–Sabin Subdivision Surfaces

Observation of the subdivision process offers the following properties of the Doo–Sabin surfaces.

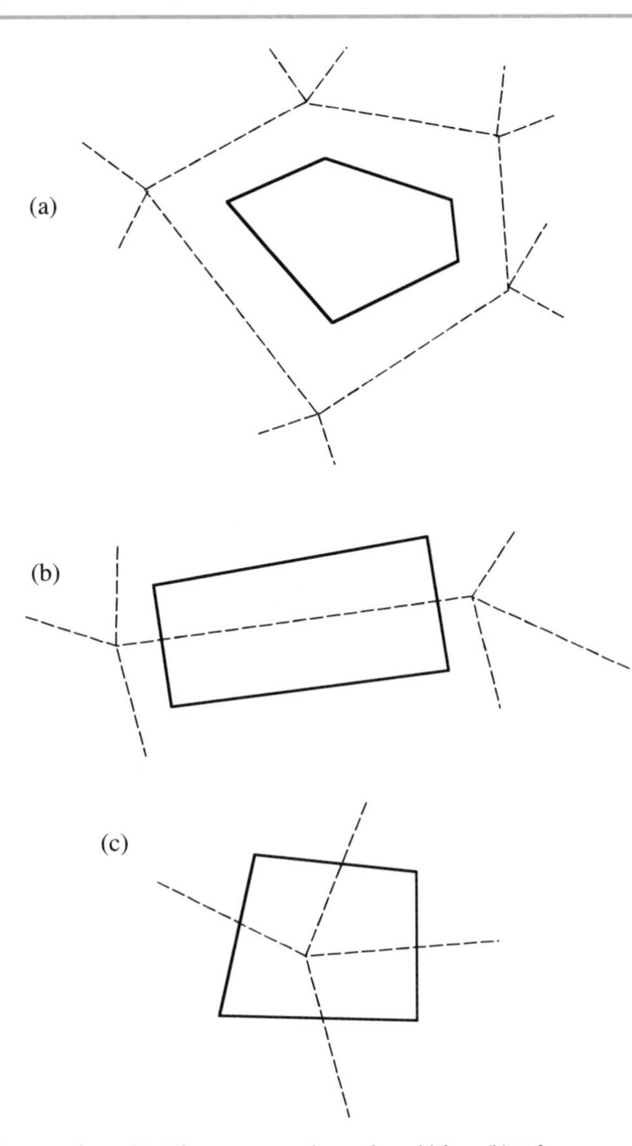

Figure 6.9 Generating new faces. (a) F-face constructed out of an old face; (b) E-face constructed out of an old edge; (c) V-face constructed out of an old vertex.

- The control mesh and the generated surface can be of any arbitrary topology.
- All vertices become 4-valent after the first subdivision. Each subdivision step creates more four-sided faces due to the increase of the E-faces which are quadrilateral.
- An n-sided face will create a new smaller n-sided face. Therefore the number of non-quadrilateral faces is fixed during the subdivision. Since a non-quadrilateral face cannot be covered using an ordinary surface model, they are referred to as *anomalous regions*. With successive subdivision of a mesh, the anomalous regions get smaller and finally converge to the extraordinary points.

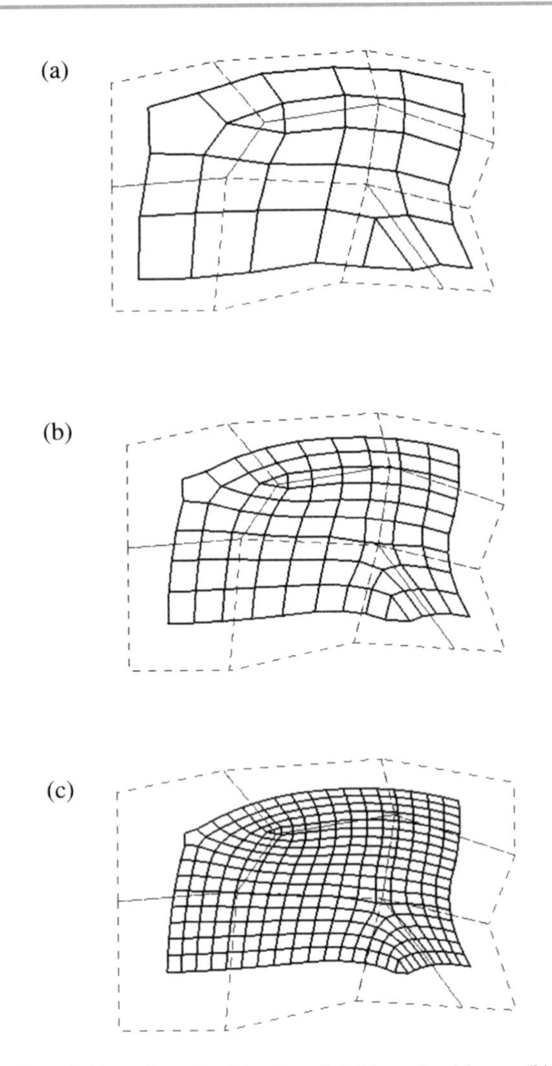

Figure 6.10 Generation of Doo–Subin surfaces. Applying the subdivision rules: (a) once; (b) twice; (c) three times. (d) Doo–Subin surface.

- As the new vertices of a face are an affine combination of the old ones, the subdivision is affinely invariant, and the limit surfaces have the convex hull property.
- The centroid of any face of each mesh is on the limit surface. This is because the faces converge towards the centroids by the subdivision rules.
- The surface produced is overall quadratic and has a well-defined tangent plane, but the curvature is not continuous (Doo and Sabin 1978). Therefore a Doo–Sabin subdivision surface is always smooth regardless of its topology.

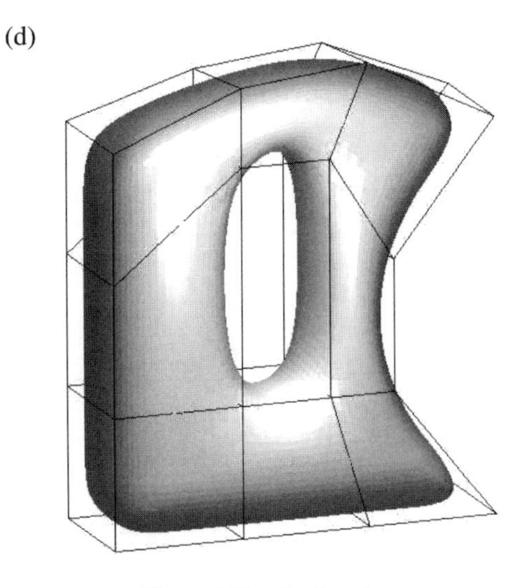

Figure 6.10 Continued.

6.3.5 Other Comments

The process of generating a Doo–Sabin subdivision surface is similar to that of a Catmull–Clark surface, except that the Doo–Sabin surface is quadratic. A surface patch is generated by applying the subdivision rules repeatedly to a mesh until a predetermined precision is reached, i.e. the size of the faces satisfies the resolution requirement. Then the surface is approximated by the polygons obtained after the subdivision steps.

However, after a few steps, most faces are quadrilateral except for a small number of anomalous faces. This suggests that standard surface schemes can be employed for the construction of the limit surface. A usual approach is to cover the regular regions with traditional surface patches, such as B-splines. The irregular regions will be further subdivided until the size is small enough for the application.

6.4 Triangular Patch-based Spline Surfaces

In this section we introduce another approach which is also capable of generating a smooth surface over an irregular mesh. The differences between this approach and the above two are: first, the final surface is not a limit surface of the mesh that merely acts as a control polyhedron; secondly, the final surface is composed of Bézier triangular patches rather than quadrilateral ones. Although quadrilaterals have many advantages, especially in rendering, triangular patches are more flexible in filling irregular areas. Unlike the above subdivision surfaces, each patch of the final surface is represented by an equation. Therefore it is easier to work out the coordinate values and normals. It is also easier to analyze the geometric properties, as Bézier triangular patches have been well studied.

6.4.1 Triangular Bézier Patches

As triangular Bézier patches are an important part of this scheme, in this section we will briefly review this subject.

Triangular Bézier patches (Choi 1991) are an extension of the ordinary quadrilateral Bézier surface patches. Their shape is determined by the control net which is of a triangular structure. To define such a triangular patch, let us first define a degree-n bivariate Bernstein polynomial as follows:

$$B^n_{ijk}(u, v, w) = \frac{n!}{i!\,j!\,k!} u^i v^j w^k \quad \text{for } 0 \le i, j, k \le n \quad \text{and} \quad i + j + k = n, \tag{6.39}$$

where u, v, w are known as the *barycentric coordinates*, which are subject to

$$u + v + w = 1 \quad \text{and} \quad 0 \le u, v, w \le 1. \tag{6.40}$$

Usually the following notation is employed:

$$\begin{aligned} \mathbf{i} &= (i, j, k) \\ |\mathbf{i}| &= i + j + k \\ |\mathbf{u}| &= u + v + w. \end{aligned} \tag{6.41}$$

Therefore an n-degree triangular Bézier patch is defined as

$$\mathbf{r}(\mathbf{u}) = \sum_{|\mathbf{i}|=n} \mathbf{V}_{\mathbf{i}} B^n_{\mathbf{i}}(\mathbf{u}), \tag{6.42}$$

where $\mathbf{V}_{\mathbf{i}}$ denotes the control points and $B^n_{\mathbf{i}}(\mathbf{u})$ denotes the bivariate n-degree Bernstein polynomials.

Owing to the use of barycentric coordinate notation, both Bernstein polynomials and the control points of such a bivariate patch are indexed using three integers. For example, in the cubic case, the control net will consist of vertices

$$\begin{array}{c} \mathbf{V}_{030} \\ \mathbf{V}_{021}\ \mathbf{V}_{120} \\ \mathbf{V}_{012}\ \mathbf{V}_{111}\ \mathbf{V}_{210} \\ \mathbf{V}_{003}\ \mathbf{V}_{102}\ \mathbf{V}_{201}\ \mathbf{V}_{300}. \end{array} \tag{6.43}$$

The corresponding cubic patch is shown in Figure 6.11.

6.4.2 Derivatives of Triangular Bézier Patches

For a tensor product surface patch, the differentiation is usually in the form of partial derivatives with respect to the parameters u and v. For a triangular patch, however, because of the baricentric parameterization, it is more appropriate to use directional derivatives.

A direction at the parametric domain of a triangular patch is defined by two points in the domain (Farin 1997), $\mathbf{u}_0$ and $\mathbf{u}_1$, where $\mathbf{u}_1 = \mathbf{u}_0 + \Delta\mathbf{u}$ and $\Delta\mathbf{u} = (\Delta u, \Delta v, \Delta w)$. The

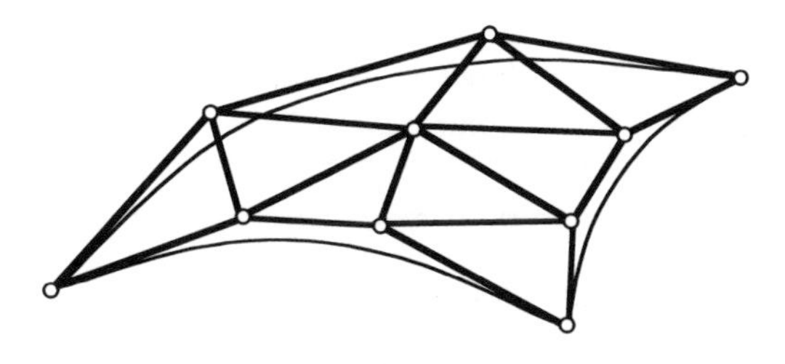

Figure 6.11 Triangular Cubic Bézier patch.

direction is specified by $\mathbf{u}_1 - \mathbf{u}_0 = \Delta\mathbf{u}$. Thus the directional derivative of a surface at $\mathbf{r}(\mathbf{u})$ with respect to $\Delta\mathbf{u}$ is defined by

$$D_d\mathbf{r}(\mathbf{u}) = \Delta u\, \mathbf{r}_u(\mathbf{u}) + \Delta v\, \mathbf{r}_v(\mathbf{u}) + \Delta w\, \mathbf{r}_w(\mathbf{u}). \tag{6.44}$$

This expression computes the tangent vector of the surface $\mathbf{r}(\mathbf{u})$ along the direction of $\Delta\mathbf{u}$. Here the partial directive $\mathbf{r}_u(\mathbf{u})$ is given by

$$\begin{aligned}
\mathbf{r}_u(\mathbf{u}) = \frac{\partial}{\partial u}\mathbf{r}(\mathbf{u}) &= \frac{\partial}{\partial u} \sum_{|\mathbf{i}|=n} \frac{n!}{i!\,j!\,k!} u^i v^j w^k\, \mathbf{V}_{\mathbf{i}} \\
&= n \sum_{|\mathbf{i}|=n} \frac{(n-1)!}{(i-1)!\,j!\,k!} u^{i-1} v^j w^k\, \mathbf{V}_{\mathbf{i}} \\
&= n \sum_{|\mathbf{i}|=n-1} \frac{(n-1)!}{i!\,j!\,k!} u^{i-1} v^j w^k\, \mathbf{V}_{i+1,j,k} \\
&= n \sum_{|\mathbf{i}|=n-1} \mathbf{V}_{\mathbf{i}+\mathbf{e}_1} B_{\mathbf{i}}^{n-1}(\mathbf{u}).
\end{aligned} \tag{6.45}$$

Owing to symmetry, the other two partial directives $\mathbf{r}_v(\mathbf{u})$ and $\mathbf{r}_w(\mathbf{u})$ can be formulated in the same way.

Substituting (6.45) into (6.44), we obtain the directional derivative

$$D_u\mathbf{r}(\mathbf{u}) = n \sum_{|\mathbf{i}|=n-1} (\Delta u\, \mathbf{V}_{\mathbf{i}+\mathbf{e}_1} + \Delta v\, \mathbf{V}_{\mathbf{i}+\mathbf{e}_2} + \Delta w\, \mathbf{V}_{\mathbf{i}+\mathbf{e}_3}) B_{\mathbf{i}}^{n-1}(\mathbf{u}), \tag{6.46}$$

where $\mathbf{e}_1 = (1, 0, 0)$, $\mathbf{e}_2 = (0, 1, 0)$ and $\mathbf{e}_3 = (0, 0, 1)$. Also note that since $|\mathbf{u}| = u+v+w = 1$, $|\Delta\mathbf{u}| = \Delta u + \Delta v + \Delta w = 0$. Comparing (6.46) with (6.42), it can be observed that the directional directives of a degree-n patch are computed with degree $n-1$ Bernstein polynomials.

Also, the first-order cross-boundary directives are determined by the control points of the boundary and those of the next row to the boundary. This property is very useful in the construction of a G^1 continuous patch composed of two or more triangular Bézier patches.

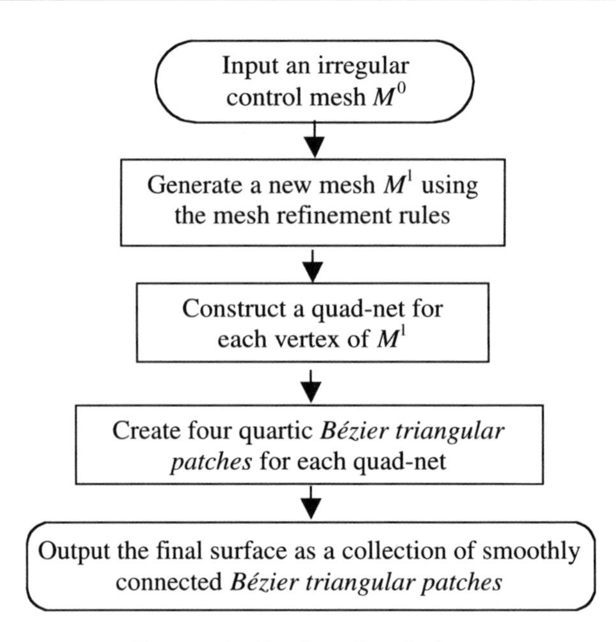

Figure 6.12 Flowchart of Loop's scheme.

6.4.3 Loop Spline Surfaces

In this section we introduce Loop's scheme (Loop 1994), which takes as input a mesh of irregular topology and generates a G^1 continuity surface consisting of triangular Bézier patches. The following flowchart (Figure 6.12) outlines the steps of Loop's scheme.

6.4.3.1 Mesh Refinement

From the above flowchart, it can be seen that Loop's scheme combines mesh subdivision with ordinary parametric surfaces. Unlike the previously introduced subdivision surfaces, this scheme has no intention of subdividing the initial mesh without limit. In fact, it is a spline surface generation scheme with a pre-processing stage. This pre-processing is a one-step subdivision (refinement).

The refinement step is to subdivide the initial mesh M^0 once, so that the irregularities are removed or isolated. The irregularities of a mesh include vertices that are not 4-valent and faces that are not quadrilateral. After the refinement, a new mesh, called M^1 is produced, which will have only 4-valent vertices. Furthermore, every non-quadrilateral face of mesh M^1 will be surrounded by four-sided faces.

The refinement step produces a vertex of M^1 for each {vertex, face} pair of the initial mesh M^0. For a face F of M^0 and its n surrounding vertices $p_0, p_1, \ldots, p_{n-1}$, the generated point p_i^1 of M^1 corresponding to a pair $\{p_i, F\}$ is given by

$$p_i^1 = \tfrac{1}{4}O + \tfrac{1}{8}p_{i-1} + \tfrac{1}{2}p_i + \tfrac{1}{8}p_{i+1}, \tag{6.47}$$

where O denotes the centroid of face F. Compared with (6.24), it is easy to see that this

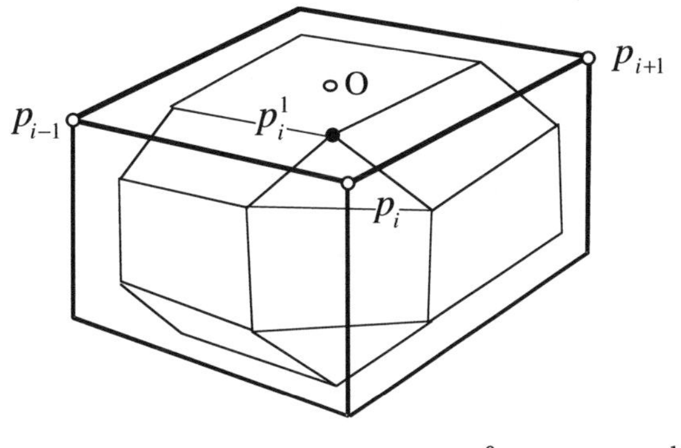

Figure 6.13 Mesh refinement. ——, mesh M^0; ——, mesh, M^1.

subdivision rule is very similar to that of the Doo–Sabin scheme. But we only need to apply it to the initial mesh once.

The faces of the refined mesh M^1 are constructed corresponding to either a vertex, a face or an edge of M^0. So there are three types of faces in the new mesh M^1.

- A face corresponding to face F of M^0. It is constructed by connecting the points p_i^1 $(i = 0, \ldots, n-1)$ of M^1. Each p_i^1 is created corresponding to the $\{p_i, F\}$ pairs, where p_i denotes all the vertices of face F.
- A face corresponding to vertex p of M^0. It is constructed by connecting the points p_j^1 $(j = 0, \ldots, m-1)$ of M^1. Each p_j^1 is created corresponding to the $\{p, F_j\}$ pairs, where F_j are all the faces that share vertex p.
- A face corresponding to an edge (p_i, p_{i+1}) of M^0. It is constructed from four points generated from four vertex–face pairs: $\{p_i, F_0\}$, $\{p_i, F_1\}$, $\{p_{i+1}, F_0\}$, $\{p_{i+1}, F_1\}$ where F_0 and F_1 are two faces sharing this edge.

The following observations can be made from this refinement:

- each k-valent vertex of M^0 will generate a k-sided face in M^1;
- each n-sided face of M^0 will generate an n-sided face in M^1;
- each edge of M^0 will generate a four-sided face in M^1.

Figure 6.13 illustrates this refinement process.

6.4.3.2 Quad-net Construction

Quad-nets are used as an intermediate structure between the refined mesh M^1 and the final surface. After the previous refinement, each vertex of mesh M^1 is 4-valent. In other words, there are exactly four faces surrounding each vertex of M^1. Like the Doo–Sabin surfaces, we hope the final surface constructed will interpolate the centroids of all faces of the new mesh and form an overall smooth surface. In achieving this, around each vertex V of mesh M^1, we construct a quadrilateral region with the centroids of the four surrounding faces being its corner points. In addition, another 12 points are placed in this region totaling 16 points, as shown in Figure 6.14. These 16 points, together with the side numbers n_0 and n_1 of the two faces corresponding to corner points A_{00} and A_{33}, respectively, are collectively referred to as a *quad-net*. And these 16 points are known as the *quad-net points*.

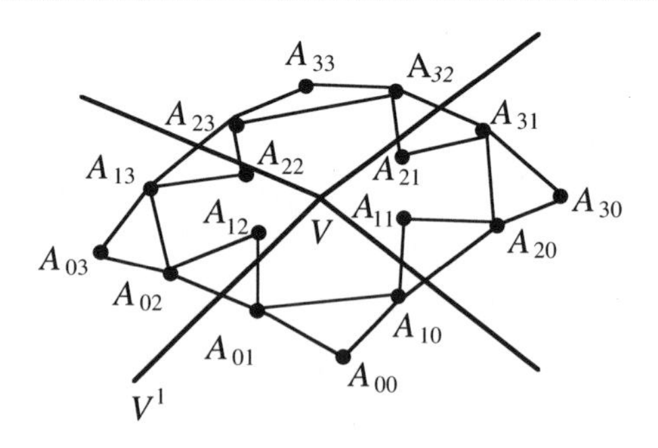

Figure 6.14 Quad-net corresponding to vertex V of M^1.

Such a quad-net is used to define a surface patch, which interpolates the centroids of the four faces joining vertex V. This surface patch will be bounded by four cubic Bézier curves, the control polygons of which are represented by $A_{00}, A_{10}, A_{20}, A_{30}$; $A_{30}, A_{31}, A_{32}, A_{33}$; $A_{33}, A_{23}, A_{13}, A_{03}$ and $A_{03}, A_{02}, A_{01}, A_{00}$, respectively. To ensure functional continuity across the patches, these four cubic boundary curves will be shared by the neighboring quad-nets. The other four quad-net points, A_{12}, A_{22}, A_{11}, A_{21}, are positioned to ensure the G^1 continuity with the adjacent surface patches.

A surface patch will be generated for each quad-net. However, apart from some special cases, each such surface patch is not a single patch, it consists of four quartic triangular Bézier patches, as illustrated in Figure 6.15. The smoothness of the final surface, therefore, relies on the functional and tangential continuity between:

- the two connected triangular patches of any two neighboring quad-nets, and
- the four triangular patches within a quad-net.

The 16 quad-net points will be generated fulfilling these two requirements.

6.4.3.3 Generation of Quad-net Points

First, let us discuss the smooth connection of two adjacent quad-nets. In essence, this is about the smooth connection of the triangular patches of each quad-net that share the quad-net points of the common boundary. To make both patches share the same tangent plane along the boundary curve, i.e. to have G^1 continuity, the first two rows of the quad-net points of both patches must satisfy certain constraints, which are (Loop 1994)

$$(1-c)A_{00} + cA_{01} = \tfrac{1}{2}A_{10} + \tfrac{1}{2}\hat{A}_{10} \tag{6.48}$$

$$\tfrac{1}{2}A_{01} + \tfrac{1}{2}A_{02} = \tfrac{1}{2}A_{12} + \tfrac{1}{2}\hat{A}_{12} \tag{6.49}$$

$$A_{03} = \tfrac{1}{2}A_{13} + \tfrac{1}{2}\hat{A}_{13} \tag{6.50}$$

where $\hat{A}_{10}$, $\hat{A}_{12}$ and $\hat{A}_{13}$ are the corresponding points of an adjacent quad-net, and c is a scalar to be determined. By solving these constraints, we will obtain the quad-net points associated with one boundary of the quad-net.

Since A_{00} is the centroid of a face of mesh M^1, constraint (6.48) indicates that all the quad-net points surrounding A_{00}, such as A_{01}, A_{10} and $\hat{A}_{10}$, must be coplanar. Therefore we can generate these quad-net points based on this observation.

Let $P_0, \ldots, P_{n-1} \in R^3$ be the vertices of the face whose centroid is A_{00}, the quad-net points $Q_0, \ldots, Q_{n-1}$ surrounding A_{00} can be found by

$$Q_i = \frac{1}{n}\sum_{j=0}^{n-1} P_j \left\{1 + \frac{2}{3}\left(1 + \cos\frac{2\pi}{n}\right)\left[\cos\frac{2\pi(j-i)}{n} + \tan\frac{\pi}{n}\sin\frac{2\pi(j-i)}{n}\right]\right\}. \tag{6.51}$$

It is easy to see that all points Q_i computed above satisfy the following expression. Hence they are coplanar.

$$\left(1 - \cos\frac{2\pi}{n}\right)O + \cos\frac{2\pi}{n}Q_i = \frac{1}{2}Q_{i-1} + \frac{1}{2}Q_{i+1}, \tag{6.52}$$

where

$$O = \frac{1}{n}\sum_{j=0}^{n-1} P_j \tag{6.53}$$

is the centroid of the face. If we choose $c = \cos(2\pi/n)$, and interpret O as the centroid A_{00}, constraint (6.48) is satisfied, which gives us three quad-net points, A_{01}, A_{10} and $\hat{A}_{10}$.

Constraint (6.50) is a special case of constraint (6.48) when the face whose centroid is A_{03} is four-sided, i.e. $n = 4$. This is because $c = \cos(2\pi/n) = 0$ when $n = 4$. As every n-sided face of M^1 ($n \neq 4$) is surrounded by four-sided faces, A_{02}, A_{13} and $\hat{A}_{13}$ are also computed using (6.51).

The inner points A_{11}, A_{12}, A_{21} and A_{22} are constructed to help maintain the cross-boundary conditions. Let V be the vertex about which the quad-net is constructed and V^1 be the vertex of the neighboring quad-net. Both quad-nets share the same boundary, as shown in Figure 6.14. If we compute A_{12} and $\hat{A}_{12}$ by

$$\begin{aligned} A_{12} &= \tfrac{1}{2}A_{01} + \tfrac{1}{2}A_{02} + \tfrac{1}{6}(V - V^1) \\ \hat{A}_{12} &= \tfrac{1}{2}A_{01} + \tfrac{1}{2}A_{02} + \tfrac{1}{6}(V^1 - V) \end{aligned} \tag{6.54}$$

constraint (6.49) is immediately satisfied by adding together these two expressions.

We have now obtained all quad-net points of the first two rows corresponding to a boundary curve. The points of the other three boundaries of a quad-net are similarly obtainable.

6.4.3.4 Generation of Triangular Bézier Patches

Once all quad-net points are obtained, we are in a position to construct the parametric patches. As stated before, each surface patch comprises four quartic triangular Bézier patches, and the key point is to make sure these four patches are connected with G^1 continuity.

For each quad-net, the Bézier triangles and the labeling scheme of the control nets are shown in Figure 6.15. The control points of these triangular patches are derived on the condition that G^1 smoothness is maintained between all four patches, and between the quad-net and the adjacent quad-nets (Loop 1994). In the following, we will take one of the Bézier triangles as an example to illustrate the construction process and the rest can be similarly constructed.

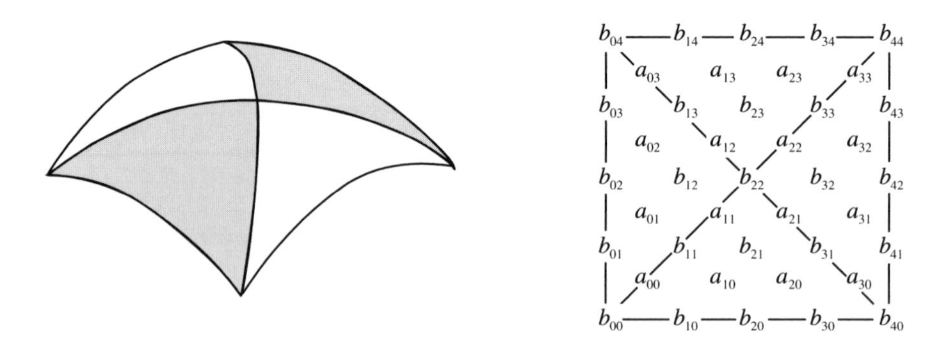

Figure 6.15 Construct four quartic Bézier triangles for each quad-net.

First of all, the boundary curves of the quad-net need to be interpolated. This leads to the boundary control points given by

$$\begin{aligned} b_{00} &= A_{00} \\ b_{01} &= \tfrac{1}{4}A_{00} + \tfrac{3}{4}A_{01} \\ b_{02} &= \tfrac{1}{2}A_{01} + \tfrac{1}{2}A_{02} \\ b_{03} &= \tfrac{3}{4}A_{02} + \tfrac{1}{4}A_{03} \\ b_{04} &= A_{03}. \end{aligned} \tag{6.55}$$

The control points in the rows next to the boundary are found by maintaining the cross-boundary continuity condition, and are defined by

$$\begin{aligned} a_{00} &= \frac{1}{2}b_{10} + \frac{1}{2}b_{01} \\ a_{01} &= \frac{c}{8}A_{00} + \frac{3-3c}{8}A_{01} + \frac{c}{4}A_{02} + \frac{1}{8}A_{10} + \frac{1}{2}A_{12} \\ a_{02} &= \frac{3-c}{8}A_{02} + \frac{c}{8}A_{03} + \frac{1}{2}A_{12} + \frac{1}{8}A_{13} \\ a_{03} &= \frac{1}{2}b_{03} + \frac{1}{2}b_{14}. \end{aligned} \tag{6.56}$$

For computing a_{01}, a_{02}, a_{10} and a_{20}, we set $c = \cos(2\pi/n_0)$, and for a_{31}, a_{32}, a_{13} and a_{23}, $c = \cos(2\pi/n_1)$. Point b_{12} has no effect on the tangential continuity conditions. Loop (1994) suggests the following equation for this point should be created:

$$b_{12} = \tfrac{7}{8}A_{12} + \tfrac{1}{8}(A_{21} - A_{11} - A_{22}) + \tfrac{3}{16}(A_{10} + A_{13}) - \tfrac{1}{16}(A_{00} + A_{03}). \tag{6.57}$$

The remaining points are constructed to ensure that the cross-boundary conditions internal to the quad-net are C^1 continuous:

$$\begin{aligned} b_{11} &= \tfrac{1}{2}a_{10} + \tfrac{1}{2}a_{01} \\ a_{11} &= \tfrac{1}{2}b_{21} + \tfrac{1}{2}b_{12} \\ b_{22} &= \tfrac{1}{2}a_{12} + \tfrac{1}{2}a_{21}. \end{aligned} \tag{6.58}$$

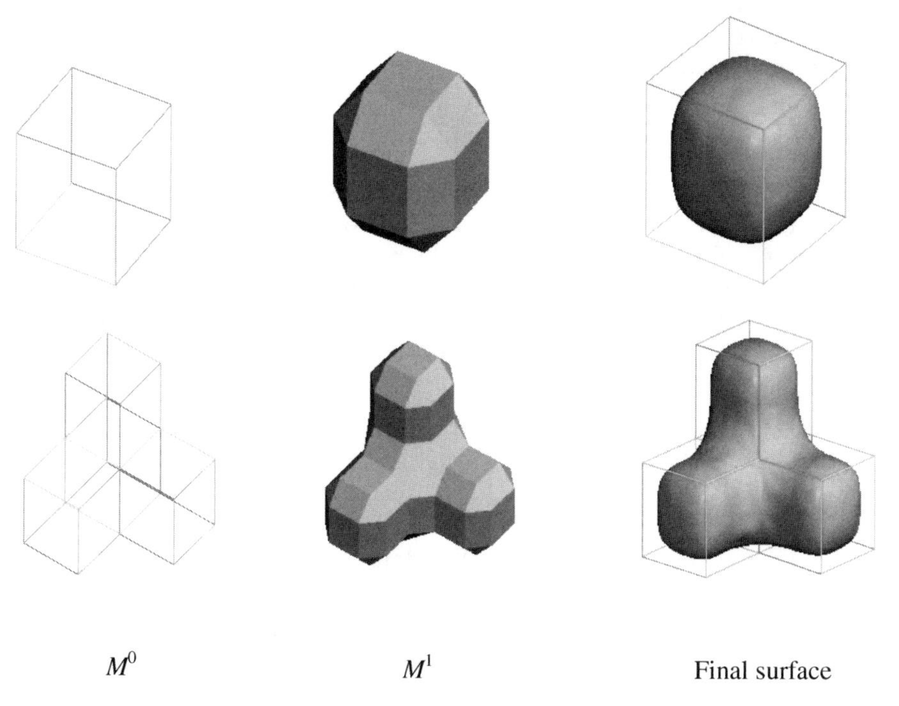

Figure 6.16 Loop spline surfaces.

The constructed quartic triangular Bézier patches are collectively output as a composite spline surface. Figure 6.16 shows two simple examples constructed using Loop's scheme.

6.4.4 Properties of Loop Spline Surfaces

The properties of Loop spline surfaces arise from the fact that they are the result of a combination of mesh subdivision and smooth connection of a network of triangular Bézier patches.

- The control mesh and hence the generated surface can be of any arbitrary topology consisting of n-sided faces and n-valent vertices.
- The new mesh comprises only 4-valent vertices after the first and only mesh subdivision step.
- An n-sided face of the initial mesh will create a new smaller n-sided face after the subdivision step.
- The centroid of any face of the new mesh will be interpolated by the generated spline surface.
- The new mesh is composed of quad-nets, each of which is usually covered by four quartic triangular Bézier patches. In some special cases, cubic triangular patches may be used instead of the quartics, and over regular regions of the mesh, a bi-quadratic Bézier patch may be used to cover a quad-net (Loop 1994).
- Overall, this scheme produces a spline surface with G^1 continuity.

- Since the surface produced is represented by an analytical formula, the surface evaluation is computationally efficient.

6.4.5 Other Comments

The refinement step is very similar to that of the Doo–Sabin subdivision, except that only one subdivision step is needed. Then the quad-nets will be constructed by generating all the quad-net points. Every such quad-net will be covered by four quartic triangular Bézier patches, once all the control points of these patches arrive.

For each quartic triangular Bézier patch, (6.42) will be used for the evaluation of the coordinates of any points of the patch. From the index scheme of a triangular Bézier patch, the control net will consist of vertices as follows:

$$\begin{array}{c} \mathbf{V}_{040} \\ \mathbf{V}_{031}\ \mathbf{V}_{130} \\ \mathbf{V}_{022}\ \mathbf{V}_{121}\ \mathbf{V}_{220} \\ \mathbf{V}_{013}\ \mathbf{V}_{112}\ \mathbf{V}_{211}\mathbf{V}_{310} \\ \mathbf{V}_{004}\ \mathbf{V}_{103}\ \mathbf{V}_{202}\ \mathbf{V}_{301}\ \mathbf{V}_{400}. \end{array} \tag{6.59}$$

Inserting these into (6.42), the patch can be given explicitly by

$$\begin{aligned} \mathbf{r}(u, v, w) = {} & \mathbf{V}_{004}w^4 + 4\mathbf{V}_{103}uw^3 + 6\mathbf{V}_{202}u^2w^2 + 4\mathbf{V}_{301}u^3w \\ & + \mathbf{V}_{400}u^4 + 4\mathbf{V}_{013}vw^3 + 12\mathbf{V}_{112}uvw^2 + 12\mathbf{V}_{211}u^2vw \\ & + 4\mathbf{V}_{310}u^3v + 6\mathbf{V}_{022}v^2w^2 + 12\mathbf{V}_{121}uv^2w + 6\mathbf{V}_{220}u^2v^2 \\ & + 4\mathbf{V}_{031}v^3w + 4\mathbf{V}_{130}uv^3 + \mathbf{V}_{040}v^4. \end{aligned} \tag{6.60}$$

The partial directives $\mathbf{r}_u(\mathbf{u}) = \frac{\partial}{\partial u}\mathbf{r}(\mathbf{u})$ and $\mathbf{r}_v(\mathbf{u}) = \frac{\partial}{\partial v}\mathbf{r}(\mathbf{u})$ can be computed directly from (6.60) by taking the partial differentiation. They can also be computed as directional directives (6.46). For example, the directives along the boundary of $u = 0$ are given by

$$D_u\mathbf{r}(\mathbf{u})|_{u=0} = n \sum_{|\mathbf{i}|=n-1} (\Delta u\, \mathbf{V}_{\mathbf{i}+\mathbf{e}_1} + \Delta v\, \mathbf{V}_{\mathbf{i}+\mathbf{e}_2} + \Delta w\, \mathbf{V}_{\mathbf{i}+\mathbf{e}_3})B_{\mathbf{i}}^{n-1}(0, v, w) \tag{6.61}$$

and the direction along which the directive is taken depends on the combination of Δu, Δv and Δw. But note that the relationship $\Delta u + \Delta v + \Delta w = 0$ must be observed.

Figure 6.17 (see Plates) are some images from the project of deformable animation characters undertaken by the National Center for Computer Animation, Bournemouth University, UK (Zhao et al. 2000) where Loop spline surfaces are employed for the surface representation.

6.5 Conclusions

Surface topology is often a restricting factor for many traditional surface modeling techniques, including the most often used non-uniform rational B-splines (NURBS). These techniques require that the control structure be quadrilateral. Although they are capable of modeling the majority of surfaces for computer animation, such limitations have proven problematic for many applications, such as branches and irregular holes. The remedies to these problems often involve patchwork trimming, which is slow and prone to numerical errors, and the smoothness at the seams is difficult to maintain.

To eradicate the problem caused by irregular topology, a new trend of using subdivision surfaces has evolved in the recent releases of animation systems. Subdivision surfaces permit models of any topology to be generated and have been broadly welcomed by the computer animation community.

In this chapter we have introduced three surface modeling schemes that are all able to create smooth surfaces over an irregular mesh: Catmull–Clark subdivision surfaces, Doo–Sabin subdivision surfaces and Loop spline surfaces. Although they all involve subdivision and are able to cope with irregular topology, each of them adopts a different construction scheme and therefore has different characteristics.

6.5.0.1 Type of Final Surfaces

Although all of the above introduced schemes undertake subdivision to some extent, not all final surfaces are the limit of subdivision. Only the surfaces generated by the Catmull–Clark and Doo–Subin schemes are genuine subdivision surfaces. Loop surfaces are actually spline surfaces.

6.5.0.2 Degree of Continuity

All three modeling schemes construct smooth surfaces as a result, meaning that the final surfaces will at least have G^1 continuity. The Catmull–Clark subdivision surfaces, however, are G^2 continuous except at the extraordinary points. Over the regular areas, meshes can be covered with cubic B-spline surface patches.

6.5.0.3 Basic Surface Patches

It is understood that the basic surface patches for the Catmull–Clark subdivision are four-sided and are generalized cubic B-spline surfaces. The basic surface patches for the Doo-Sabin subdivision are generalized quadratic B-spline surfaces (also four-sided). Loop surfaces are, however, composed of triangular patches. Usually they are quartic triangular Bézier patches, although in some regions lower-degree four-sided patches are used.

6.5.0.4 Sharp and Semi-sharp Features

Sharp and semi-sharp features are important for modeling of animation characters. By modifying the subdivision rules or rearranging the control points, these schemes all have potential for producing such features. Currently, however, the Catmull–Clark subdivision surfaces are ahead of the other two schemes owing to the development by DeRose et al. (1998). Such features have made the subdivision surfaces even more powerful for computer animation.

6.5.0.5 Surface Expressions

Many traditional workers in the community of surface modeling are used to producing their surfaces with a given expression. On this grounds people may find the subdivision surfaces (Catmull–Clark's and Doo–Sabin's) hard to master. This is because these surfaces do not have an explicit expression for their representations. On the other hand, the Loop surface does provide such expressions, as the whole spline surface is made up of triangular Bézier patches whose evaluation is straightforward.

Acknowledgment

I would like to gratefully acknowledge Mr Yunfeng Zhao, Dr Jinjin Zheng and Mr Mark Hodgkin for the preparation of the illustrations. Gratitude also goes to Professor John Vince for his valuable suggestions.

Bibliography

Ball, A.A. and Storry J.T., Conditions for tangent plane continuity over recursively defined B-spline surfaces. *ACM Transactions on Graphics*, 7(2), 83–102, 1988.

Catmull, E. and Clark J., Recursively generated B-spline surfaces on arbitrary topological meshes. *Computer Aided Design*, **6**, 350–355, 1978

Choi, B., *Surface modeling for CAD/CAM*, Elsevier Science, 1991

DeRose, T., Kass, M. and Truong, T., Subdivision surfaces in character animation. *Computer Graphics (SIGGRAPH'98)*, pp. 85–94, 1998

Doo, D. and Sabin, M., Behaviour of recursive division surfaces near extraordinary points. *Computer-Aided Design*, **6**, 356–360, 1978

Farin, G., *Curves and Surfaces for Computer Aided Geometric Design*, 3rd edn, Academic Press, 1997

Halstead, M., Kass, M. and DeRose T., Efficient, fair interpolation using Catmull–Clark surfaces. *Computer Graphics (SIGGRAPH'93)*, pp. 35–44, 1993

Joy, K., *On-line Geometric Modeling Notes*, Computer Science Department, University of California, Davis, 1999

Loop, C., Smooth spline surfaces over irregular meshes. *Computer Graphics (SIGGRAPH'94)*, pp. 303–310, 1994

Nasri, A.H., Polyhedral subdivision methods for free-form surfaces. *ACM Transactions on Graphics*, **6**(1), 29–73, 1987

Zhao, Y., Zhang, J.J. and Comninos, P., A deformable leg model for computer animation, *The 5th World Conference on Integrated Design & Process Technology*, Dallas, Texas, 2000

7 Rendering and Shading

Ian Stephenson

7.1 Introduction

Computer animation takes place in a virtual 3D world. However it is normally visualized through a 2D screen made up of pixels. The rendering process is the computer animator's camera, which records the virtual world in a format that can be broadcast. While the rendering process involves many well-documented stages, these have often been considered in isolation, and it is not always clear how these techniques can be combined into a practical renderer.

This chapter takes a systems approach to rendering by considering the implementation of a simple renderer, and showing how the stages interact. Though by its nature this forces us to give preference to the approaches used by one particular renderer, note will be given, where possible, to the issues and implications of other design choices.

In order to provide a solid basis for discussion, we will base our renderer around the RenderMan standard (Upstill 1989; Apodaca and Gritz 1999; Pixar 2000), as this forces us to address practical problems rather than constructing some imaginary system. In particular, RenderMan includes procedural shading, which must be integrated with the geometry stages of the renderer. Procedural shading is a requirement for any production renderer and has serious implications for a renderer's construction, which only become apparent once a full system is considered. It is also presupposed that most surfaces will be curved rather than implementing a system to handle polygons and then forcing more interesting surfaces to conform to that model.

7.2 Input

Any rendering system must begin with the parsing of a scene description. The RenderMan standard defines both a C API and an RIB (RenderMan interface bytestream) file format, either of which allows a scene to be fed into the renderer. Any renderer that can support these forms of input may be used interchangeably. A modeling package simply generates an RIB, which allows an animator to select any compliant renderer, based on the requirements of the project. It also allows us to develop a renderer, which may be used for "real" work without concern as to how the scene is to be produced.

7.2.1 RIBs

Although the C API and the RIB interface are functionally almost identical, the RIB interface is simpler to discuss, as it avoids the lexical complexities of C. However in practice, our renderer would be written to implement the C API internally. A simple lexical analyzer (written in lex) can easily read in an RIB file and call the equivalent C call for each command with little regard to the semantics of the commands.

A complete description can be found in the RenderMan Specification (Pixar 2000), so we will restrict our discussion to the most interesting and relevant sections. The C API includes two versions of many commands in the forms *RiXxxx* and *RiXxxxV*. Although the *RenderMan Companion* (Upstill 1989) refers primarily to the former style, the second form is more useful. A simple library can convert *Ri* calls to the *RiV* form.

An RIB file consists of basically two parts: the setup phase which commences at the beginning of the file, and the world description, which contains the actual scene to be rendered, enclosed in *WorldBegin* and *WorldEnd* commands. Prior to the world description, the setup phase is essentially describing the virtual camera that will be used, both in terms of its position and its operation.

The nature of the camera is defined by setting a number of *Options*. These are parameters which apply to the whole scene, and include the *Projection* type, *Output Resolution*, and other standard options along with a mechanism (the *Option* command) to allow renderer-specific extensions. BMRT (Gritz and Hahn 1996), for example, uses such an extension to control the use of radiosity.

Any transformations prior to *BeginWorld* are taken to define the position of world space relative to the camera.

7.2.2 GState

One of the main tasks of the front end to a renderer is to manage the graphics state (*GState*). This consists of the *current transformation matrix*, and the *current attributes*.

Transformations are applied in RenderMan such that each transformation applies to all objects, which follow. For example,

```
Sphere 1 -1 1 360
Scale 0.5 0.5 0.5
Sphere 1 -1 1 360
Translate 1 0 0
Sphere 1 -1 1 360
```

creates a sphere of unit radius at the origin; a sphere of 0.5 at the origin; and a sphere of radius 0.5 at (0.5, 0, 0). To achieve this, each transformation is converted to matrix form and multiplied by the *current transformation matrix* (CTM), which it replaces ($\text{newCTM} = \text{oldCTM} \times M$).

In order to facilitate hierarchical modeling the current transformation can be stored and restored later using *TransformBegin* and *TransformEnd*. This is easily implemented as a linked list of matrices. All operations are performed on the final element on the list. *TransformBegin* copies the last element

```
newM->value=CTM->value;
newM->previous=CTM;
CTM=newM;
```

while *TransformEnd* removes it

```
CTM=CTM->previous;
```

Attributes are more general than transforms, which they include, along with such things as the current surface color and surface shader. These are managed in a similar way using *AtttributeBegin* and *AttributeEnd.* As for options, a renderer may add its own specific attributes using the *Attribute* command (for example *VectorMan* uses this to control the number of wires drawn for wire-frame renders). Similarly to transformations, attributes apply to all objects, which follow unless they are overwritten with a new value or popped from the attribute stack. In most cases the renderer need take no action other than to record the attribute at the point it is actually encountered.

7.2.3 Parameter Lists

The RenderMan API allows arbitrary variables to be passed into the renderer in the form of parameter lists. These are either standard, renderer-specific or more typically relate to the parameters of a shader. At this stage of the rendering system they simply need to be recorded for future use.

7.3 Objects

The primitive geometry types a renderer is required to implement have already been covered in other chapters, but each form of geometry provides different challenges to the renderer, as well as the animator. Having parsed its input, a renderer must therefore convert the geometry into a common form, which can be handled by the later stages. This is typically a regular mesh of quadrilaterals known as micro-polygons (MPs).

7.3.1 Quadrics

RenderMan supports a number of standard quadric objects such as a sphere, torus, cone, disk, hyperboloid and paraboloid. All of these can be represented simply by a function, which takes u and v surface coordinates and returns a point P in object space. These surfaces can therefore be reduced to a mesh of squares at arbitrary resolution – as will be seen in the following sections, this is ideal for shading, and hence these objects may be rendered trivially.

7.3.2 Patches

Patches within RenderMan may be linear, cubic (with arbitrary basis functions) or rational (NURBS). In addition, uniform patches may be specified individually or as part of meshes. Furthermore, a *PatchMesh* may be periodic or non-periodic. Regardless of these complexities patches are parametric, and hence can be handled identically to quadrics.

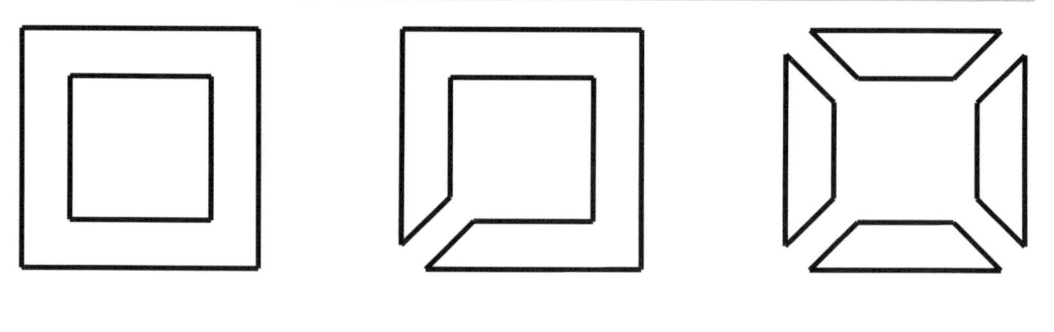

Figure 7.1 Splitting polygons.

7.3.3 Polygons

Although polygons are generally regarded as being trivial to render, they are in fact one of the hardest primitives within the original RenderMan standard to handle. While many texts deal with Phong shading large triangles, we need to reduce polygons to quads or triangles, which can be passed to the shading stage as patches.

Polygons come in three forms: convex, concave and concave with holes. A convex polygon may be trivially converted to quads (one of which may be degenerate, i.e. a triangle), which may then be shaded. Holes may be removed from a polygon by splitting along a line from the outside edge to a point on the hole. This is shown in Figure 7.1. Concave polygons may be converted into two or more convex polygons by repeatedly splitting them between two points, such that the new edge is inside the original shape and does not intersect any existing edges.

7.3.4 Subdivision Surfaces

Being non-parametric, subdivision surfaces (as discussed in Chapter 6) inherit some of the problems of polygons, in that they lack globally consistent texture coordinates. Repeated subdivision produces a polygonal mesh, which could be used to approximate the surface, but this would be difficult to shade well.

Fortunately, in the case of Catmull–Clark surfaces (Catmull 1974), following subdivision all faces become quadrilaterals, most of which are simply B-spline surfaces, and hence locally parametric. Those patches near extraordinary vertices (whose valence is other than 4, and hence are not B-splines) can still be evaluated parametrically, as shown in Stam (1998). We can therefore split a surface into a number of parametric patches for passing to the shading engine.

7.3.5 Points and Curves

Lightweight primitives are intended to be minimally shaded, either consisting of a single shading sample or a strip of samples (in the case of a curve). We can therefore pack these together into a pseudo-patch for shading. While this will mean that shading calculations based on derivatives are incorrect, such operations are meaningless for these primitives, as they have no width.

7.3.6 Blobby Objects

Pixar's Render Man (PRMan) 3.9 introduced implicit surfaces (Bloomenthal et al. 1997) in the form of the *Blobby* (Pixar 1999) command. While a number of techniques exist for polygonalizing isosurfaces (Duff 1992; Watt and Watt 1992; Heckbert 1994), simple polygonalization is a poor technique when procedural shading is being used. Standard techniques such as *Marching Cubes* produce a large number of small irregular polygons rather than the regular mesh, which the shading engine prefers, but this is a penalty that must currently be accepted, in exchange for the flexibility of implicit surfaces.

Normals for all object types may be calculated symbolically by providing a function that calculates **N** from u and v. If no such function may be simply derived (or simply has not yet been written) a surface normal may be calculated as

$$\mathbf{N} = \frac{dP}{du} \times \frac{dP}{dv}. \tag{7.1}$$

A renderer may use a mix of symbolic and analytical normals.

7.4 Transforms and Coordinate Systems

Objects are defined in their own object space. All points, vectors and normals must be converted into common world space for shading, but care must be taken in their application within the rendering pipeline.

The calculations of shading and texturing can be performed in any coordinate system. When coordinate systems are not explicitly specified within the shader code the renderer references a default space known as *current space*. This space is renderer dependent, but is usually equivalent to either the camera or object space. Object space may be efficient when the rendering is being done by ray tracing (ray–object intersections being difficult to calculate in any other space). World space is in principle the most appealing space in which to operate, but in practice camera space is most commonly used for scan-line renderers, as it allows all the linear transforms to be performed prior to shading. The first task of the back-end of our renderer is therefore to transform all the micro-polygons of our surface into camera space.

So far we have considered only points. However, we must distinguish between vectors used to represent points, free vectors and vectors that represent surface normals. These correspond to the shading language (SL) types *point*, *vector* and *normal*, and must be handled differently when transformed both by SL code and by the rendering pipeline. Points respond as expected to scaling, rotation, shear and translations. However, as vectors represent simply an offset to an arbitrary point they are invariant under translation. This can be handled simply in homogeneous coordinates by representing the vector $[x\ y\ z]$ as $[x\ y\ z\ 0]$ and the corresponding point as $(x, y, z, 1)$. When a translation matrix is applied, the vector is unchanged while the point is moved.

The case for normals is somewhat more complex. Consider the 2D case of a sine wave along the x-axis (Figure 7.2). If the graph is scaled in the x-direction (becoming flatter)

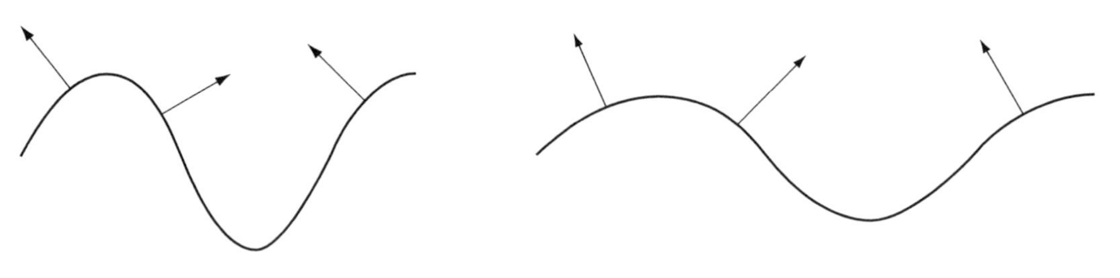

Figure 7.2 Transforming normals.

then the normals become smaller in x (more vertical). Mathematically for a transform M, normals must be scaled by the inverse of the transpose of M. In practice this is often equivalent to M, so the error may go unnoticed for some time – particularly as we rarely care about the magnitude of the normal, merely its direction.

7.5 Shading

The *shading engine* shades a patch in a single pass, rather than shading each point in turn. While certain simple shaders can be evaluated a point at a time, more complex operations require information about the surface, rather than just the point. For this reason the previous stages of the pipeline have reduced the geometry to a mesh. Provided that each facet of the mesh is smaller than one pixel in the output image, the resulting surface should look perfectly smooth.

The required resolution of the mesh to be shaded can be estimated by calculating the bounding box for the patch, and projecting it into screen space. Alternatively, a low-resolution mesh can be generated, and the micro-polygons in this mesh can be measured in screen space. These are only approximations, and degenerate geometry or poorly written shaders can give rise to artifacts. The RenderMan attribute *ShadingRate* allows the user to scale the resolution of the shaded mesh. A finer mesh gives a better approximation to the underlying surface, particularly when displacement is used, at the expense of render time, while a coarser mesh can dramatically speed up test renders.

For certain objects it may be found that certain parts of the mesh require very fine dicing while others require coarse dicing (for example, a ground plane extending from the foreground off into the distance). Making the micro-polygons too big leads to a lack of detail in the image, while choosing MPs that are too small leads to wasted effort and potential aliasing problems. For this reason patches may be split at this stage and each part shaded separately. Patches may also be split if the size of the mesh required to shade them correctly were too large, and hence would require more memory than is available.

7.5.1 The Shading Process

RenderMan supports a number of shader types. However, the most important for the shading of surfaces are *Displacement*, *Surface* and *Atmosphere*. The relationship between these is shown in Figure 7.3. Displacement may modify the position and normals of points on the mesh, while Surface calculates the color of the points, taking into account the observer

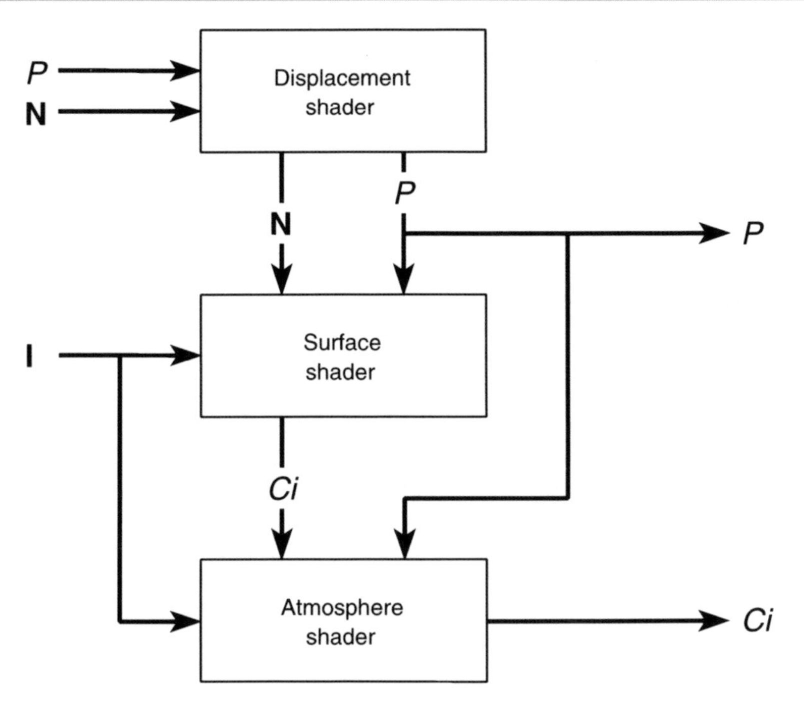

Figure 7.3 The shading process.

(viewing along a vector **I**) and the lights in the scene.[1] Atmosphere modifies this color to simulate the effect of the space between the surface and the observer.

Each of these are written as a function, which are run one after the other. Communication between the shaders is limited to the sharing of global variables. Following shading we will have final versions of *P* and *Ci* (the *Position* and *Color* of each micro-polygon) ready for drawing.

7.5.2 Shading Language

RenderMan shaders are written using SL – a high-level programming language designed for describing shaders. Though much like C, extended data types, operator overloading and special rendering functions allow shaders to be written far more succinctly than in more general programming languages. Conversely this narrow focus does present some limits to the operations that can be implemented.

A shader describes shading calculations in terms of a single point, and these are duplicated across the surface, and hence all features must be described implicitly rather than explicitly. For example, to draw a disk we must consider whether each point is inside the disk and color it appropriately rather than drawing the disk onto the surface. Rather than shading each point to completion, each operation in the code is executed for the whole surface before moving to the next operation.

[1] In practice certain renderers (particularly PRMan) allow displacement to be applied in surface shaders, though this can give rise to artifacts, and is non-portable.

This approach of defining a set of operations, and then performing these operations in parallel across a large data set is known as SIMD computing (single-instruction multiple data, see Flyn 1972). In addition to fitting well within the shading paradigm (where a surface must be colored), it can be implemented very efficiently, is ideal for hardware acceleration (Olano 2000) and allows operations such as the calculation of surface normals to be implemented far more easily than could be done if each point were shaded in turn.

7.5.3 The Shading Engine

The basic data structure on which the shader engine operates is a 2D array of points, each point being logically connected to its neighbors to form a regular patch. By this stage all geometry has been converted into this format. For each point we need to store shading variables such as P (position) and **N** (surface normal). Hence, we have an array of memory addresses associated with each point (or in SIMD terminology, node) in which these variables can be stored.

Though the logical layout of the array is `machine[upos][vpos][memloc]`, it is more efficient to store these as `machine[memloc][upos+vpos*width]`, as shown in Figure 7.4. This keeps all instances of each variable in one place which will give better cache performance on most processors – to execute `ss = s*2` we only access memory areas `machine[saddress]` and `machine[ssaddress]` rather than running through the whole of memory. Using a 1D array for the actual points simply makes indexing easier.

The SIMD Virtual Machine

Shaders are typically compiled to a virtual machine code, which is interpreted at run time. This simplifies the compilation process and allows compiled shaders to be used across a range of available hardware (in a renderfarm, for example).

While this approach may appear to be slow (as it is in the case of the Java virtual machine (VM) or the BCPL INTCODE system (Richards and Whitby-Stevens 1979)) this is in fact not the case when the technique is used for procedural shading. Any overhead in the interpretation of code is incurred due to the instruction decode – an instruction is read in, and the interpreter must decide how it should be executed. If it takes nine CPU cycles to

float machine[u][v][mem]

u v P N	u v P N
u v P N	u v P N

float machine[mem][u+uwidth*v]

u: 0 1 2 3
v: 0 1 2 3
N: 0 1 2 3
P: 0 1 2 3

Figure 7.4 The shading machine data structure.

work out that the instruction is an add operation, and then one cycle to perform the add, such a system would indeed be extremely slow. However in the case of shading, having decoded the add instruction we must apply it to (say) a 100×100 patch – 10 000 cycles. The instruction overhead has dropped from 90% of CPU time to 0.01%.

In addition, many shading operations can be added to the instruction set. So, for example, we add a noise function as a single instruction. This will take a significant time to execute, (noise accounts for up to 50% of the CPU time used by many typical shaders) again reducing the significance of the instruction decode.

The code of our shading engine is therefore simply a loop, which reads an instruction from the shader (often simply a text file), identifies the instruction and then performs that operation on all of the points on the surface.

```
while(fscanf(opstream,"%s",operation))
{
...
if(strcmp(operation,"add")==0)
     {
     fscanf(opstream,"%s",source1Name);
     fscanf(opstream,"%s",source2Name);
     fscanf(opstream,"%s",destName);
     s1 = lookup(source1Name);
     s2 = lookup(source2Name);
     dest = lookup(destName);
     for(i=0;i<nodeCount;i++)
          {
          machine[dest][i]= machine[s1][i] + machine[s2][i];
          }
     }
...
}
```

7.5.3.1 Derivatives

Within the shading language a number of operations (known as derivative or area functions) depend not just on the point being shaded but on the way variables change across the surface. The most obvious of these is *calculatenormal*, which finds the normal of a new displaced surface. Though at first sight challenging to implement, once a SIMD approach is adopted, providing a numerical approximation to the new surface normal becomes trivial.[2]

```
if(strcmp(operation,"calculatenormal")==0)
     {
     for(i=0;i<maxnodes;i++)
          {
          dPdu=(machine[pAddr][i+1] - machine[pAddr][i])
              /(machine[uAddr][i+1] - machine[uAddr][i]);
```

[2]This code is somewhat simplified as it assumes we are operating on *P* and **N**. It also ignores the case where the point for which the normal is being calculated is on the edge of the grid. Great care must be taken throughout the renderer to ensure that grid edges are handled correctly.

```
        dPdv=(machine[pAddr][i+uwidth] - machine[pAddr][i])
            /(machine[vAddr][i+uwidth] - machine[vAddr][i]);
        machine[nAddr][i]= crossProduct(dPdu,dPdv);
        }
    }
```

7.5.3.2 Branches and Loops

So far we have applied each SIMD instruction to every node of the virtual machine. In order to implement the shading language we need to implement conditionals (*if–then–else*) and loops (*for*, *while* and *illuminance*).

When a conditional statement is encountered in a SIMD context it is generally not a branch, in that both paths of the code must be executed, each path being applied to a different set of nodes. This is managed by an array of flags indicating which nodes are currently active. When an if statement is encountered, all nodes which fail the test are turned off, and the code in the body of the conditional is executed.

At the end of the conditional we need to wake up the sleeping nodes. This can be achieved by marking the nodes with the value of a label to which they have "branched". When a label instruction is encountered, any nodes sleeping on that label are woken. Note that no branch has actually taken place – the interpreter simply keeps reading instructions. In the case of an *else* clause, all active nodes are suspended, and a label wakes up all nodes, which should execute the else. Finally, at the end of the conditional all the nodes are woken by a further label.

SL	VM code	Node1	Node2
if(s>0.5)	gt s 0.5	true	false
	jmpIfFalse 1		sleep on 1
{do this}	....	active	sleep
else	jmp 2	sleep on 2	sleep
	label 1	sleep	wakeup
{do that}	....	sleep	active
done	label 2	wakeup	active

In addition to conditional forward branching, we simply need unconditional backward branches to be able to implement all SL looping constructs. These are implemented by changing the program counter, which indicates which instruction the interpreter will process next – we actually do need to branch. Though backward branches are unconditional they are only taken if there is an active node. If no nodes are active then the loop is complete, and the interpreter can be allowed to progress naturally to the next instruction.

One complication to this scheme is the use of a shared stack pointer. Shading VMs often use a stack-based architecture for reasons of simplicity. The stack data itself is unique to each node, but due to the large numbers of pushes and pops encountered in typical code it is attractive to share the stack pointer among all active nodes. A potential problem would be if a set of nodes with one value of the stack pointer was put to sleep, the stack pointer changed, and the nodes woke to an incorrect stack pointer.

This cannot occur for most types of branch, as the stack is always empty between statements, and most branches are statements, so SP = 0 at the end of the *if* clause, and should be zero at the start of the *else* clause. The two exceptions to this are the conditional operator COND?A:B, which executes a branch within an expression, and in the case of a function call where there may be data on the stack from the calling function while statements are executed in the child function.

The stack pointer problem can be handled by recording the correct value of the stack pointer for each label. When a branch is first made to a label, the current stack pointer is recorded. When a label is encountered, if there is a recorded stack pointer then it is restored. This operates correctly.

7.5.3.3 Lighting

So far we have ignored the subject of lighting. The reason for this is that in the case of local illumination, lighting is simply a shading calculation – lights encountered in the RIB stream are simply recorded and passed to the shading engine. Instructions such as diffuse and specular can be added to our VM, which take their parameters of position, normal, roughness and observer from the stack, and loop over all lights within the scene calculating the lighting using standard lighting equations (Hall 1989; Upstill 1989; Cook and Torrance 1982).

If non-standard light shaders are applied these must also be executed using the shading VM. This can either use a separate instance of the machine, or the shading calculations may be performed on the same array as the nodes, which actually require the data. In either case the results of these calculations should be cached, as the result of the light shader will be used by both the diffuse and specular calculations.

7.5.3.4 Uniform vs. Varying Data

Not all calculations need be repeated for every point. Many values are uniform across the shaded surface, and hence SL provides a mechanism for identifying this, allowing repeated calculations to be avoided.

The most obvious strategy for implementing such an optimization would be to add qualifiers to each instruction specifying whether it operates in uniform or varying data. The compiler then tracks this information through the parse tree and generates code that uses the new instructions. At run-time operations such as *addUU* need only perform a single addition. Instructions such as *addUV* may also be severely optimized compared with *addVV*.

A more interesting approach is to monitor the type of expressions at run time. A push operation can record whether the value at the top of the stack value is now uniform or varying. An add operation can check the types of data on which it is being asked to operate and optimize accordingly. Though not stack-based the VEX shading language provided in the Mantra renderer uses such an approach (SideFX 2000).

7.5.4 Standard Shading Functions

Most of the functions required by shading are translated into a single instruction. The code for the operation is then implemented as part of the shader engine. Though most of these instructions are relatively standard, and self-evident, a small number are unusual and worthy of comment.[3]

7.5.4.1 Fresnel

Fresnel reflection (refraction) occurs when light bounces off (passes into) a non-conducting surface – typically glass or ceramic. While the use of this function is well documented, its actual implementation is little known. A full discussion may be found in Hall (1989), but the results can be summarized in the following code:

```
float F_d_pl (normal N, vector L, vector T, float eta)
    {
    float amplitude = (N.L+ eta * N.T) /(N.L - eta* N.T);
    return amplitude * amplitude;
}

float F_d_pp (normal N, vector L, vector T, float eta)
    {
    float amplitude = (eta*N.L + N.T) /
        (eta*N.L - N.T);
    return amplitude * amplitude;
    }

/* ***********************************************************
* double F_d_R (N, L, T, eta)
* DIR_VECT *N, *L, *T (in) - N, L, T vectors
*
* Returns the average reflectance for a dielectric.
* N and L are assumed to be to the same side of the
* surface.
*/
float F_d_R (normal N, vector L, vector T, float eta)
    {
    return (F_d_pl (N, L, T, eta) +
        F_d_pp (N, L, T, eta)) / 2.0;
    }
```

Reflection and refraction directions may be calculated as follows:

```
vector reflect(vector I, N)
{
    return I - 2*(I.N)*N;
}
```

[3] Further useful shader functions can be found in Ebert et al. (1998).

```
vector refract(vector I, N; float eta)
{
    float IdotN = I.N;
    float k = 1 - eta*eta*(1 - IdotN*IdotN);

    return k < 0 ? (0,0,0): eta*I - (eta*IdotN + sqrt(k))*N;
}
```

7.5.4.2 Noise

Interesting shaders require surface features to behave in an unpredictable fashion. The use of a simple random function would lead to unrepeatable renders, and frame-to-frame inconsistencies due to popping and aliasing. Perlin noise provides a function that is locally smooth but globally random.

The approach to producing noise is to assign a random gradient to each *lattice point*, i.e. integer. These gradients are then interpolated, producing a value between +1 and −1, which is zero at lattice points (Figure 7.5).

This is implemented by summing wavelets. A wavelet is defined by a weighting function:

$$1 - 3|t|^2 + 2|t|^3 \qquad (\text{for } -1 \leq t \leq 1), \tag{7.2}$$

which is multiplied by a linear function kt, where k is the gradient at the lattice point. These functions are show in Figure 7.6.

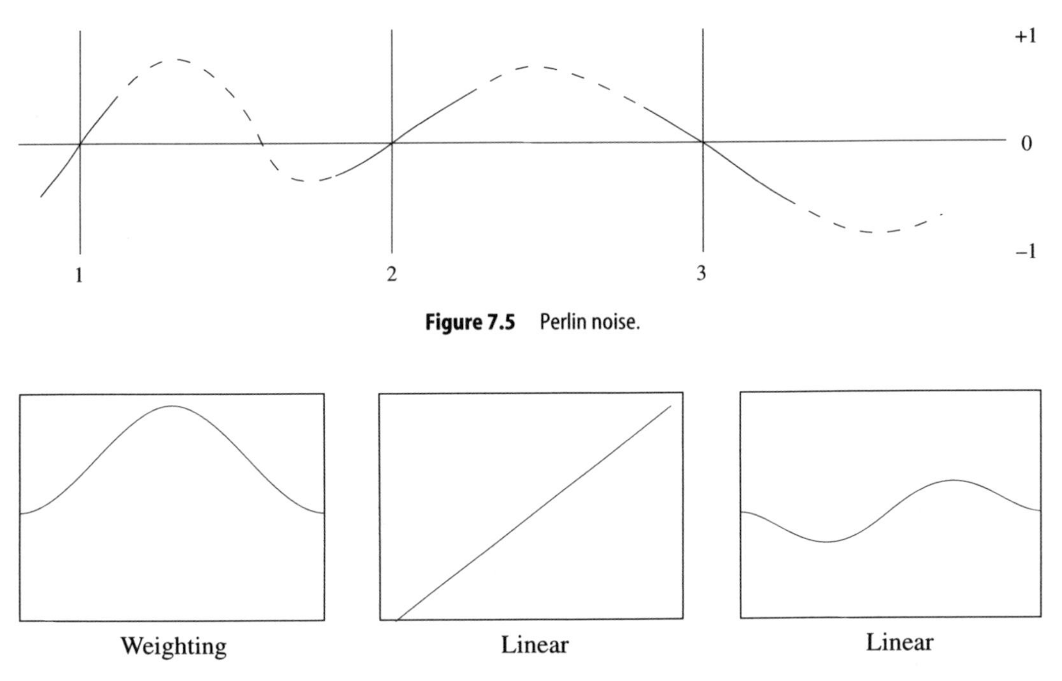

Figure 7.5 Perlin noise.

Figure 7.6 The noise wavelet.

For a point x, the value of $noise(x)$ is found by summing the two wavelets at $floor(x)$ and $floor(x) + 1$:

```
float noise(float x)
{
    int w=floor(x);
    float t=x-w;
    float k=gradientTable[w];

    float wavelet1=(-3|t|^{2}+2|t|^{3}*k*t

    w=w+1;
    t=t-1;
    k=gradientTable[w];

    float wavelet2=(-3|t|^{2}+2|t|^{3})*k*t
    return wavelet1+wavelet2;
}
```

A hash table is used to store pre-calculated gradients and assign them to lattice points in a fashion which minimizes repetition. Perlin noise is discussed in depth in Ebert et al. (1998), and a full implementation is provided.

7.5.4.3 Area

Though simple to implement, the definition of the RenderMan *Area* function provided by Upstill (1989) is somewhat misleading. A more accurate description can be found in Pixar (2000), which defines area as

$$length\left(\frac{dP}{du}du \times \frac{dP}{dv}dv\right). \tag{7.3}$$

This corresponds to the area in space which the current micro-polygon is occupying, and can be used to ensure any features smaller than this size are blended out by the shader to avoid aliasing.

7.6 Scan Conversion

We now have a geometric primitive reduced to a mesh of quadrilaterals, each correctly shaded. These must now be assembled into an image.

The points must first be transformed from camera space (an orthogonal space with the camera at the origin) to screen space, typically by using a perspective projection. This can be broken down into two stages: the point is first divided by its own z-coordinate and transformed into the viewing plane. The unit-square of the viewing plane is then scaled to the output resolution. The exact details of these transformations are controlled by *RiDisplay*, *RiProjection* and *RiScreenWindow*.

Alternative projections may be available, as RenderMan allows custom projections to be written in SL. These are simply executed as the final stage of the shading pipeline, rather than using a standard hard-coded projection. Few renderers implement this facility, as it becomes impossible to invert the process, which may be required by some shaders.

7.6.1 Z-buffer

The simplest method of generating an image from micro-polygons is through the use of a Z-buffer. For each pixel in the final image its distance from the camera is stored. As data is written to the frame-buffer, the depth of the new pixel is compared with the old pixel, and the new value recorded only if it is in front of the old pixel.

The micro-polygons may be flat or Gouraud shaded (depending on the *ShadingInterpolation* attribute), though this should have little effect, as the polygons should all be of the order of one pixel in size. This should be considered when selecting a plotting algorithm.

The main limitation of this approach is that it handles transparency very badly. It is hard to represent a transparent object in front of another, as we can store only the depth of one of them, even though both may be visible. One option is to place sub-pixel holes in transparent objects – a 50% transparent object would be rendered at only 50% of pixels, allowing any objects behind to be seen through the gaps. Though in principle this is a valid approach the results are very poor unless a high sampling rate is used.

The key advantage of this approach is its simplicity, and hence speed. Z-buffering is often available in hardware, and as scan conversion can take up a significant amount of time, it may be beneficial to pass off this task to a graphics card (even though the final image will be written to disk). It is also relatively memory efficient, as only four extra bytes are needed per pixel (a floating-point z-channel in addition to the RGB values).

7.6.2 REYES

In order for transparency to be correctly handled, we need to record all of the micro-polygons that intersect with a particular pixel. Only once all primitives have been rendered can we reduce this list to a single RGB value. This handles opacity well, as we store both foreground and partially obscured objects. We also have information about how much of the pixel a polygon is covering, which will allow us to produce a higher-quality image when the image is resampled.

The main drawback of this approach is its memory usage. For each pixel in the final image we are storing a linked list of micro-polygons, representing every object in the scene. This can be improved slightly by throwing away any MPs that are completely obscured, but the system is still impractical.

To use this approach for practical rendering the output image must be "bucketed" into squares of a more manageable size. Each bucket is processed in turn. Any objects that intersect the current bucket[4] are rendered – micro-polygons that fall into adjacent buckets are held over, while the current bucket is fully processed. At any time only a small number of pixels are being processed, and hence the data structures remain of manageable size.

[4]The calculation of this intersection must take into account displacement. As we have not yet shaded the object at the point of bucketing, this can only be done on the basis of user-supplied displacement bounds. Incorrect bounds lead to objects being clipped at bucket boundaries.

7.7 Resampling

In order to avoid sampling artifacts ("*jaggies*"), a Z-buffer image will be generated at a higher resolution than the final output image (typically two to three times in each direction, controlled by *RiPixelSamples*). This high-resolution image must be resampled to produce an image at the resolution required. While the pixels could simply be averaged, a considerably better image can be produced by using a weighted average of samples from both inside and outside the area covered by the output pixel.

Similarly for a REYES approach we need to collapse the list of micro-polygons to a single value. Rather than just summing them, the coverage can be sampled at several points in and around the pixel. The points may be randomly perturbed (a technique known as jittering) to further reduce aliasing.

7.7.1 Fourier and Nyquist

To appreciate the problem of resampling we need to consider a little signal processing theory. This is most easily thought about in terms of an audio signal, but applies equally to images.

Sound is a rapid change in air pressure. An audio signal displayed on an oscilloscope shows how the pressure changes over time (Figure 7.7). When a computer records sound it stores the amplitude of the waveform (typically) 44 000 times per second. In a similar way a digital image stores a color sample at many points across the page. Provided these samples are close enough together, the sound can be played back or the image reproduced accurately.

An alternative way to think of sound is as a set of frequencies. Musical notes produce pressure changes, which repeat – the note *A*, for example, goes from high pressure to low pressure 440 times a second. However a simple 440 Hz sine wave will sound very dull, and interesting sounds are generated by mixing together sine waves of all frequencies at varying intensities. We can therefore represent sound by plotting amplitude against frequency as shown in Figure 7.8. The human ear can typically hear sounds in the range of 20 Hz to 20 kHz.

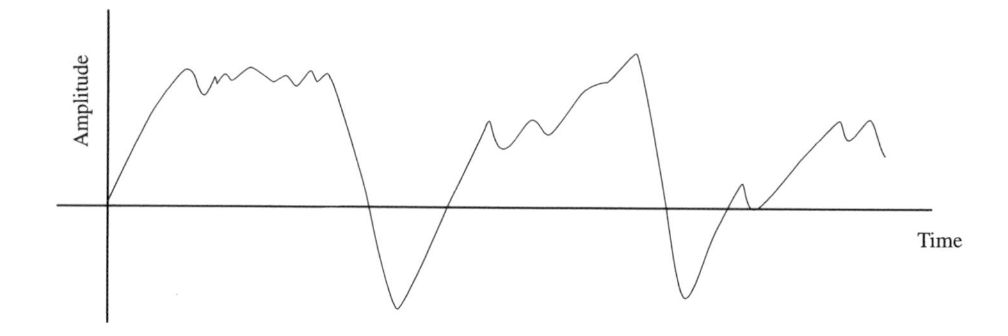

Figure 7.7 A time domain signal.

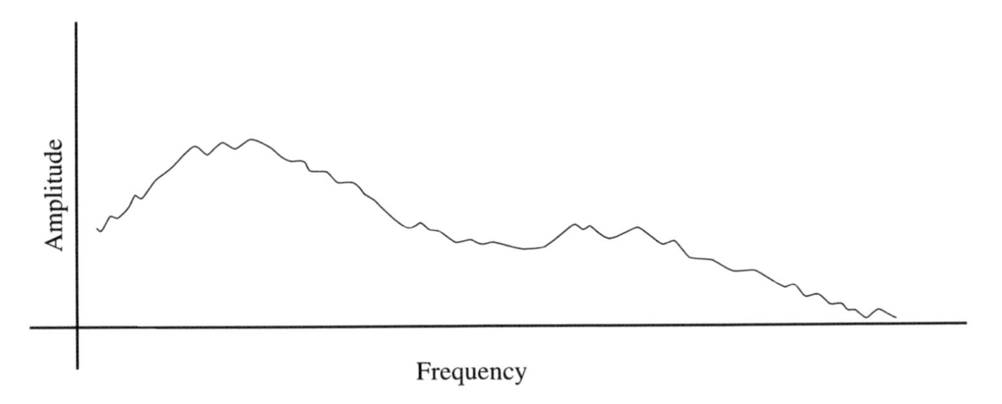

Figure 7.8 A frequency domain signal.

These two representations of the audio signal contain equivalent information. We can therefore convert from the first amplitude–time (time domain) format to the second frequency–time (frequency domain) representation. This conversion is known as a Fourier transform, and may be performed efficiently by a computer using an algorithm known as a FFT (fast[5] Fourier transform). The Fourier transform is a self-inverse function in that it can also be used to covert from the frequency domain back to the time domain.

Nyquist theory says that in order to represent a signal accurately we must sample it at twice its highest frequency. If this condition is not maintained we will see aliasing when the signal is replayed. In the case of audio, we are interested in signals of up to 20 kHz, and hence we record at 44K samples per second – slightly more than double the highest frequency.

An image behaves in exactly the same fashion. If we have a checkerboard pattern we must have one sample in each back square, and one in each white square to get any resemblance between the original scene and the final image. If we want to accurately reproduce the sharp edges to the checkerboard squares we would need an even higher sampling rate.

7.7.2 Convolution

The interim image we have following Z-buffering is at a higher resolution than the final image we require. This will accurately reproduce the scene up to a frequency of half the sampling rate. Any higher-frequency components will be present as aliasing errors, but by choosing a high enough resolution (and carefully written shaders), this should have been minimized. However, in reducing the image to the final output resolution we are reducing the sampling frequency. We must therefore remove any high-frequency components from the interim image during the reduction process.

Removing the high-frequency components from a signal in the frequency domain is conceptually simple – consider a graphic equalizer which might be found in an audio system. The audio signal at each frequency is multiplied by the value of the slider (Figure 7.9). To remove signals above 4 kHz we simply set those sliders to zero.

[5] Something of an optimistic name – the amount of data involved ensures that the process is still painfully slow.

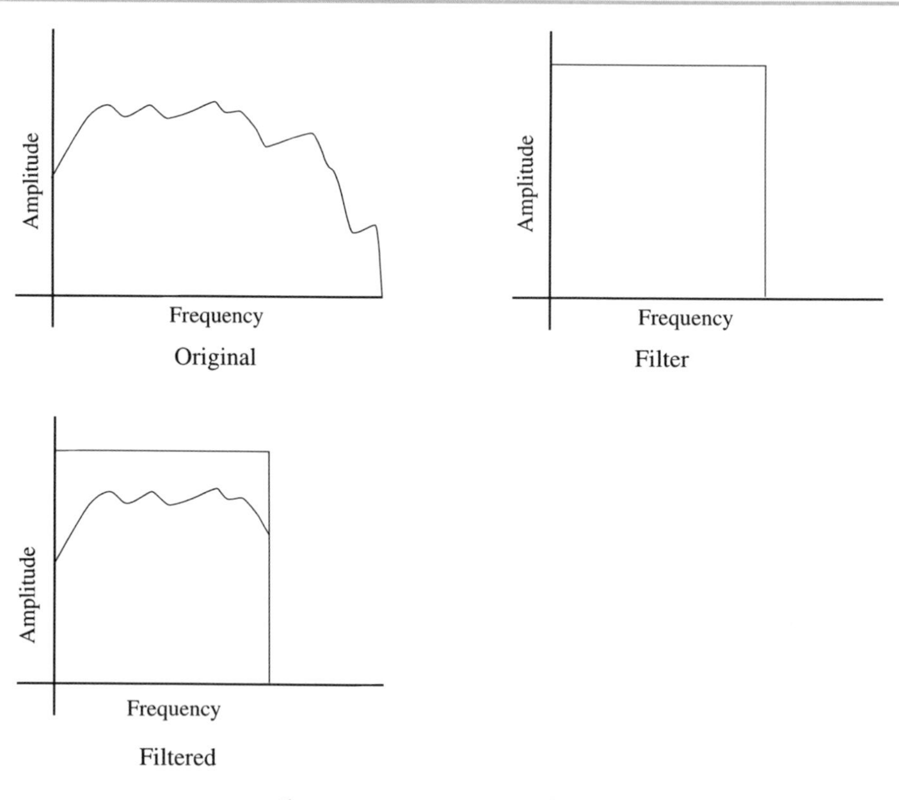

Figure 7.9 Frequency domain filtering.

Filtering a signal in the time domain is not quite so simple. Multiplication in the frequency domain is equivalent to a process known as convolution in time (and vice versa). In digital form convolution is performed as a weighted average of the signal in the area of the pixel:

```
for(i=-w;i<w;i+=delta)
    sum+=f(i)*p(x+i)*delta;
```

where x is the sample being processed, w is the width of the filter, f is the filter weight function and p is the original signal being filtered. This is repeated for each point in the image.

The filter weighting function f should be the Fourier transform of the filter F applied by multiplication in the frequency domain. As F is ideally a step function, this suggests that f should be the Fourier transform of F. This relationship is shown in Figure 7.10. Unfortunately as will be demonstrated this is not practical.

7.7.3 Resampling Kernels

We now have the theory necessary to evaluate some of the filter functions commonly used (and available in RenderMan through the *RiFilterFunction* call).

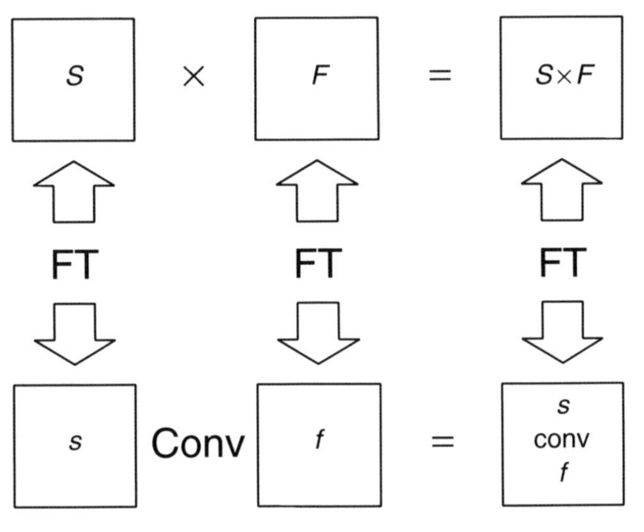

Figure 7.10 Convolution.

Box

A box filter (Figure 7.11) is the simplest to implement in that it requires all samples within the pixel area to be averaged with equal weight. Unfortunately if we look at the Fourier transform of the box filter it starts removing frequencies far below the required cutoff point (leading to blurring), and still allows many higher frequencies to get through (aliasing).

Triangle

A significant improvement may be achieved by the use of a triangle (or tent) filter function (Figure 7.12). Here we use samples from points outside the strict pixel area. Although this may seem counter-intuitive, the results are a filter that has a much sharper cutoff than the box filter, resulting in an image which is significantly better.

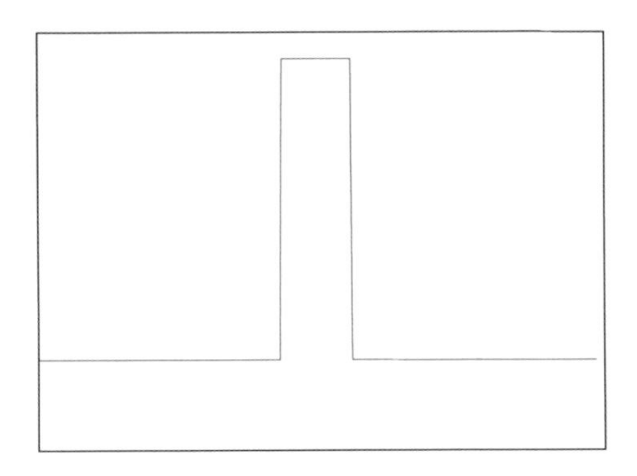

Box

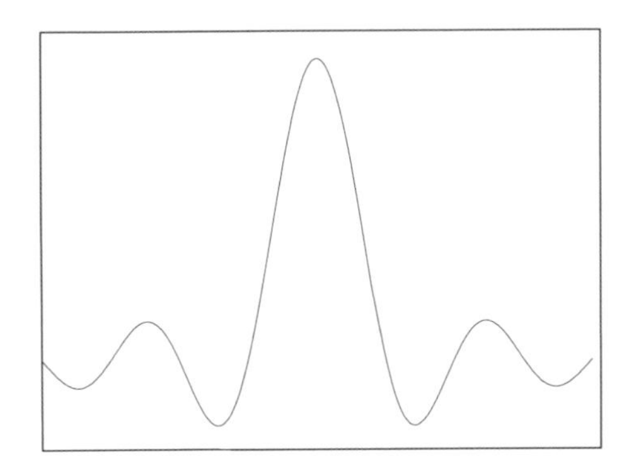

FT(Box) = sync

Figure 7.11 The box filter.

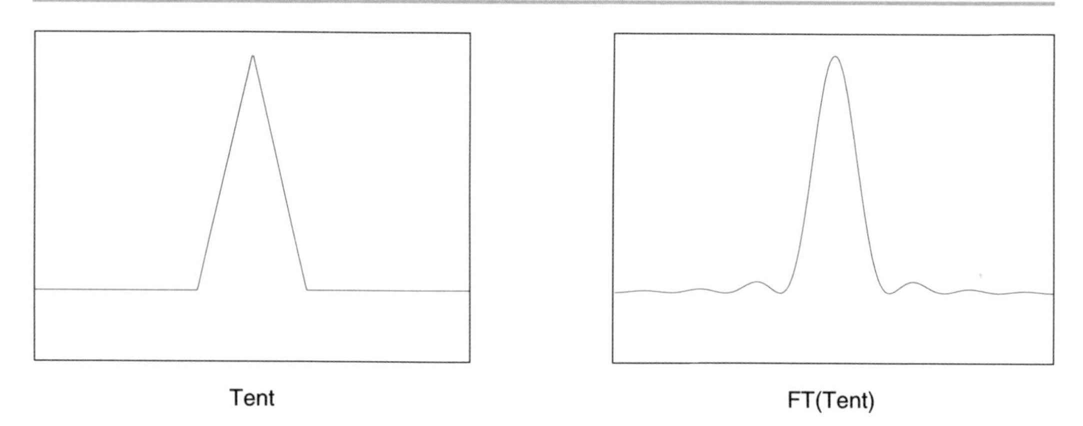

Figure 7.12 The tent filter.

Gaussian

The Fourier transform of a Gaussian is itself a Gaussian curve (Figure 7.13), and hence the resultant filter behaves well in the frequency domain. Unfortunately, as it never quite reaches zero, each pixel is in principle dependent on the whole image. In practice sample values from more than a few pixels away can be ignored.

Sinc

Finally, in the case of the sinc function (Figure 7.14) we find a filter that produces the box step in the frequency domain that we would ideally like. Unfortunately, like the Gaussian, the sinc function requires pixel samples from a large (theoretically infinite) area. It achieves its sharp cut-off by subtracting samples from surrounding pixels. Once again, though this may seem strange, justification for this can be found by examining the frequency domain representation.

The choice of *PixelSamples* and *PixelFilter* is basically a tradeoff between speed and quality. More samples will produce a better image (by increasing the Nyquist frequency of the interim image) at the cost performance. The more complex filters will require samples

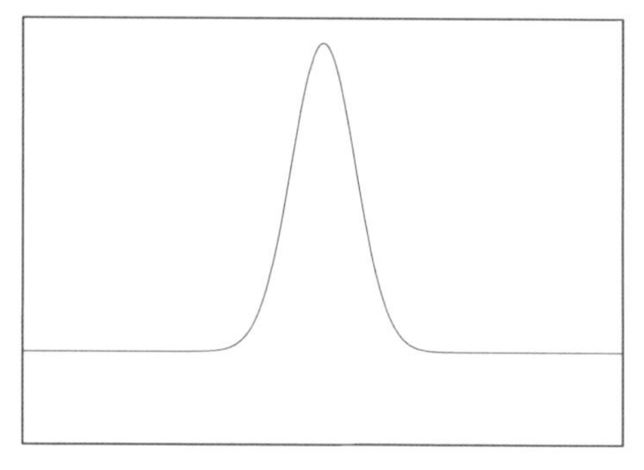

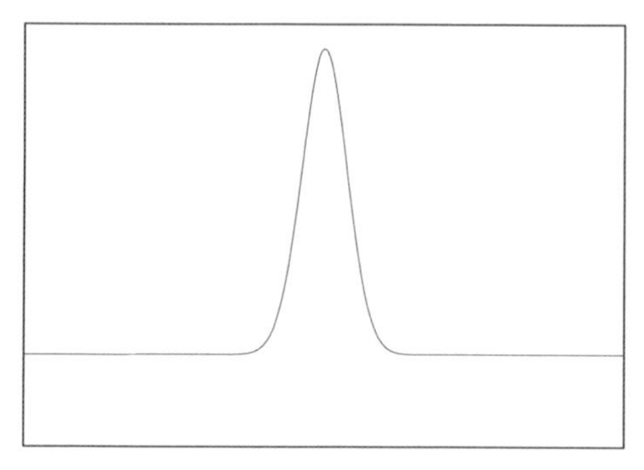

Figure 7.13 The Gaussian filter.

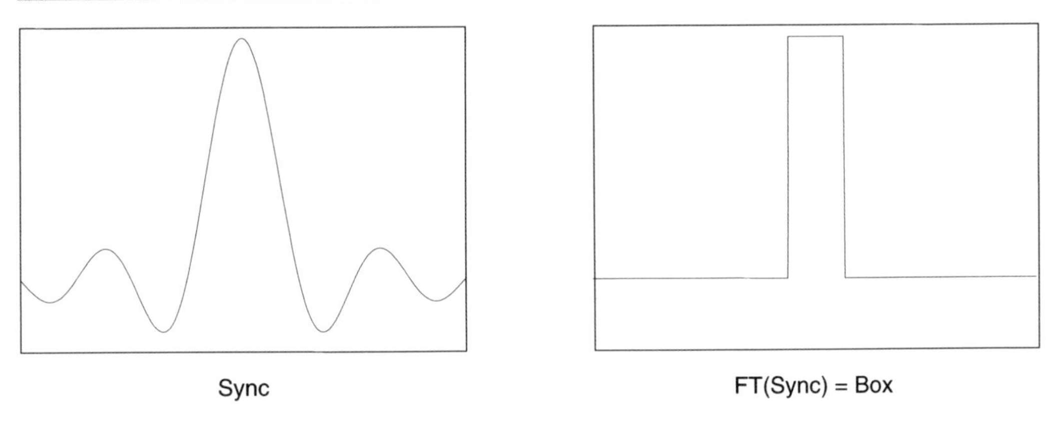

Figure 7.14 The sinc filter.

from larger areas of the image, again slowing down the render. Care must be taken if the sample area is simply truncated, as this can also lead to artifacts in the final image (Blinn 1998).

7.7.4 Mip-mapping

During rendering the shading engine will frequently access pre-drawn textures stored on disk. While these can be accessed on a point-by-point basis, it should be remembered that this also is a form of resampling, controlled by the same laws of signal processing as previously discussed. Using convolution, the correct area sample can be calculated, though this will be slow. Performance is significantly improved by storing textures at a range of resolutions – a technique known as Mip-mapping.

The original version (often scaled up so its resolution is a power of 2) of the image is used as the base of a pyramid. Copies of the image are then stored at half, quarter, eighth resolution, and so on until the image is a single pixel. This can be performed off-line, and the resultant images stored in special texture files (which will be four-thirds as big as the original image). The renderer can then directly access pre-filtered versions of the image at any resolution it requires at minimal cost.

It should be remembered when generating the MipMap that even though the image is being reduced by powers of two, it is still being resampled, and best results can be obtained by using one of the higher-order filtering functions (rather than the simple box as seems tempting). As this process is only done once per texture rather than for every frame, the additional cost is insignificant.

7.8 Exposure, Imaging and Quantization

Following resampling, the image is subjected to three final image-processing steps. The response of most displays is non-linear – a pixel value of 200 is not twice as bright as a value of 100. However to correctly perform lighting calculations within the renderer depends on

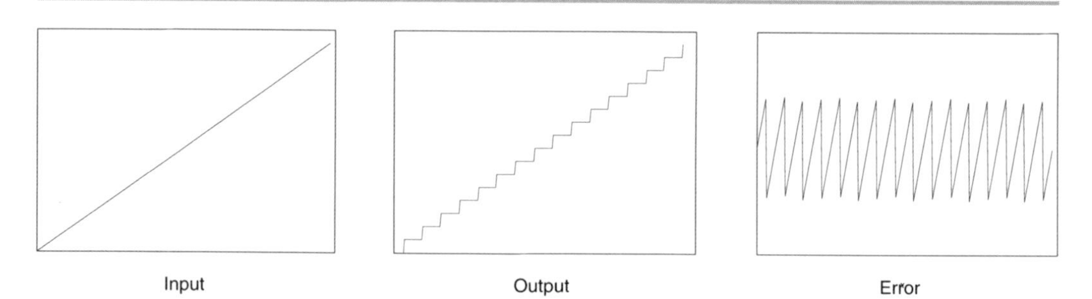

Figure 7.15 Quantization error.

the color space being linear. At some stage between scan conversion and final viewing the image must be "gamma corrected" using the equation: $C_{out} = C_{in}^{1/\gamma}$, where γ is the amount of correction required. *RiExposure* allows this to be done before the image is written to disk.

Unfortunately, this approach has two drawbacks. First, the exact amount of gamma correction required is dependent on the display being used. Different computer types will require different values, and if precise calibration is required, variation may be found from machine to machine. Film or video require further correction. Secondly, once the image has been converted to a non-linear color space it cannot be composited using the standard compositing mathematics (Porter and Duff 1984). Prior to use a compositor will have to remove the gamma correction and re-apply it to the final image.

It is therefore generally recommended that images are generally written to disk without gamma correction, and correction applied during the display/video/film transfer process. However, at the *RiExposure* stage of the pipeline pixels are still in floating-point format, while they are likely to have been quantized when they reach the transfer stage. Hence if the image is being generated in a single pass, and sent directly to a device of known attributes applying gamma at the *exposure* stage may produce better results.

Following *exposure*, *RiImager* allows a shader to be applied to the image as a whole. Note that this is an *option* rather than an *attribute* like all other shader types. This is intended to allow custom image-processing operations to be applied to the pixels.

The final stage of the renderer must be to output the image. However most output formats require pixels of limited bit depth (typically 8–16 bits per channel), while internally the renderer will manipulate floating-point values. Quantization converts from the internal format to the output format.

If the floating-point value is simply rounded (or truncated) to the nearest integer we get an error value which (for linear color gradients) is both periodic and discontinuous (Figure 7.15). Unfortunately the eye is designed to be sensitive to both of these artifacts, as in real scenes they typically convey important information. This is avoided by applying a random value to the signal with an amplitude equal to half of one quantization level.

The resultant error in the output signal (Figure 7.16) is still of the same amplitude when measured analytically, but that error is now spread evenly over all the samples. The error is no longer periodic and hence is visually less intrusive. Over areas of greater than a few pixels, the error averages to zero.

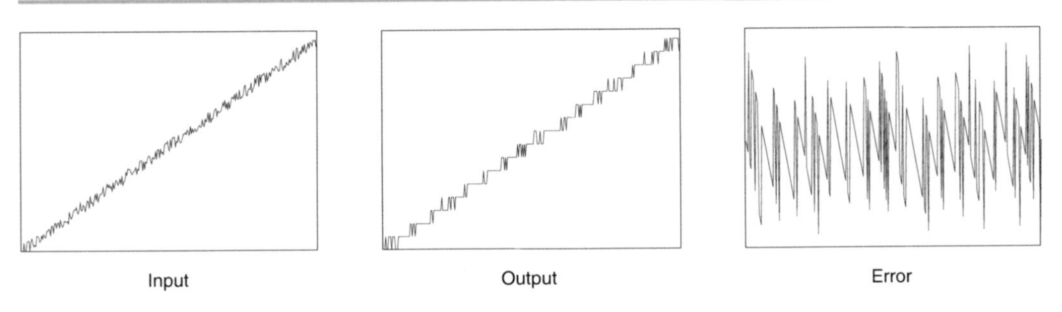

Figure 7.16 Dithering.

7.9 Advanced Features

7.9.1 Spectral Colors

Light in computer graphics is typically represented by three components representing red, green and blue. However this representation is neither necessary nor even sufficient for the generation of images under all circumstances (Hall 1989). The RI specification supports more complex color models, though to date no renderer in common use supports this feature.

7.9.1.1 Light and Vision

Like sound, light consists of a spectrum of frequencies (though the waves in question are electromagnetic rather than pressure). A color represents a mix of many frequencies, and so to accurately represent it we should record the strength of each frequency.

Human eyes are not sensitive to most of the electromagnetic spectrum, containing cells that respond to three specific frequency bands. Though these bands are sometimes referred to as R, G and B they are in fact very broad, overlap and correspond poorly to our typical ideas of these colors. We perceive the color of a continuous spectrum of light based on the extent to which it stimulates these three sets of cells. As a result there are many spectra that will be perceived as the same color (these are known as metamers).

7.9.1.2 Color Spaces

By picking (almost) any three basis colors, in theory any visible color could be reproduced. However, in practice physical sources are limited to a range in intensity between 0 (completely off) and 1 (as bright as is possible). A TV screen may, in fact, be dark gray rather than black when turned off, and will never be as bright as a 10K spotlight. Any physical device therefore has a limited range of colors it can reproduce – this is known as the *gamut* of the device. Note however, that the gamut is a function of a device rather than a color space which mathematically has no such limitations.

While the utility of color values greater than 1 is obvious, the need for negative values of a basis color can be seen by considering mixing a color from primaries Rb (Red with a little blue mixed in), Gb (green with blue) and B (pure blue): having set the correct level of R and

G with Rb and Gb we may already have too much B, so the required value for B is less than zero.

In theory negative color values are not a problem. They are however unwelcome. For this reason a standard color space is defined known as CIExyz. The three stimuli X, Y and Z can be mixed using only positive weights to produce any visible color. In addition, the spectra of X, Y and Z are defined by the standard, and hence a CIExyz color may be used as a reference.

The RGB color space is not so well defined in that R, G and B typically refer to the colors of the phosphors on the monitor at which the user happens to be sitting. RGB on one system will not typically correspond to RGB on another system. To address this the NTSC standard defines a particular set of R, G and B colors in terms of CIExyz. A calibrated display system can store colors internally as NTSCrgb, and convert them to the RGB of the monitor for display.

7.9.1.3 Spectral Colors

Just as the RGB color space is not as standard as we would like to think, even the use of three channels of color should be questioned. While such a model is adequate for display of the final image it may not be appropriate for performing calculation.

Consider the case of a transparent object, which lets a narrow band of yellow light through, but blocks red and green completely. This scenario is shown in Figure 7.17. Light from a

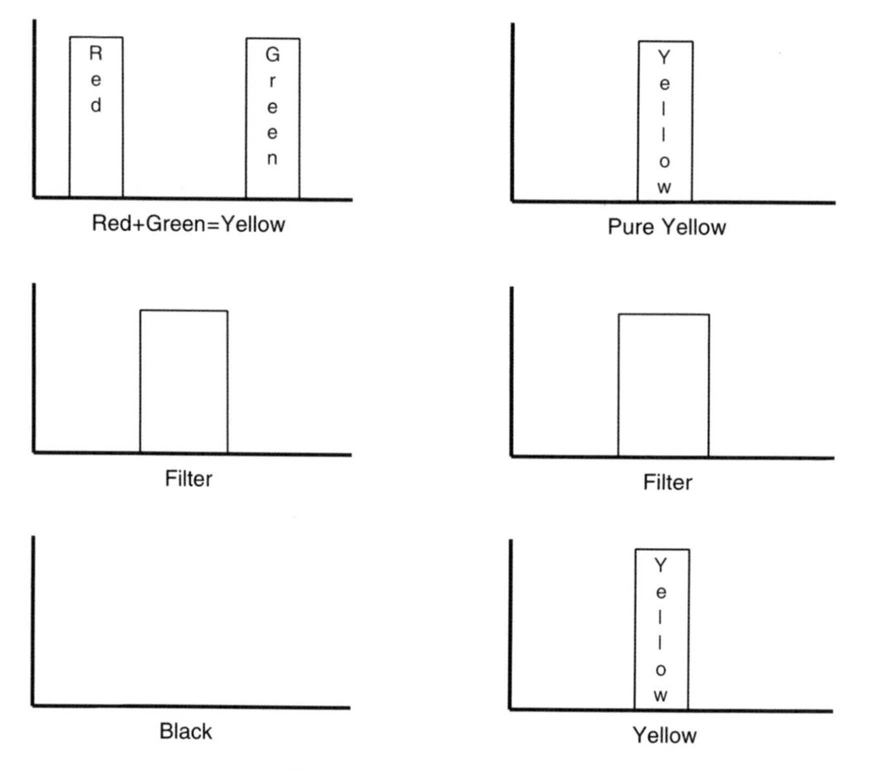

Figure 7.17 Color quantization.

sodium lamp would pass unimpeded, yet a mix of red and green light set to exactly match the sodium yellow to the human eye would be blocked completely.

While such degenerate cases are rare, more subtle interactions are common. The problem can be reduced by increasing the number of samples used internally for color calculations. Up to ten color samples might be used to produce a better approximation to the full color spectrum.

7.9.2 Motion Blur

Just as point sampling an image produces a poor quality image due to spatial aliasing, sampling a scene at a single point in time can lead to temporal aliasing. A real camera's shutter is open for a finite amount of time, and this leads to motion blur in the captured image. This must be simulated in the digital image if the results are to appear natural.

PRMan implements motion blur in the REYES architecture by calculating the position of each micro-polygon at both the start and end of the shutter time. During super-sampling each sample is taken at different times, and the position of each MP is interpolated. This adds little overhead to the rendering process, though it has the limitation that MPs are shaded only once (the lighting at the beginning of the motion may be very different from that at the end), and subtleties in the motion path are lost.

A renderer based around a Z-buffer may implement motion blur by rendering different super-samples at different times (Kirk 1992). In the simplest case where two time samples are required, alternate lines may be masked and the motion-blurred objects rendered in two positions. Non-moving objects are rendered as normal, and the interlaced images are combined during the filtering stage to produce a single motion-blurred image.

7.9.3 Deferred Shading

The approach adopted here (and in most scan-line renderers supporting the RenderMan standard) is one of shading before hiding. That is, we fully shade each object, prior to determining which parts of it are visible. This approach is essential if displacement is to be supported, as only post-shading do we actually know where the surface is. If displacement is not required it is clearly more efficient to avoid redundant shading calculations by determining visible surfaces first.

Scan conversion may be performed on the unshaded geometry by storing the shading parameters (P, $\mathbf{N}$, s, t, Cs, etc.) at each pixel. Shading calculations can then be performed using simply the values in the frame-buffer (Perlin 1985). However, simply shading output pixels prohibits derivative operations, as there is no connectivity information available.

By retaining the original scene database, and storing u, v and an object reference, the local geometry of the scene can be reconstructed. Visible points may be determined, and the area of objects around the visible pixels may be shaded in the normal fashion.

Deferred shading allows the scene to be rapidly re-shaded without repeating geometry calculations. It is therefore appropriate for use in interactive preview rendering systems where users can directly manipulate the appearance of the scene (including the location of lights, which are simply parameters to the shading process). The extraction of u, v coordinates for a scene also forms the basis of 3D paint systems.

7.9.4 Occlusion Culling

Though it has not been discussed for reasons of clarity, throughout the rendering pipeline any geometry (object, mesh or micro-polygon) that is not visible should be discarded as soon as possible. Detecting that an object is not visible avoids the creation of thousands of micro-polygons. While this has little effect on the final image, it can dramatically reduce rendering times.

A piece of geometry may be invisible because it is outside the viewing fustrum, back-facing (for closed objects – *RiSides*) or fully obscured by other objects. The chances of an object being fully occluded can be dramatically increased by sorting the scene database so closest objects are rendered first. This avoids rendering objects that are subsequently painted over.

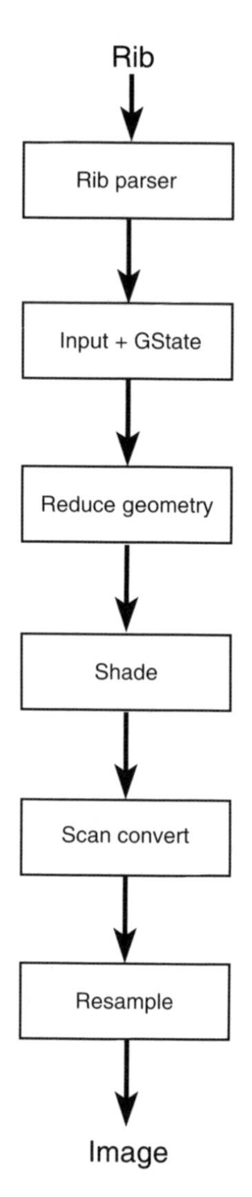

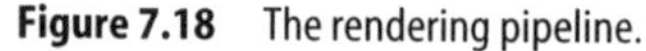

Figure 7.18 The rendering pipeline.

7.10 Conclusions

The rendering pipeline of a simple scan-line renderer based around a procedural shading engine is summarized in Figure 7.18. The input is read and fed into a front end which tracks the graphics state (attributes and transformations). When geometry is encountered it is passed to the geometry reduction stage which converts all input forms to a patch mesh of such resolution that artifacts will not be visible. This mesh of micro-polygons can then be shaded, and then added to the image. Once complete the image will be resampled to produce the final output.

This pipeline in Figure 7.18 represents one of the simplest forms possible. For reasons of efficiency these stages may be interleaved in a more complex fashion. A renderer incorporating more complex lighting techniques would require a more complex construction. However in either case these blocks form the basis of the work that must be done.

Bibliography

Apodaca, A.A. and Gritz, L., *Advanced RenderMan: Creating CGI for Motion Pictures.* Morgan-Kaufmann, 1999.

Blinn, J., *Dirty Pixels.* Morgan-Kaufmann, 1998.

Bloomenthal, J., et al., *Introduction to Implicit Surfaces.* Morgan-Kaufmann, 1997.

Catmull, E., *A Subdivision Algorithm for Computer Display of Curved Surfaces,* UTEC, CSc, pp. 4–133, University of Utah, 1974.

Cook, R.L. and Torrance, K.E., A reflectance model for computer graphics, *ACM Transactions Graphics,* **5** (1), 1982.

Duff, T., Interval arithmetic and recursive subdivision for implicit surfaces and CSG, *SIGGRAPH,* 1992.

Ebert, D. et al., *Texturing and Modeling.* AP Professional, 1998.

Flyn, M.J., Some computer organizations and their effectiveness. *IEEE Trans. Computing,* **C-21** 948–960, 1972.

Gritz, L. and Hahn, J. BMRT: a global illumination implementation of the Renderman standard VIN3, 1996.

Hall, R., *Illumination and Color in Computer Generated Imagery.* Springer-Verlag,, 1989.

Heckbert, P.S., *Graphics Gems IV,* AP Professional, 1994.

Kirk, D., *Graphics Gems III.* AP Profesional, 1992.

Olano, M., Hardware accelerated rendering, *SIGGRAPH* 2000, Course Notes.

Perlin, K., An image synthesizer. *Computer Graphics* **19** (3), 287–96 (*SIGGRAPH*) 1985.

Pixar, *Application note #31,* 1999.

Pixar, *RenderMan Interface Specification v3.2,* 2000.

Porter, T. and Duff, T., Compositing digital images, *Computer Graphics,* **18** (3) 1984.

Richards, M. and Whitby-Stevens, C. *BCPL – the Language and its Compiler.* Cambridge University Press, 1979.

SideFX Software, *The VEX Shading Language,* 2000.

Stam, J., Exact evaluation of Catmull–Clark subdivision surfaces at arbitrary parameter values, *SIGGRAPH* 1998. Addison-Wesley.

Upstill, S., *The RenderMan Companion: a Programmer's Guide to Realistic Computer Graphics.* Addison-Wesley, 1989.

Watt, A. and Watt, M. *Advanced Animation and Rendering Techniques.* Addison-Wesley, 1992.

Index